Springer Series in
MATERIALS SCIENCE 84

Springer Series in
MATERIALS SCIENCE

Editors: R. Hull R. M. Osgood, Jr. J. Parisi H. Warlimont

The Springer Series in Materials Science covers the complete spectrum of materials physics, including fundamental principles, physical properties, materials theory and design. Recognizing the increasing importance of materials science in future device technologies, the book titles in this series reflect the state-of-the-art in understanding and controlling the structure and properties of all important classes of materials.

71 **Dissipative Phenomena in Condensed Matter**
Some Applications
By S. Dattagupta and S. Puri

72 **Predictive Simulation of Semiconductor Processing**
Status and Challenges
Editors: J. Dabrowski and E.R. Weber

73 **SiC Power Materials**
Devices and Applications
Editor: Z.C. Feng

74 **Plastic Deformation in Nanocrystalline Materials**
By M.Yu. Gutkin and I.A. Ovid'ko

75 **Wafer Bonding**
Applications and Technology
Editors: M. Alexe and U. Gösele

76 **Spirally Anisotropic Composites**
By G.E. Freger, V.N. Kestelman, and D.G. Freger

77 **Impurities Confined in Quantum Structures**
By P.O. Holtz and Q.X. Zhao

78 **Macromolecular Nanostructured Materials**
Editors: N. Ueyama and A. Harada

79 **Magnetism and Structure in Functional Materials**
Editors: A. Planes, L. Manõsa, and A. Saxena

80 **Ion Implantation and Synthesis of Materials**
By M. Nastasi and J.W. Mayer

81 **Metallopolymer Nanocomposites**
By A.D. Pomogailo and V.N. Kestelman

82 **Plastics for Corrosion Inhibition**
By V.A. Goldade, L.S. Pinchuk, A.V. Makarevich and V.N. Kestelman

83 **Spectroscopic Properties of Rare Earths in Optical Materials**
Editors: G. Liu and B. Jacquier

84 **Hartree–Fock–Slater Method for Materials Science**
The DV–X Alpha Method for Design and Characterization of Materials
Editors: H. Adachi, T. Mukoyama, and J. Kawai

85 **Lifetime Spectroscopy**
A Method of Defect Characterization in Silicon for Photovoltaic Applications
By S. Rein

86 **Wide-Gap Chalcopyrites**
Editors: S. Siebentritt and U. Rau

Volumes 20–70 are listed at the end of the book.

H. Adachi · T. Mukoyama · J. Kawai (Eds.)

Hartree-Fock-Slater Method for Materials Science

The DV-Xα Method for Design and Characterization of Materials

With 132 Figures and 33 Tables

Springer

Editors:

Professor Hirohiko Adachi
Kyoto University
Center of Fundamental Chemistry
34-4,Takano-Nishihiraki-cho
Sakyo-ku, Kyoto 606-8103, Japan
E-mail: adachi@fukui.kyoto-u.ac.jp

Professor Dr. Jun Kawai
Kyoto University
Dept. of Materials Science and Engineering
Sakyo-ku, Kyoto, 606-8501, Japan
E-mail: jun.kawai@materials.mbox.media.kyoto-u.ac.jp

Professor Takeshi Mukoyama
Kansai Gaidai University
16-1 Nakamiya-Higashinocho
Hirakata, Osaka 573-1001, Japan
E-mail: mukoyama@kansaigaidai.ac.jp

ISSN 0933-033X

ISBN-10 3-540-24508-1 Springer Berlin Heidelberg New York
ISBN-13 978-3-540-24508-7 Springer Berlin Heidelberg New York

Library of Congress Control Number: 2005924367

Springer is a part of Springer Science+Business Media
springer.com

Printed in Germany

Typesetting by the editors
Production: LE-TEX Jelonek, Schmidt & Vöckler GbR, Leipzig
Cover concept: eStudio Calamar Steinen
Cover production: *design & production* GmbH, Heidelberg

Printed on acid-free paper 57/3141/YL SPIN: 11383833 5 4 3 2 1 0

Springer Series in

MATERIALS SCIENCE

Editors: R. Hull   R. M. Osgood, Jr.   J. Parisi   H. Warlimont

20 **Microcluster Physics**
By S. Sugano and H. Koizumi
2nd Edition

21 **The Metal-Hydrogen System**
By Y. Fukai   2nd Edition

22 **Ion Implantation in Diamond, Graphite and Related Materials**
By M. S. Dresselhaus and R. Kalish

23 **The Real Structure of High-T_c Superconductors**
Editor: V. Sh. Shekhtman

24 **Metal Impurities in Silicon-Device Fabrication**
By K. Graff   2nd Edition

25 **Optical Properties of Metal Clusters**
By U. Kreibig and M. Vollmer

26 **Gas Source Molecular Beam Epitaxy**
Growth and Properties of Phosphorus Containing III–V Heterostructures
By M. B. Panish and H. Temkin

27 **Physics of New Materials**
Editor: F. E. Fujita   2nd Edition

28 **Laser Ablation**
Principles and Applications
Editor: J. C. Miller

29 **Elements of Rapid Solidification**
Fundamentals and Applications
Editor: M. A. Otooni

30 **Process Technology for Semiconductor Lasers**
Crystal Growth and Microprocesses
By K. Iga and S. Kinoshita

31 **Nanostructures and Quantum Effects**
By H. Sakaki and H. Noge

32 **Nitride Semiconductors and Devices**
By H. Morkoç

33 **Supercarbon**
Synthesis, Properties and Applications
Editors: S. Yoshimura and R. P. H. Chang

34 **Computational Materials Design**
Editor: T. Saito

35 **Macromolecular Science and Engineering**
New Aspects
Editor: Y. Tanabe

36 **Ceramics**
Mechanical Properties, Failure Behaviour, Materials Selection
By D. Munz and T. Fett

37 **Technology and Applications of Amorphous Silicon**
Editor: R. A. Street

38 **Fullerene Polymers and Fullerene Polymer Composites**
Editors: P. C. Eklund and A. M. Rao

39 **Semiconducting Silicides**
Editor: V. E. Borisenko

40 **Reference Materials in Analytical Chemistry**
A Guide for Selection and Use
Editor: A. Zschunke

41 **Organic Electronic Materials**
Conjugated Polymers and Low Molecular Weight Organic Solids
Editors: R. Farchioni and G. Grosso

42 **Raman Scattering in Materials Science**
Editors: W. H. Weber and R. Merlin

43 **The Atomistic Nature of Crystal Growth**
By B. Mutaftschiev

44 **Thermodynamic Basis of Crystal Growth**
P–T–X Phase Equilibrium and Non-Stoichiometry
By J. Greenberg

45 **Thermoelectrics**
Basic Principles and New Materials Developments
By G. S. Nolas, J. Sharp, and H. J. Goldsmid

46 **Fundamental Aspects of Silicon Oxidation**
Editor: Y. J. Chabal

Springer Series in
MATERIALS SCIENCE

Editors: R. Hull R. M. Osgood, Jr. J. Parisi H. Warlimont

49 **Alkali Halides**
A Handbook of Physical Properties
By D. B. Sirdeshmukh, L. Sirdeshmukh, and K. G. Subhadra

50 **High-Resolution Imaging and Spectrometry of Materials**
Editors: F. Ernst and M. Rühle

51 **Point Defects in Semiconductors and Insulators**
Determination of Atomic and Electronic Structure from Paramagnetic Hyperfine Interactions
By J.-M. Spaeth and H. Overhof

52 **Polymer Films with Embedded Metal Nanoparticles**
By A. Heilmann

53 **Nanocrystalline Ceramics**
Synthesis and Structure
By M. Winterer

54 **Electronic Structure and Magnetism of Complex Materials**
Editors: D.J. Singh and D. A. Papaconstantopoulos

55 **Quasicrystals**
An Introduction to Structure, Physical Properties and Applications
Editors: J.-B. Suck, M. Schreiber, and P. Häussler

56 **SiO_2 in Si Microdevices**
By M. Itsumi

57 **Radiation Effects in Advanced Semiconductor Materials and Devices**
By C. Claeys and E. Simoen

58 **Functional Thin Films and Functional Materials**
New Concepts and Technologies
Editor: D. Shi

59 **Dielectric Properties of Porous Media**
By S.O. Gladkov

60 **Organic Photovoltaics**
Concepts and Realization
Editors: C. Brabec, V. Dyakonov, J. Parisi and N. Sariciftci

61 **Fatigue in Ferroelectric Ceramics and Related Issues**
By D.C. Lupascu

62 **Epitaxy**
Physical Principles and Technical Implementation
By M.A. Herman, W. Richter, and H. Sitter

63 **Fundamentals of Ion-Irradiated Polymers**
By D. Fink

64 **Morphology Control of Materials and Nanoparticles**
Advanced Materials Processing and Characterization
Editors: Y. Waseda and A. Muramatsu

65 **Transport Processes in Ion-Irradiated Polymers**
By D. Fink

66 **Multiphased Ceramic Materials**
Processing and Potential
Editors: W.-H. Tuan and J.-K. Guo

67 **Nondestructive Materials Characterization**
With Applications to Aerospace Materials
Editors: N.G.H. Meyendorf, P.B. Nagy, and S.I. Rokhlin

68 **Diffraction Analysis of the Microstructure of Materials**
Editors: E.J. Mittemeijer and P. Scardi

69 **Chemical-Mechanical Planarization of Semiconductor Materials**
Editor: M.R. Oliver

70 **Applications of the Isotopic Effect in Solids**
By V.G. Plekhanov

Preface

The aim of this book is to present and explain the discrete variational (DV) Xα molecular orbital method for calculations of the electronic structure and properties of molecules. With recent advance of computers, several self-consistent molecular orbital methods have been developed. The DV-Xα method is one such method.

The approximation for the exchange potential, now denoted by the Xα potential, was first introduced in 1951 by J.C. Slater. The Xα method is usually called as the Hartree–Fock–Slater method for atoms. J. Korringa (1947), W. Kohn and N. Rostoker (1954) proposed the Korringa–Kohn–Rostoker (KKR) method to calculate electronic states of solids with a translational symmetry by the use of the multiple scattering approach. Though this approach was believed to be suitable for energy band calculations, K.H. Johnson (1967) demonstrated that this approach is also applicable to isolated molecules, using the Xα potential. Johnson's method was called the multiple scattering (MS) Xα or scattered wave (SW) Xα method. After the work of Johnson, complicated molecules or surfaces, without any periodic boundary conditions, could be calculated with a similar method of solid state physics.

The DV-Xα method first appeared in 1970 in Chicago and was developed by D.E. Ellis, H. Adachi (one of the editors and authors of this book), and other young researchers at that time. In contrast, to multiply scattered spherical waves used for the MS-Xα or SW-Xα method, the wave functions in the DV-Xα method are expressed by a linear combination of atomic orbitals (LCAO). The atomic wave functions are given in numerical form and the matrix elements in the secular equation are calculated numerically with the DV-integration method. The advantage of the DV-Xα method is the possibility to obtain realistic molecular potentials and wave functions with relatively small basis set. This method is also applicable to molecules or clusters both for ground and excited states, which have no translational symmetry.

Based on this merit, the DV-Xα method has been used for design of alloys and ceramics, surface and interface chemistry, and core level spectroscopies, because, in such systems, the added atoms, surface, or core hole are regarded as an impurity to break the periodic boundary condition. Recently the DV-Xα method has been applied to design of various kinds of industrial materials, such as electric batteries, catalysts, etc.

The reason for such wide applications is in the fact that by using the DV-Xα method it is easy to converge the electronic structure of clusters that contain any atomic number elements up to $Z = 115$; any kind of unstable valence states, such as Fe^{5+}; any bond length, such as an iron in the earth's core; or a very short-lived transition state, such as the 1s core hole state. The DV-Xα method is still advancing including the relativistic effect, spin multiplet, and spin-orbital interaction.

This book presents various topics of the DV-Xα method from the basic concept to important applications and is divided into three parts. The first part is fundamental and the basic principle of the DV-Xα method is explained in Chap. 1. The second part is devoted to materials science. Chapters 2 to 7 describe alloy design, lattice imperfections, ceramics, magnetic properties, optical materials and heavy elements. The third part covers applications for spectroscopy. X-ray spectroscopy, core hole spectroscopy and Auger spectroscopy are discussed in Chaps. 8 to 10.

We wish to thank Dr. I. Kishida for his efforts to transform many complicated mathematical formulae into TeX. It is also a pleasure to thank Claus Ascheron of Springer for his encouragement and patience over the course of preparing this book.

Kyoto, Oct. 2005

Hirohiko Adachi
Jun Kawai
Takeshi Mukoyama

Contents

List of Contributors

Hiroһiko Adachi
Fukui Institute
for Fundamental Chemistry
Kyoto University
Takano-Nishihiraki-cho 34-4
Sakyo-ku, Kyoto 606-8103
Japan

Toshiaki Enoki
Graduate School
of Science and Engineering
Tokyo Institute of Technology
2-12-1 Ookayama, Meguro-ku
Tokyo 152-8550
Japan

Kimichika Fukushima
Advanced Energy Design
and Engineering Department
Isogo Engineering Center
Toshiba Corporation
8, Shinsugita-cho
Isogo-ku, Yokohama 235-8523
Japan

Tomohiko Ishii
Faculty of Engineering
Kagawa University
2217-20, Hayashicyo, Takamatsu
Kagawa 761-0396
Japan

Toshio Kamiya
Materials and Structures Laboratory
Tokyo Institute of Technology
Mailbox R3-4, 4259 Nagatsuta
Midori-ku, Yokohama 226-8503,
Japan

Jun Kawai
Department of Materials Science
and Engineering
Kyoto University
Sakyo-ku, Kyoto 606-8501
Japan

László Kövér
Institute of Nuclear Research
of the Hungarian Academy
of Sciences
P.O. Box 51
H-4001, Debrecen
Hungary

Masataka Mizuno
Center for Atomic
and Molecular Technologies
Graduate School of Engineering
Osaka University
2-1 Yamadaoka
Suita, Osaka 565-0871
Japan

Masahiko Morinaga
Department of Materials Science
and Engineering
Graduate School of Engineering
Nagoya University, Furo-cho
Chikusa-ku, Nagoya 464-8603
Japan

Takeshi Mukoyama
Kansai Gaidai University
16-1 Nakamiya-Higashinocho
Hirakata, Osaka, 573-1001
Japan

Yoshinori Murata
Department of Materials Science
and Engineering
Graduate School of Engineering
Nagoya University
Furo-cho, Chikusa-ku
Nagoya 464-8603
Japan

Kazuyoshi Ogasawara
Department of Chemistry
Kwansei Gakuin University
2-1 Gakuen, Sanda
Hyogo 669-1337, Japan

Naoki Ohashi
National Institute
for Materials Science (NIMS)
1-1 Namiki, Tsukuba
Ibaraki 305-0044
Japan

Rika Sekine
Department of Chemistry
Faculty of Science
Shizuoka University
Suruga-ku, Shizuoka 422-8529
Japan

Junzo Tanaka
National Institute
for Materials Science (NIMS)
1-1 Namiki, Tsukuba
Ibaraki 305-0044
Japan

Masahiro Yamashita
Graduate School of Science
Tohoku University
Aoba, Aramaki, Aoba-ku
Sendai, 980-8578, Japan

Hiroshi Yukawa
Department of Materials Science
and Engineering
Graduate School of Engineering
Nagoya University
Furo-cho, Chikusa-ku
Nagoya 464-8603
Japan

Part I

Fundamental

1

DV-Xα Method and Molecular Structure

Hirohiko Adachi

1.1 Molecular Orbital Theory

In a molecule, the potential acting on an electron can be expressed as a summation of the potentials of the constituent atoms. In such a model, the electron moves about in the attracting field of their nuclei; in other words, the valence electron can move all over the molecule, because the potential barrier is depressed midway between the atoms where the atomic potentials overlap (see Fig. 1.1).

Molecular orbital theory addresses such an electron system. The molecular orbital corresponds to the wave function of the electron moving about in the molecule. In order to obtain the molecular orbital ϕ_l, we need to solve the Schrödinger equation for the molecule, written as

$$h\phi_l(r) = \varepsilon_l \phi_l(r) , \tag{1.1}$$

where h is the one-electron Hamiltonian for the molecule, given by

$$h(\boldsymbol{r}) = -\frac{1}{2}\nabla^2 + V(\boldsymbol{r}) . \tag{1.2}$$

Because it is impossible to solve the partial differential equation of (1.1) rigorously, we need to make an approximation of the solution to the equation. In most molecular orbital (MO) methods, an MO wave function ϕ_l is expressed by a linear combination of atomic orbitals (LCAO) written as

$$\phi_l = \sum_l C_{il}\chi_i , \tag{1.3}$$

where C_{il} is the coefficient and χ_i the atomic orbital as a basis function. In order to find a solution to the Schrödinger equation, the Rayleigh–Ritz variational method is usually employed to obtain a secular equation. The

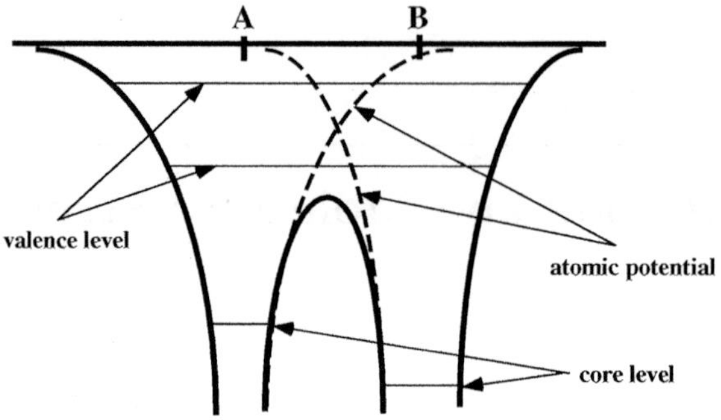

Fig. 1.1. Schematic diagram of the molecular potential resulting from superimposing atomic potentials

molecular orbital can be obtained by solving the secular equation written in a matrix formula as

$$(\underset{\sim}{\boldsymbol{H}} - \underset{\sim}{\varepsilon} \cdot \underset{\sim}{\boldsymbol{S}}) \cdot \underset{\sim}{\boldsymbol{C}} = 0 , \tag{1.4}$$

where $\underset{\sim}{\boldsymbol{H}}$ and $\underset{\sim}{\boldsymbol{S}}$ are the matrices whose elements are the so-called resonance and overlap integrals, respectively. Solving the equation yields the eigenvalue $\underset{\sim}{\varepsilon}$ and eigenvector $\underset{\sim}{\boldsymbol{C}}$, which correspond to the one-electron energy and the coefficient of the LCAO wave function for the molecular orbital, respectively. The matrix elements H_{ij} and S_{ij} in the secular equation are given by integrals of atomic orbitals as

$$S_{ij} = \int \chi_i \chi_j \, \mathrm{d}v , \tag{1.5}$$

$$H_{ij} = \int \chi_i H \chi_j \, \mathrm{d}v , \tag{1.6}$$

and are referred to as the "resonance integral" and "overlap integral," respectively, as mentioned above. For the case of $i = j$, we sometimes call the integral H_{ii} the "Coulomb integral." In the MO calculation, we should first evaluate the matrix elements H_{ij} and S_{ij} of (1.5) and (1.6) to construct the secular equation and then solve it by diagonalizing the matrix to obtain $\underset{\sim}{\varepsilon}$ and $\underset{\sim}{\boldsymbol{C}}$ as the eigenvalue and eigenvector. Thus, we obtain the eigenvalue ε_l, which corresponds to the one-electron MO energy and the eigenvector to the coefficient C_{il} for the atomic orbital i in the LCAO wave function of the lth molecular orbital.

Next, we try to understand the spatial distribution of electronic charge. We can analyze the result of the MO calculation with the aid of the Mulliken population analysis [1]. In this method, the total number of electrons in the system can be written as

$$n = \sum_l \sum_i \sum_j Q_{ij}^l , \quad Q_{ij}^l = f_l C_{il} C_{jl} S_{ij} , \tag{1.7}$$

where f_l is the occupation number of the lth MO level. Q_{ij}^l and Q_i^l are related by

$$Q_i^l = \sum_j Q_{ij}^l \,, \tag{1.8}$$

are the contributions of the lth MO state to the orbital population Q_i and the overlap population Q_{ij}, defined as

$$Q_i = \sum_l \sum_j Q_{ij}^l \,, \tag{1.9}$$

and

$$Q_{ij} = \sum_l Q_{ij}^l \,. \tag{1.10}$$

They are regarded as the effective number of electrons in the ith atomic orbital, and the strength of the covalent interaction between atomic orbitals i and j, respectively. When the value of Q_{ij} is positive, the interaction between atomic orbitals i and j is bonding, while an antibonding interaction occurs in the case of a negative overlap population. The sum of Q_i over all i belonging to atom A,

$$Q_{\mathrm{A}} = \sum_{i \in \mathrm{A}} Q_i \,, \tag{1.11}$$

is the effective charge of the atom. Thus, the net charge of atom A, N_{A}, is given by the difference between the atomic number Z_{A} and Q_{A}; thus,

$$N_{\mathrm{A}} = Z_{\mathrm{A}} - Q_{\mathrm{A}} \,. \tag{1.12}$$

The bond overlap population (BOP), Q_{AB}, which is defined as the sum of Q_{ij} over all i belonging to atom A and all j belonging to atom B – namely,

$$Q_{\mathrm{AB}} = \sum_{i \in \mathrm{A}} \sum_{i \in \mathrm{B}} Q_{ij} \,, \tag{1.13}$$

is considered as a measure of strength of covalent bonding between atoms A and B.

1.2 Discrete Variational (DV) Xα Molecular Orbital Method

In the conventional first-principle molecular orbital method, an analytical function such as the Gaussian-type orbital (GTO) or the Slater-type orbital (STO) is utilized as the basis functions of the LCAO of (1.3), so that the analytical integrations for $\boldsymbol{H}$ and $\boldsymbol{S}$ are feasible. In such a case, a large number of basis functions is necessary to represent an atomic orbital. This means that a large-scale computation for the preparation of the matrix elements is inevitable with this method.

In the discrete variational Xα (DV-Xα) method [2–4], we adopt an LCAO MO function given by (1.3) similar to the usual MO method, but we use the real atomic orbital for the basis function χ_i instead of the analytical function, so that one can prepare basis functions, each of which corresponds to one atomic orbital. This drastically reduces the required computational memory and time. The atomic orbital used as the basis function can be obtained by numerically solving the Schrödinger equation for an atom. The atomic orbital wave function is given by

$$\chi = R_{\mathrm{n}l}(r) \cdot \boldsymbol{y}_{l\mathrm{m}}(\theta, \phi) \ , \tag{1.14}$$

where $\boldsymbol{y}_{l\mathrm{m}}$ are the well-known spherical harmonics, and $R_{\mathrm{n}l}$ is the radial wave function, which we can calculate numerically with the following equation for a given atom:

$$\left\{ -\frac{1}{2}\frac{\mathrm{d}^2}{\mathrm{d}r^2} - \frac{1}{r}\frac{\mathrm{d}}{\mathrm{d}r} + V(r) \right\} R_{\mathrm{n}l}(r) = \varepsilon_{\mathrm{n}l} R_{\mathrm{n}l}(r) \ . \tag{1.15}$$

The numerical computation of the above equation is practicable by the use of a computer code similar to that introduced by Herman and Skillman [5]. Figure 1.2 shows the radial wave functions $R_{\mathrm{n}l}$ for C and O atoms, which can be used for MO calculations of a molecule that contains these atoms.

The atomic orbital χ is the product of $R_{\mathrm{n}l}$ and the spherical harmonics $\boldsymbol{y}_{l\mathrm{m}}$, as shown by (1.14). We use the numerical atomic orbital thus obtained for the

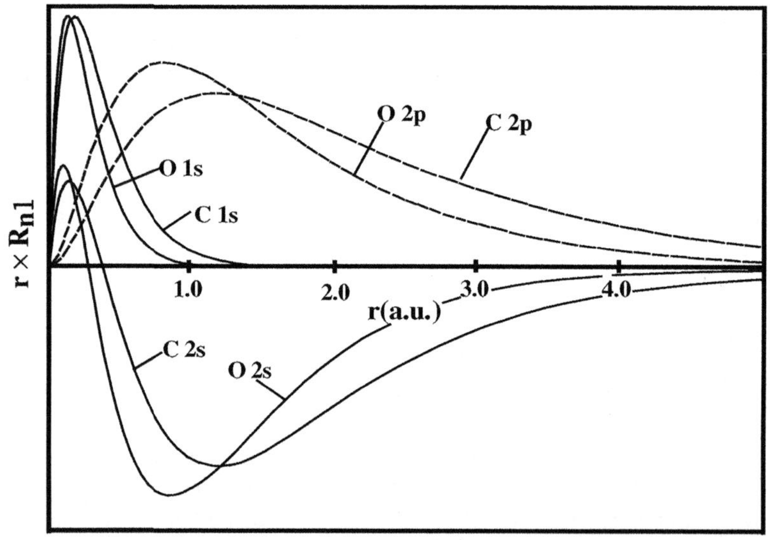

Fig. 1.2. Radial wave function of atomic orbitals for C and O atoms

basis function in the DV-Xα MO method; naturally, analytical integrations for the resonance and overlap matrix elements, H_{ij} and S_{ij}, are impossible to carry out. Instead, a numerical integration [4, 6] should be adopted for the calculation of H_{ij} and S_{ij}. In this method, the integration is replaced by a summation; namely,

$$\int \mathrm{d}\boldsymbol{r} \cdot g(\boldsymbol{r}) \Rightarrow \sum_{k=1}^{N} \omega(\boldsymbol{r}_k) \cdot g(\boldsymbol{r}_k) \,. \tag{1.16}$$

Here $g(\boldsymbol{r})$ is the integrand, and $\omega(\boldsymbol{r}_k)$ is the weight of the sampling point $\boldsymbol{r}_k$. For the integration, we first take a total of N sampling points in three-dimensional real space to calculate the integrand values at each sampling point, and then sum up the integrand values multiplied by the weight of the point over all sampling points. The matrix elements are thus given by

$$H_{ij} = \sum_{k=1}^{N} \omega(\boldsymbol{r}_k)\chi_i(\boldsymbol{r}_k)h(\boldsymbol{r}_k)\chi_j(\boldsymbol{r}_k) \,, \tag{1.17}$$

$$S_{ij} = \sum_{k=1}^{N} \omega(\boldsymbol{r}_k)\chi_i(\boldsymbol{r}_k)\chi_j(\boldsymbol{r}_k) \,, \tag{1.18}$$

where $h(\boldsymbol{r}_k)$ is the one-electron Hamiltonian given by

$$h(\boldsymbol{r}_k) = -\frac{1}{2}\nabla^2 + V(\boldsymbol{r}_k) \,. \tag{1.19}$$

$V(\boldsymbol{r}_k)$ is the effective molecular potential and is written by

$$V(\boldsymbol{r}_1) = -\sum_{\nu} \frac{Z_\nu}{|\boldsymbol{r}_1 - \boldsymbol{R}_\nu|} + \int \frac{\rho(\boldsymbol{r}_2)}{r_{12}} \mathrm{d}\boldsymbol{r}_2 - 3\alpha \left\{ \frac{3}{8\pi}\rho(\boldsymbol{r}_1) \right\}^{1/3} \,, \tag{1.20}$$

where Z_ν and $\boldsymbol{R}_\nu$ are the atomic number and position of atom ν, respectively. The molecular charge density is given by

$$\rho(\boldsymbol{r}) = \sum_{l} \rho_l(\boldsymbol{r}) = \sum_{l} f_l |\phi_l(\boldsymbol{r})|^2 \,. \tag{1.21}$$

Thus, the first and second terms of (1.20) are the attractive potential from the nuclei and the repulsive potential of the electron cloud of the whole system, respectively. The third term is the exchange-correlation potential proposed by Slater [7, 8], and α is the scaling parameter, the only parameter used in this method.

In performing the DV numerical integration described above, we have to generate the N sampling points. We first take three random numbers, a_k, b_k, and c_k, which are distributed between 0 and 1, to set the kth sampling point $\boldsymbol{r}_k$. We can choose these numbers, for example, to be the decimal parts of $a_k = k \times 2^{1/2}$, $b_k = k \times 3^{1/2}$, and $c_k = k \times 5^{1/2}$, for $k = 1$ to N. In order to determine the point $\boldsymbol{r}_k(r_k, \theta_k, \phi_k)$ in polar coordinates using a_k, b_k, and c_k, we set

$$\left.\begin{aligned} \cos\theta_k &= 2b_k - 1 , \\ \phi_k &= 2\pi c_k , \\ r_k &= \ln\frac{\exp(a_k/A)}{1 + \{1 - \exp(a_k/A)\}\exp(-R_0)} , \end{aligned}\right\} \tag{1.22}$$

where A is an adjustable parameter and R_0 is chosen to be a value near the atomic radius. Both are parameters of the Fermi-like distribution function for the sampling point (see Fig. 1.3).

Then, the density of the sampling point is given by

$$d(r) = \frac{A}{4\pi r^2\{1 + \exp(r - R_0)\}} . \tag{1.23}$$

The sampling points are distributed around a given atom according to the above equations. For a molecular calculation, we take these points around all constituent atoms. The density of the sampling point $\boldsymbol{r}$ can be given by the sum of contributions from all atoms in the molecule; thus,

$$D(\boldsymbol{r}) = \sum_{\nu} t_\nu \cdot d_\nu(\boldsymbol{r}_\nu) , \quad \sum_{\nu} t_\nu = 1 . \tag{1.24}$$

Consequently, the weight of the sampling point is

$$\omega(\boldsymbol{r}) = \frac{1}{N \times D(\boldsymbol{r})} . \tag{1.25}$$

The distribution of the sampling points for a diatomic molecule is displayed in Fig. 1.4. The sampling points are relatively condensed near the nucleus where the variations of the wave functions and the potential are rapid, while the points are dilute far from the nucleus since these functions change smoothly.

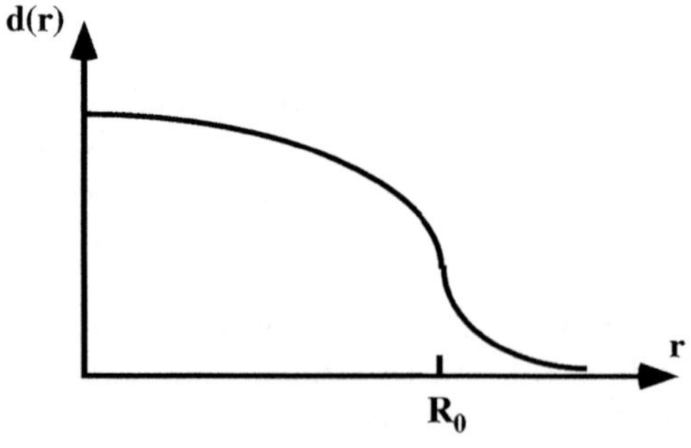

Fig. 1.3. Density of the sampling point of a Fermi-like function

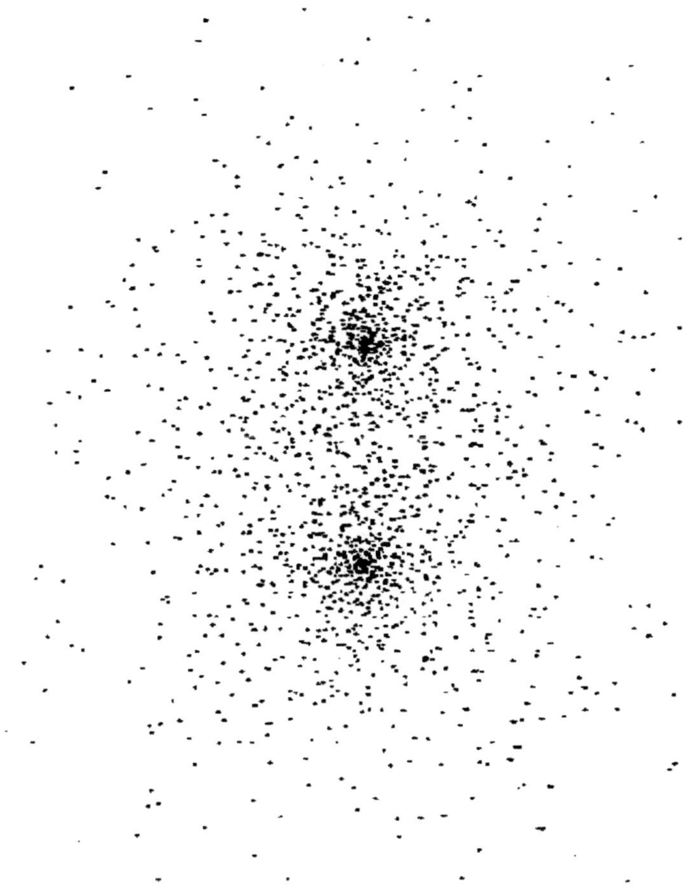

Fig. 1.4. Projection of 2000 sampling points around the C and O atoms in a CO molecule

Next, we should evaluate the integrand at a given sampling point constructed by the atomic orbital functions and potential function, as given in (1.17) and (1.18). The radial part of the atomic orbital given by (1.14) is previously calculated using (1.15). The molecular potential is also expressed by (1.20); thus, we are able to evaluate the integrand value at the point.

Figure 1.5 illustrates how to estimate the values of atomic orbital functions at a sampling point $\boldsymbol{r}$ locating around the nuclei. The integrand is multiplied by the weight $\omega(\boldsymbol{r})$ for each sampling point and then summed up over all the sampling points to obtain the integral values of H_{ij} and S_{ij}. The secular equation thus composed can be solved by an ordinary matrix diagonalization to obtain the eigenvalues and eigenvectors.

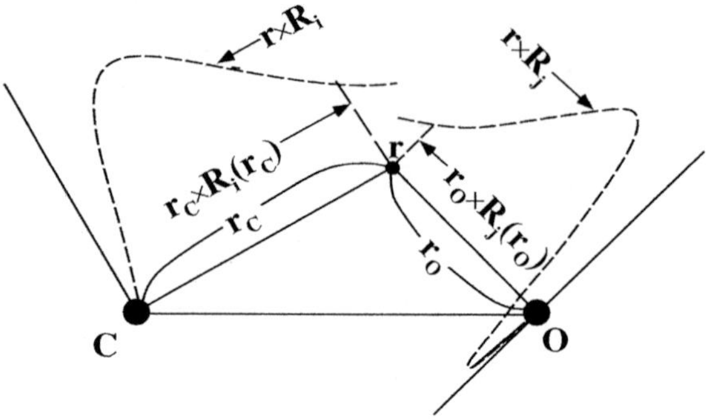

Fig. 1.5. Evaluation of the radial wave functions of atomic orbitals for C and O atoms at a sampling point $\boldsymbol{r}$

1.3 Molecular Orbital Calculation of H_2

Next we are going to perform an actual calculation of the molecular orbital. The simplest case of a molecular problem is a hydrogen molecule. Consider two hydrogen atoms A and B approaching each other to form a hydrogen molecule. In this case, we have an LCAO molecular orbital given by

$$\phi = C_{\mathrm{A}}\chi_{\mathrm{A}} + C_{\mathrm{B}}\chi_{\mathrm{B}} \,, \tag{1.26}$$

where χ_{A} and χ_{B} are $H \times 1s$ orbitals of atoms A and B. The secular equation (1.4) can thus be expressed as

$$\begin{matrix} (H_{\mathrm{AA}} - \varepsilon)C_{\mathrm{A}} + (H_{\mathrm{AB}} - \varepsilon S_{\mathrm{AB}})C_{\mathrm{B}} = 0 \\ (H_{\mathrm{BA}} - \varepsilon S_{\mathrm{BA}})C_{\mathrm{A}} + (H_{\mathrm{BB}} - \varepsilon)C_{\mathrm{B}} = 0 \end{matrix} \,, \tag{1.27}$$

where we can set

$$\varepsilon_0 = H_{\mathrm{AA}} = H_{\mathrm{BB}} \,, \quad S_{\mathrm{BA}} = S_{\mathrm{AB}} \,, \quad H_{\mathrm{BA}} = H_{\mathrm{AB}} \,, \tag{1.28}$$

since the atoms A and B are equivalent. Then, the eigenvalue ε can be obtained by solving the determinantal equation

$$\begin{vmatrix} \varepsilon_0 - \varepsilon & H_{\mathrm{AB}} - \varepsilon S_{\mathrm{AB}} \\ H_{\mathrm{AB}} - \varepsilon S_{\mathrm{AB}} & \varepsilon_0 - \varepsilon \end{vmatrix} = (\varepsilon_0 - \varepsilon)^2 - (H_{\mathrm{AB}} - \varepsilon S_{\mathrm{AB}})^2 = 0 \,. \tag{1.29}$$

Then, we have two values

$$\varepsilon_{\pm} = \frac{\varepsilon_0 \pm H_{\mathrm{AB}}}{1 \pm S_{\mathrm{AB}}} \tag{1.30}$$

for eigenvalues; namely, the orbital energies for the H_2 molecule. Because the molecular wave function ϕ_l is normalized, we have

$$\int |\phi_l|^2 \,\mathrm{d}v = C_{\mathrm{A}}^2 + C_{\mathrm{B}}^2 + 2C_{\mathrm{A}}\, C_{\mathrm{B}} S_{\mathrm{AB}} = 1 \,. \tag{1.31}$$

From (1.27) and (1.31), we obtain

$$C_{\mathrm{A}} = C_{\mathrm{B}} = \sqrt{\frac{1}{2(1+S_{\mathrm{AB}})}} \tag{1.32}$$

for $\varepsilon = \varepsilon_+$, and we also have

$$C_{\mathrm{A}} = -C_{\mathrm{B}} = \sqrt{\frac{1}{2(1-S_{\mathrm{AB}})}} \tag{1.33}$$

for $\varepsilon = \varepsilon_-$. As a consequence, the two MO functions

$$\begin{aligned} \phi_+ &= \sqrt{\frac{1}{2(1+S_{\mathrm{AB}})}}(\chi_{\mathrm{A}} + \chi_{\mathrm{B}}) \\ \phi_- &= \sqrt{\frac{1}{2(1-S_{\mathrm{AB}})}}(\chi_{\mathrm{A}} - \chi_{\mathrm{B}}) \end{aligned} \tag{1.34}$$

are given for these two energy states. These states are referred to as the "bonding state" (+) and the "antibonding state" (−), respectively.

The values of S_{AB} and H_{AB}, as well as ε_0, are the functions of the interatomic distance R between A and B, and are calculated as shown in Fig. 1.6. The overlap integral S_{AB} takes a value between 0 and 1, as expected. The

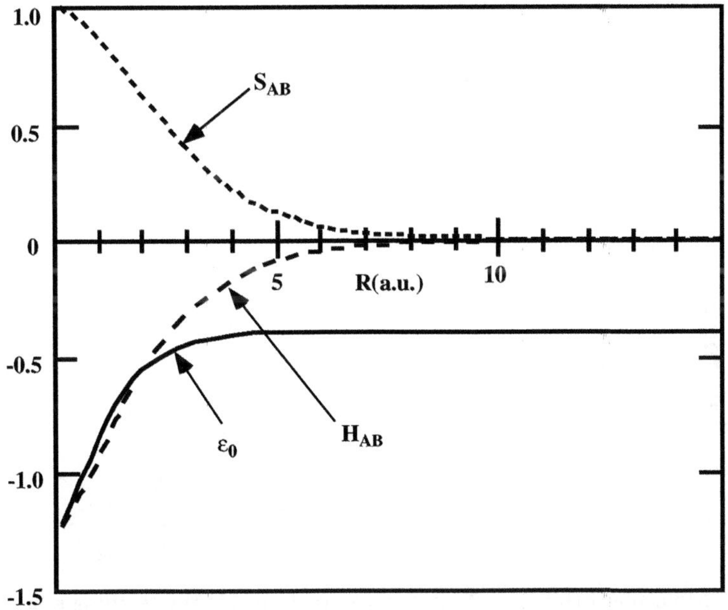

Fig. 1.6. Variation of S_{AB}, H_{AB} and ε_0 with R. H_{AB} and ε_0 are given in Ry

Coulomb integral ε_0 is the atomic energy of hydrogen perturbed by the potential of another atom, and the resonance integral H_{AB} is approximately the product of ε_0 and S_{AB}.

At the equilibrium distance of the H_2 molecule of 0.74 Å (1.4 a.u.), we obtain the values of 0.783, −0.765, and −0.703 for S_{AB}, ε_0, and H_{AB} (in units of Ry, where 1 Ry = 13.6 eV), respectively, from the DV-Xα calculation. The MO energies and wave functions are calculated using these values. The level of the bonding state is found to be depressed compared with ε_0, while that of the antibonding state is elevated, as shown in Fig. 1.7.

Figures 1.8a,b are the plots of the wave functions and the contour maps for these MO states. For a deeper understanding of these states, we look into the charge densities and the difference charge densities, which are elucidated in Fig. 1.9. The molecular charge density can be obtained by the square of the MO function and summing up over the occupied states; it is expressed by

$$\rho(\boldsymbol{r}) = \sum_l \rho_l(\boldsymbol{r}) , \quad \rho_l(\boldsymbol{r}) = |\chi_l(\boldsymbol{r})|^2 . \tag{1.35}$$

The difference density is defined to be the difference between the charge density of the molecule and the sum of atomic charge densities, and it indicates the variation of charge density during the formation of the molecule. In the figures, we note an increase in the charge density for the bonding state and a decrease for the antibonding state midway between the two nuclei of A and B. In the bonding state, the increase in the density of negative charge located around the midpoint of the A–B bond attracts these two positively charged nuclei,

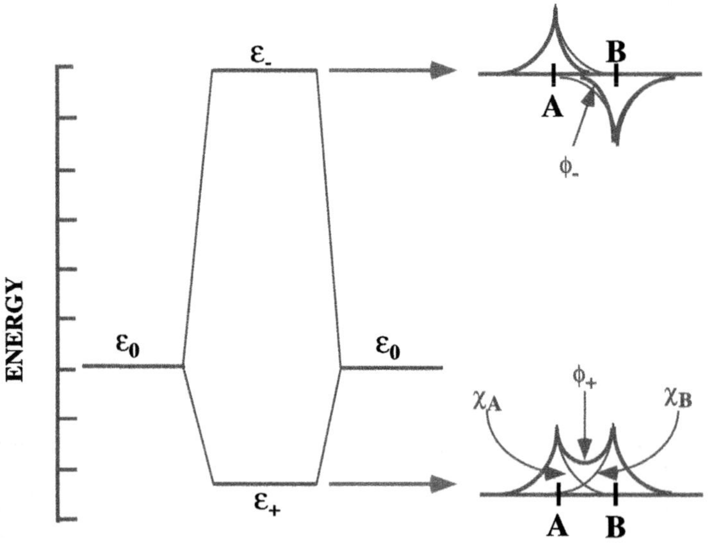

Fig. 1.7. Molecular orbital levels and wave functions for the bonding state (ε_+ and ϕ_+) and antibonding state (ε_- and ϕ_-) of an H_2 molecule

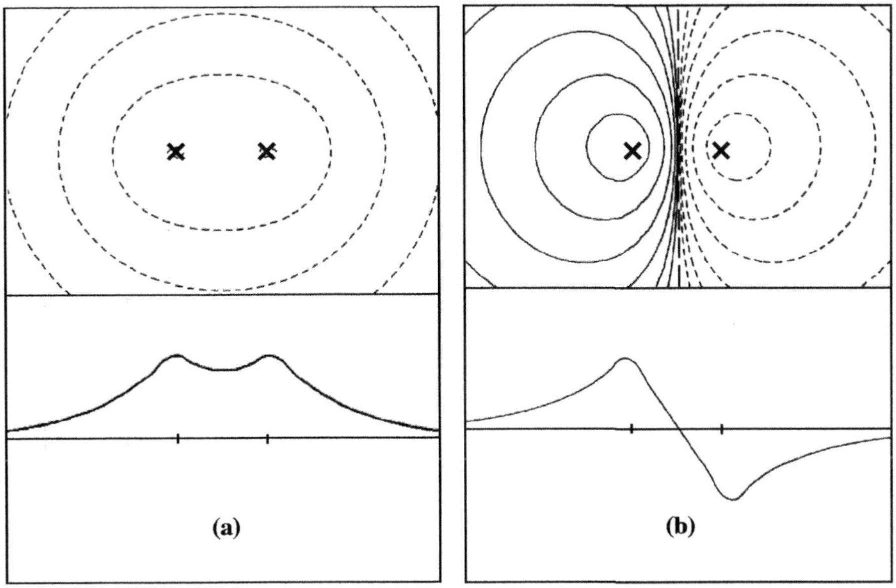

Fig. 1.8. Plots of MO wave functions and contours for (**a**) the bonding and (**b**) the antibonding state

and causes these two atoms to unite to form a molecule. This force can be regarded as the essence of covalent bonding. By contrast, the increases of electronic charge in the antibonding state are located outside the A–B bond, and serve to pull apart these two atoms.

1.4 Covalency and Ionicity

In addition to the covalent bonding described above, we should consider another important concept – the ionicity – which is deduced from molecular orbital theory. Consider the case of a diatomic molecule similar to the H_2 molecule, but where the two atoms are of different elements. Then, unlike the case of the H_2 molecule, the Coulomb integral H_{AA} in (1.27) is not equal to H_{BB}. In this case, we have two eigenvalues given by

$$\varepsilon_{\pm} = \frac{(\varepsilon_A + \varepsilon_B - 2S_{AB}H_{AB}) \pm \sqrt{(\varepsilon_A + \varepsilon_B - 2S_{AB}H_{AB})^2 - 4(1 - S_{AB}^2)(\varepsilon_A\varepsilon_B - H_{AB}^2)}}{2(1 - S_{AB}^2)} ,$$

where we set $\varepsilon_A = H_{AA}$ and $\varepsilon_B = H_{BB}$. In evaluating the eigenvalue, we need four parameters – ε_A, ε_B, H_{AB} and S_{AB} – in this case. However, having to vary all of these parameters complicates the calculation of the eigenvalues. Thus, we fix three parameters for simplicity; we set $\varepsilon_B = -0.5$, $H_{AB} = -0.5$ and

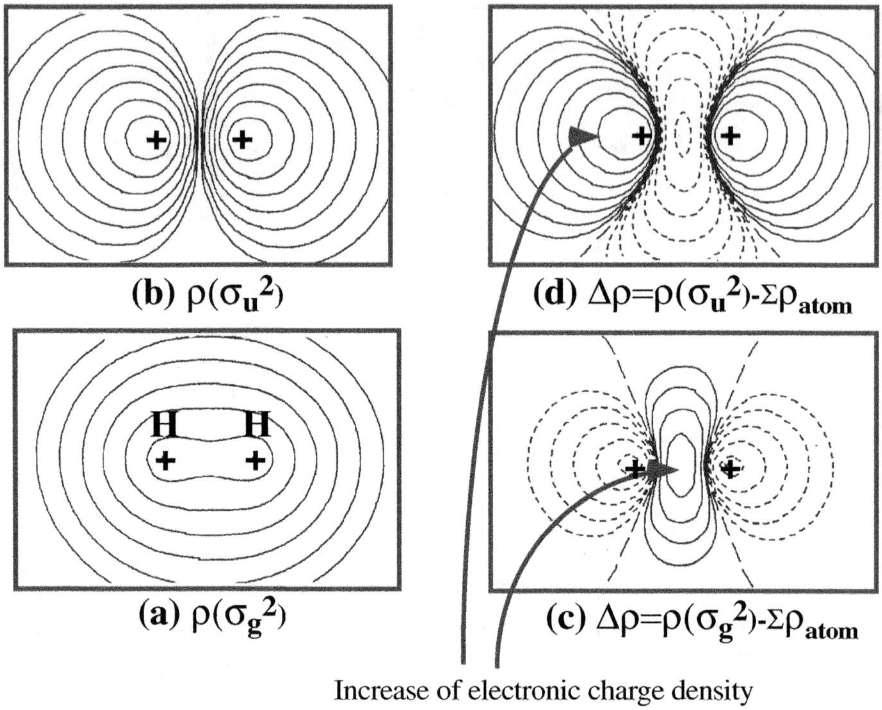

Fig. 1.9. Electronic charge density (ρ) and difference charge density ($\Delta\rho$) for bonding (σ_g) and antibonding (σ_u) electronic states

$S_{AB} = 0.3$, and then vary only ε_A. The MO wave functions can be calculated by (1.27) and (1.31). Then, we implement Mulliken population analysis given by (1.7)–(1.13) to clarify the charge density quantitatively. When we take $\varepsilon_A = 0, -0.5, -1.5$, and -2.0, the results shown in Table 1.1 are obtained.

Figure 1.10 demonstrates the changes in MO wave functions, eigenvalues, orbital populations, and overlap populations. It can be noted from the result that the chemical bonding between different elements consists of both covalency and ionicity.

In case (a), $Q_A = 0.486$ and $Q_B = 1.514$ are obtained. This means that approximately half an electron is transferred from atom A to atom B. It is also

Table 1.1. Variations of MO wave functions, eigenvalues, and charge densities

	ε_A	ε_+	C_A	C_B	ε_-	C_A	C_B	Q_A	Q_B	Q_{AB}
(a)	0	−0.645	0.363	0.814	0.425	0.975	−0.660	0.487	1.514	0.377
(b)	−0.5	−0.769	0.620	0.620	0.000	0.845	−0.845	1.000	1.000	0.462
(c)	−1.0	−1.059	0.876	0.286	−0.259	0.575	−1.009	1.687	0.313	0.300
(d)	−2.0	−2.007	0.978	0.067	−0.415	0.247	−1.045	1.952	0.048	0.078

seen that the electronic charge transfers from the atom with a higher atomic level to that with a lower level, by comparing cases (a) and (c). This effect corresponds to the charge transfer from an electropositive to an electronegative atom. The larger the level separation of two atoms, the more amount of charge transfer occurs; namely, the bonding becomes more ionic, as seen in case (d). From the change in the overlap population Q_{AB}, the covalent bonding is stronger when the level separation is smaller with maximum at $\varepsilon_{\mathrm{A}} = \varepsilon_{\mathrm{B}}$.

1.5 DV-Xα Molecular Orbital Calculation for CO Molecule

Next, we try to apply the DV-Xα method to a slightly more complicated system like carbon monoxide, i.e., CO molecule. We assume that the C and O atoms lie on the z-axis, and the internuclear distance is 1.128 Å. In this case, we take into account the $1s$, $2s$, and $2p$ orbitals for both

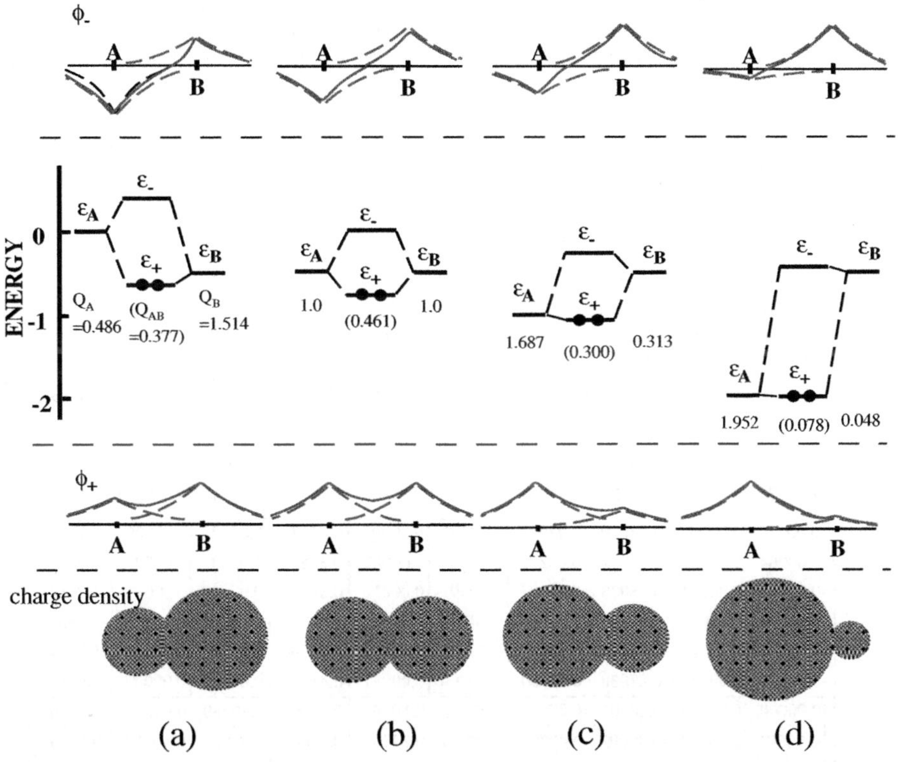

Fig. 1.10. Changes in MOs, levels, charge densities and bonding nature as ε_{A} is varied

C and O atoms in the basis functions. Then the molecular orbital ϕ_l is given by

$$\phi_l = C_{1l}\chi_{C1s} + C_{2l}\chi_{C2s} + C_{3l}\chi_{C2p_z} + C_{4l}\chi_{O1s} + C_{5l}\chi_{O2s} + C_{6l}\chi_{O2p_z} + C_{7l}\chi_{C2p_x} + C_{8l}\chi_{O2p_x} + C_{9l}\chi_{C2p_y} + C_{10l}\chi_{O2p_y} \,. \quad (1.36)$$

The molecular orbital calculation is carried out according to the procedure described in Sect. 1.2. First, the atomic orbital calculations for C and O are

Table 1.2. Matrix elements of S_{ij}, H_{ij}, and C_{il} in the secular equation for the CO molecule

$\tilde{S}$ =

1.00000									
0.00000	1.00000								
0.00000	0.00001	1.00000							
0.00022	-0.05249	-0.08946	1.00000						
-0.05629	0.44001	0.51559	0.00000	1.00000					
0.10867	-0.40043	-0.17823	0.00000	0.00001	0.99999				
0.00000	0.00002	-0.00002	0.00000	-0.00001	0.00000	1.00000			
0.00000	-0.00001	-0.00002	-0.00002	-0.00005	0.00004	0.34846	1.00000		
-0.00002	-0.00003	0.00004	0.00000	0.00000	0.00000	-0.00001	0.00000	1.00001	
0.00000	-0.00001	0.00000	0.00000	-0.00003	0.00001	0.00000	-0.00004	0.34847	1.00000

$\tilde{H}$ =

-19.97042									
0.01273	-1.34577								
0.03295	-0.59983	-1.23273							
-0.00879	1.98593	3.38317	-37.75324						
1.13662	-1.11027	-1.26833	0.00640	-2.06514					
-2.18855	0.82106	0.59308	-0.03064	0.46516	-1.36072				
0.00001	0.00000	0.00003	0.00000	0.00002	0.00001	-0.46074			
0.00000	0.00003	0.00005	0.00031	0.00008	-0.00005	-0.50237	-0.82941		
0.00020	0.00002	-0.00001	0.00000	-0.00001	0.00003	0.00000	0.00001	-0.46074	
0.00002	0.00002	0.00002	0.00008	0.00003	-0.00001	0.00001	0.00004	-0.50237	-0.82941

l=	1	2	3	4	5	6	7	8	9	10
ε_l=	-37.75336	-19.97046	-2.20137	-1.08878	-0.90714	-0.90713	-0.67008	-0.16282	-0.16282	0.96624
$\tilde{C}$ =	0.00015	1.00000	-0.07310	-0.04612	0.00001	-0.00006	0.04677	0.00001	0.00000	0.15975
	-0.00064	-0.00090	-0.17371	-0.35206	0.00001	0.00002	-0.83823	0.00001	0.00002	-1.06367
	-0.00038	-0.00204	-0.18036	-0.10197	-0.00001	0.00012	0.56824	-0.00003	0.00001	-1.21191
	-1.00007	-0.00047	-0.02566	-0.02815	0.00001	-0.00003	0.00716	0.00000	0.00000	-0.16370
	0.00067	0.00071	-0.74768	0.59682	0.00001	-0.00023	0.01200	0.00002	0.00002	1.17369
	-0.00118	0.00030	0.27388	0.72700	-0.00006	0.00051	-0.41667	-0.00001	0.00002	-0.87143
	0.00000	0.00000	-0.00001	0.00002	0.32881	0.10365	0.00002	-0.80259	0.61254	-0.00001
	-0.00001	0.00000	-0.00002	0.00004	0.78799	0.24839	0.00002	0.53654	-0.40952	-0.00002
	0.00000	0.00001	0.00000	-0.00002	-0.10363	0.32881	-0.00001	0.61253	0.80259	0.00000
	0.00000	-0.00004	-0.00001	-0.00002	-0.24847	0.78800	0.00002	-0.40948	-0.53657	0.00001

performed using (1.15) with neutral charge densities as the initial step, to obtain the numerical atomic orbitals previously shown in Fig. 1.2. Then, the sampling points are randomly generated according to (1.22)–(1.25). In order to evaluate the integrals S_{ij} and H_{ij}, we use numerical the integration scheme given by (1.17) and (1.18). The integrand values in the equations are evaluated in the manner shown in Fig. 1.5. Thus, the matrix elements S_{ij} and H_{ij} are obtained. The results are shown in Table 1.2. In this case, 1 000 000 sampling points are used for DV integrations.

The matrices $\underset{\sim}{\boldsymbol{H}}$ and $\underset{\sim}{\boldsymbol{S}}$ are symmetric; that is, $S_{ij} = S_{ji}$ and $H_{ij} = H_{ji}$. Therefore, only the lower triangular parts are given in the table. The values of the matrix elements in the shaded parts should be zero if the integration is rigorously accurate. Then, the 10×10 matrices can be divided into one 6×6 matrix and two 2×2 matrices, which correspond to the irreducible σ and doubly degenerate π representations.

The next step in the molecular orbital calculation is to solve the secular equation (1.4). The eigenvalue $\underset{\sim}{\varepsilon}$ and eigenvector $\underset{\sim}{\boldsymbol{C}}$, which give the molecular orbital energies and the coefficients of the LCAO wave function in (1.3) or (1.36), respectively, are obtained by solving the matrix equation (1.4). Thus, ten eigenvalues ε_l are equivalent to MO energies, six of which are of the σ-type and four are of the doubly degenerate π-type. The matrix elements C_{il}

Table 1.3. Orbital eigenvalues (ε_l) and occupation numbers (f_l) for a CO molecule

l		ε_l (Ry)	ε_l (Hr)	ε_l (eV)	f_l
1	1σ	−37.753	−18.877	−513.579	2.000
2	2σ	−19.970	−9.985	−271.669	2.000
3	3σ	−2.201	−1.100	−29.946	2.000
4	4σ	−1.089	−0.544	−14.811	2.000
5	1π	−0.907	−0.454	−12.340	4.000
7	5σ	−0.670	−0.335	−9.115	2.000
8	2π	−0.163	−0.081	−2.215	0.000
10	6σ	0.966	0.483	13.144	0.000

Table 1.4. Mulliken population analysis for the CO molecule

	Initial	Final
C-1s	2.00000	2.00034
C-2s	2.00000	1.59647
C-2p	2.00000	1.81375
O-1s	2.00000	2.00019
O-2s	2.00000	1.83141
O-2p	4.00000	4.75783

of the eigenvector, as well as ε_l, are also given in the table. The eigenvalues ε_l are aligned from left to right in the order of increasing energy. The first four levels are of σ-type and can be labeled 1σ–4σ. The fifth and sixth levels are doubly degenerate 1π. The seventh is 5σ, the eighth and ninth are 2π, and the tenth is 6σ. For the ground electronic state, 14 electrons occupy from the lowest level 1σ up to 5σ, according to the Pauli principle, as indicated in Table 1.3.

The coefficients of the MO wave function are given by matrix $\underset{\sim}{\boldsymbol{C}}$ in Table 1.2. For example, the seventh MO (5σ) energy ε_7 is -0.67008 (Ry) as shown in the table. The coefficients for the seventh MO are given in the seventh column of matrix $\underset{\sim}{\boldsymbol{C}}$. Then the MO can be written as

$$\begin{aligned}\phi_7 = {} & 0.0468\chi_{\mathrm{C}1s} - 0.8382\chi_{\mathrm{C}2s} + 0.5682\chi_{\mathrm{C}2p_z} \\ & + 0.0072\chi_{\mathrm{O}1s} + 0.0120\chi_{\mathrm{O}2s} - 0.4167\chi_{\mathrm{O}2p_z} \, .\end{aligned}$$

The MO wave function can be visually illustrated in the form of a contour map. Figure 1.11 represents the contour map of the 5σ MO wave function and its plot on the molecular axis.

The Mulliken population analysis is carried out along the (1.7)–(1.13), to get the orbital populations for C and O atoms. The result is listed in Table 1.4. The calculation is made assuming the charge density constructed by neutral atomic charge as the first self-consistent field (SCF) iteration. Then,

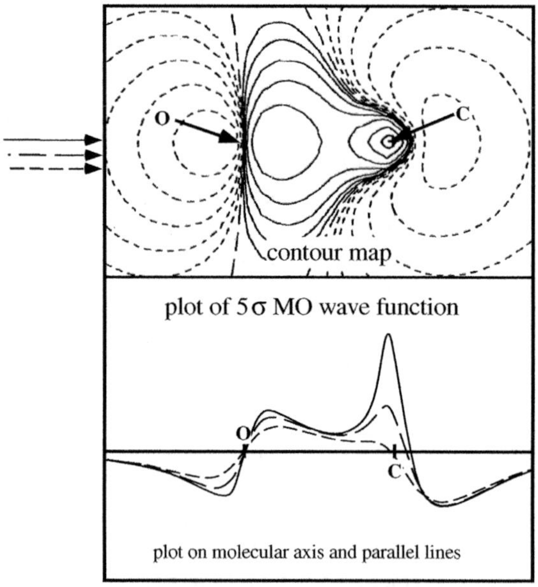

Fig. 1.11. Plots of the wave function and contour map of the 5σ MO

the initial orbital populations correspond to the electronic configurations for neutral atoms. The result of the MO calculation shows that both intra- and interatomic charge transfer take place, as seen in the table. In order to achieve an SCF, the final orbital populations are fed back to construct the atomic charge densities for the next SCF iteration. In practice, the initial charge density is admixed with the final charge density; that is,

$$\rho_{\text{next}} = (1 - x)\rho_{\text{old}} + x\rho_{\text{new}} ,$$

where ρ_{old} and ρ_{new} are initial and final charge densities of the MO calculation, respectively, and ρ_{next} is the charge density for the next iteration. The

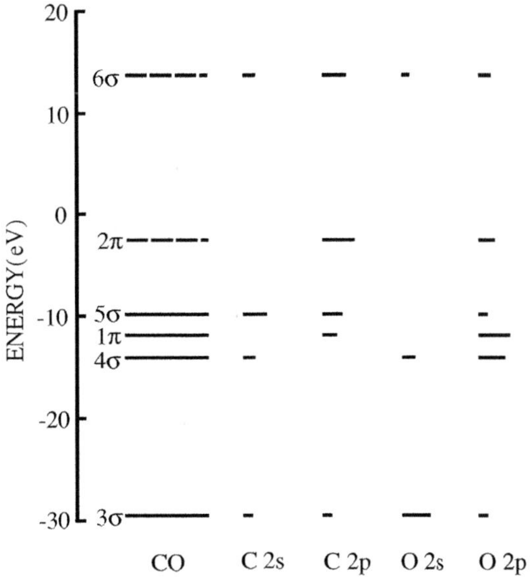

Fig. 1.12. Molecular orbital levels and atomic orbital components for CO

Table 1.5. Molecular orbital energies and atomic orbital components for CO molecule as obtained via an SCF calculation

	ε_l (eV)	f_l	C		O	
			$2s$	$2p$	$2s$	$2p$
1σ	−512.432	2.000				
2σ	−272.659	2.000				
3σ	−29.486	2.000	0.1167	0.1186	0.6541	0.1105
4σ	−14.025	2.000	0.2033	−0.0065	0.2521	0.5511
1π	−11.819	4.000	0.0	0.2830	0.0	0.7170
5σ	−9.812	2.000	0.4700	0.3921	0.0099	0.1279
2π	−2.405	0.000	0.0	0.7170	0.0	0.2830
6π	13.763	0.000	0.2100	0.4959	0.0840	0.2104

Table 1.6. Result of the Mulliken population analysis for CO molecule in an SCF

	C	O
$1s$	2.0003	2.0002
$2s$	1.5801	1.8321
$2p$	2.1403	4.4470
Q_A	4.4470	8.2793
Net charge	+0.2793	−0.2793

MO calculation is repeated with the charge density ρ_{next}. The iterations are repeated until an SCF, where the final charge density coincides with the initial charge density, is achieved. Figure 1.12 displays the valence MO levels obtained using the SCF calculation.

The atomic orbital components for each MO are indicated by the length of the horizontal line. In the figure, the occupied levels are denoted by solid lines and the empty levels are denoted by dotted lines. Thus, the highest occupied MO (HOMO) is 5σ, and the lowest unoccupied MO (LUMO) is 2π. The level energies and the orbital components are also summarized in Table 1.5, and the orbital populations and effective charges are listed in Table 1.6.

References

1. R.S. Mulliken, J. Chem. Phys., **23**, 1833, 841, 2339, 2343 (1995)
2. D.E. Ellis, G.S. Painter, Phys. Rev., **B2**, 2887 (1970)
3. H. Adachi, M. Tsukada, C. Satoko, J. Phys. Soc. Jpn., **45** 875 (1978)
4. H. Adachi, *Introduction to Quantum Materials Chemistry*, Sankyo Press, 1991
5. F. FHerman, S. Skillman, *Atomic Structure Calculations*, Printice–Hall, 1963
6. F.W. Averill, D.E. Ellis, J. Chem. Phys., **59**, 6412 (1973)
7. J.C. Slater, Phys. Rev., **81**, 385 (1951)
8. J.C. Slater, *Quantum Theory of Molecules and Solids*, vol. 4, McGraw–Hill, 1974

Part II

Materials Science

2

Alloy Design Based on the DV-Xα Cluster Method

Masahiko Morinaga, Yoshinori Murata, and Hiroshi Yukawa

Recently, there has been great progress made in a molecular orbital approach to alloy design. This approach is based on electronic structure calculations using the DV-Xα cluster method. It is applicable not only to structural alloys, but also to functional alloys. For example, new high-Cr ferritic steels for turbine rotors have recently been developed using this approach. The hydrogen storage alloy, which is an important functional alloy, has also been investigated in a fundamental manner. The hydrogen absorption and desorption characteristics of the alloy are well-understood according to this approach. Even in oxides, this method is applicable, and we briefly present one example of a perovskite-type oxide, $SrTiO_3$, which shows protonic conduction when the oxide is doped with acceptor ions.

2.1 Introduction

A theoretical method for alloy design has been developed on the basis of the molecular orbital calculation of electronic structures. Employing this method, we can consistently model a variety of materials (e.g., metals, semiconductors, and ceramics) by making use of the common ground of electron theory. The phase stability of nickel-based superalloys can be treated quantitatively using this method [1,2]. In addition, even mechanical properties are predictable for both aluminum alloys and magnesium alloys with multiple components [3,4]. Furthermore, new structure maps have been proposed to predict the crystal structures of intermetallic compounds [5].

Employing this method, various materials have been developed – for example, heat-resistant single-crystal Ni-based superalloys [6], high-strength titanium alloys [7], implant titanium alloys [8], and heat-resistant high-Cr ferritic steels for power plants [9]. It is applicable not only to these structural materials, but also to functional materials (e.g., hydrogen storage alloys). In this chapter, recent progress in this method will be reviewed, with a focus on Ni-based superalloys, high-Cr ferritic steels, and hydrogen storage alloys.

2.2 DV-Xα Molecular Orbital Method

The DV-Xα cluster method is a molecular orbital calculation method assuming the Hartree–Fock–Slater approximation [10,11]. The exchange-correlation between electrons, V_{XC}, is expressed using Slater's Xα potential [12],

$$V_{XC} = -3\alpha \left[\frac{3}{8\pi} \rho(r) \right]^{1/3} , \tag{2.1}$$

where $\rho(r)$ is the electron density at position r, the parameter α is fixed at 0.7, and the self-consistent charge approximation is used in the calculation. The matrix elements of the Hamiltonian and the overlap integrals are calculated by a random sampling method. The molecular orbitals are constructed by a linear combination of numerically generated atomic orbitals (LCAO).

Local electronic structures can be calculated using this method even for large molecules. Depending on the crystal structure, appropriate cluster models are employed in this calculation. For example, a cluster model for bcc Fe is shown in Fig. 2.1. Here, "cluster" means a hypothetical molecule representing the crystal structure of material.

As shown in Fig. 2.2, the density of states of electrons obtained from the cluster calculation resembles the result of the band calculation despite the use of a relatively small Fe_{15} cluster in the calculation. In either case, the Fermi energy level lies at the peak position of the 3d band. The similarity is simply interpreted as being due to the spatial localization of the d-orbitals around each atomic site in bcc Fe. Thus, the cluster approach is a good approximation to the electronic structure of transition metals like iron. In order to get alloying parameters, a variety of alloying elements, M, are placed at the center of the cluster shown in Fig. 2.1. Further explanation of the calculation method is given elsewhere [13].

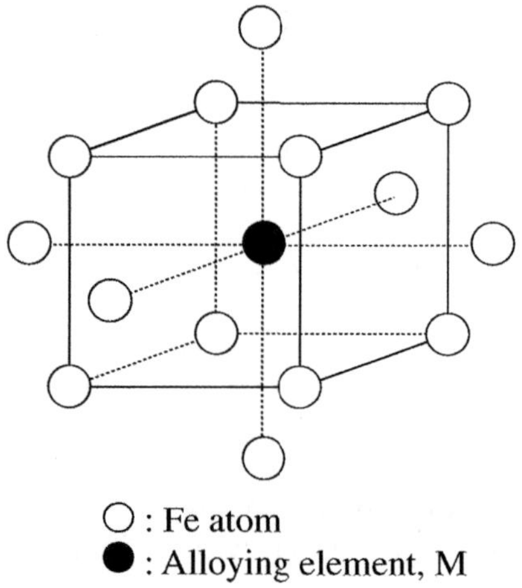

Fig. 2.1. Cluster model of bcc Fe employed in the calculation

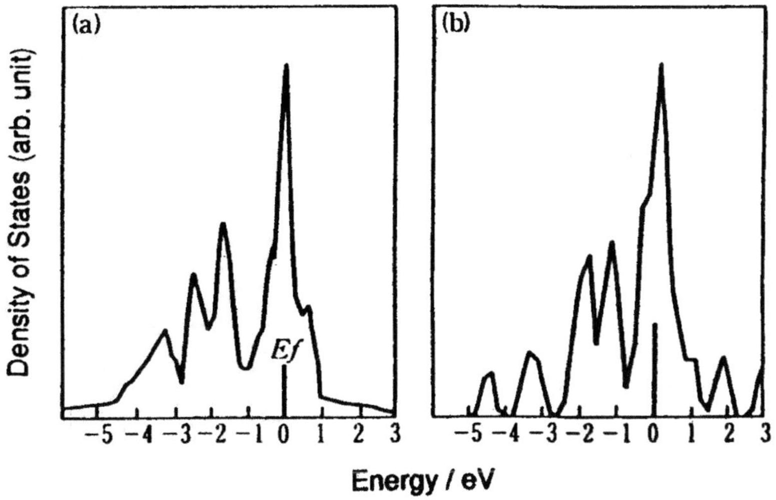

Fig. 2.2. Comparison of the electron density of states for bcc Fe between (**a**) the band calculation and (**b**) the cluster calculation

2.3 Alloying Parameters

2.3.1 *d*-Orbital Energy Level, Md

Two alloying parameters are obtained from the cluster calculation and used for alloy design [14]. One is the *d*-orbital energy level of alloying transition metal, M, in a base metal, X. For example, the results of the level structure of bcc Fe alloyed with various 3*d* alloying elements are shown in Fig. 2.3. In the case of a pure Fe cluster (i.e., M = Fe) the levels of $8e_g$ to $16t_{2g}$ originate mainly from the Fe 3*d* orbitals, and form the Fe 3*d* band where the Fermi energy level lies (indicated by the arrows in Fig. 2.3). In the case of an alloyed Fe cluster, new energy levels due mainly to the *d*-orbitals of the alloying transition metal appear above the Fermi energy level. For instance, the $13e_g$ and $16t_{2g}$ levels (drawn as broken lines) correspond to these new levels. These levels are called Md levels, and their energy height changes systematically with the order of elements in the periodic table.

The Md levels correlate with the electronegativity and the atomic radius of elements. In fact, it is well-known that the energy eigenvalue determined by the Xα method itself the negative of the Mulliken electronegativity. The Md levels increase as the electronegativity of the alloying element decreases, as shown in Fig. 2.4. In addition, as shown in Fig. 2.5, the Md levels increase with increasing atomic radius of the element. The average energy of these two *d*-orbital levels is hereafter referred to as Md. The Md values are listed in Table 2.1 for various alloying elements in bcc Fe.

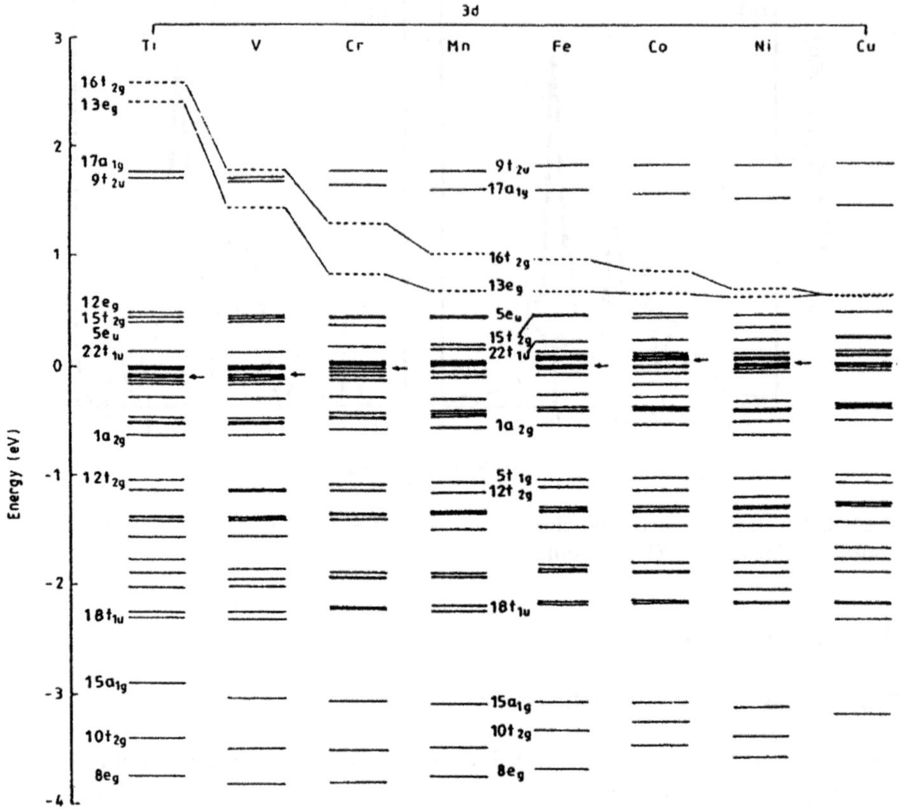

Fig. 2.3. Energy level structures of pure and alloyed Fe with $3d$ transition elements. The Fermi energy level is indicated by an *arrow* (←)

2.3.2 Bond Order, Bo

The other alloying parameter is the bond order. Here, the overlap population $Q_{\nu\nu'}$ of electrons between two atoms ν and ν' is defined as

$$Q_{\nu\nu'} = \sum_{l}\sum_{ij} C_{il}^{\nu} C_{jl}^{\nu'} \int \Psi_i^{\nu} \Psi_j^{\nu'*} \, \mathrm{d}V \, . \tag{2.2}$$

Here, Ψ_i^{ν} and $\Psi_j^{\nu'}$ are the wave functions of the ith and jth orbitals of atoms ν and ν', respectively. C_{il}^{ν} and $C_{jl}^{\nu'}$ are the coefficients that show the magnitude of the linear combination of atomic orbitals in the lth molecular orbital. In this paper, $Q_{\nu\nu'}$ is called the bond order and it is used as a measure of the strength of the covalent bond between M and X atoms.

For example, the values of the bond order (hereafter referred to as Bo) are listed in Table 2.1 for a variety of alloying elements in bcc Fe. This bond order also changes according to the position of elements in the periodic table.

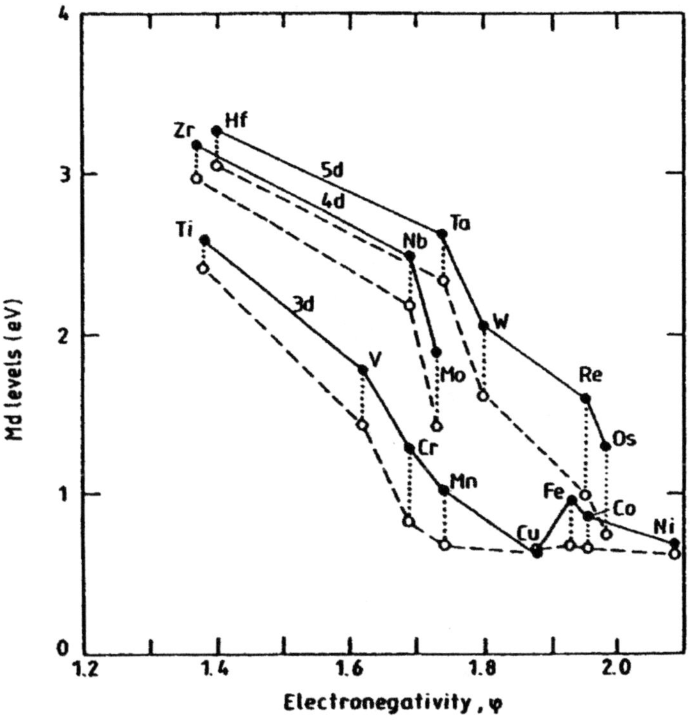

Fig. 2.4. Md levels versus electronegativity. The *empty circles*, ○, refer to e_g, while the *filled circles*, •, refer to t_{2g}

A number of alloying parameters have been proposed for describing alloying behavior (e.g., the electrons-per-atom ratio (e/a), the atomic radius, the electronegativity, and the electron vacancy number). However, most of these parameters are used to describe pure metals and cannot properly describe alloying effects. In contrast, both the Md and the Bo are new alloying parameters obtained for the first time from the molecular orbital calculation of alloyed clusters. Both the Md and the Bo have been calculated for various metal systems – for example, Ni [15], Cr [16], Ti [7], Zr [17], Nb [18], and Mo [18]. Such calculations have recently been extended to heavy metals (e.g., W [19]) using a relativistic DV method. Here, for convenience, the values of Md and Bo relevant to nickel alloys are listed in Table 2.2.

2.3.3 Average Parameter Values for Typical Alloys

For an alloy, the average values of Md and Bo are defined simply by taking the compositional average, and $\overline{\mathrm{Md}}$ and $\overline{\mathrm{Bo}}$ are defined as

$$\overline{\mathrm{Md}} = \sum \mathrm{X}_i \cdot (\mathrm{Md})_i \ , \tag{2.3}$$

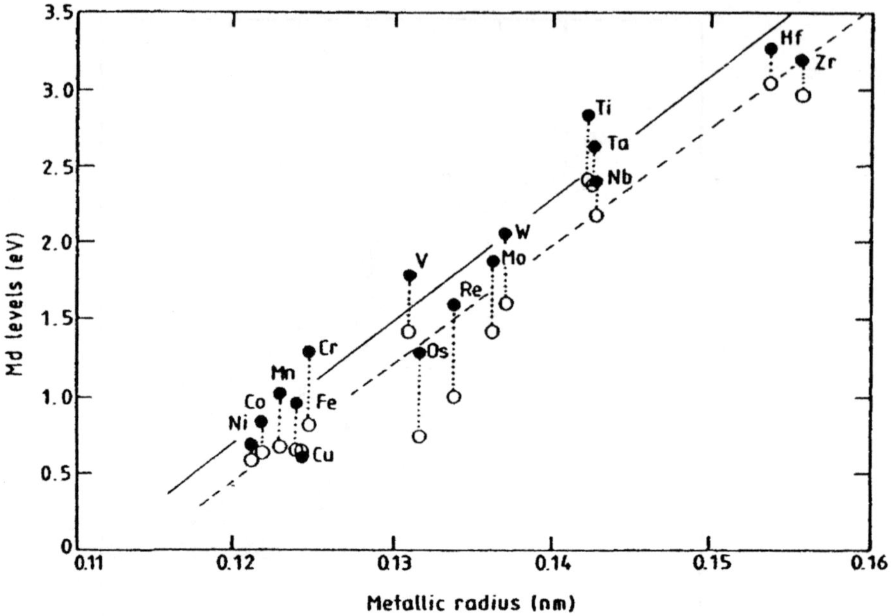

Fig. 2.5. Correlation of Md levels with metallic radius. The *empty circles*, ∘, refer to e_g, while the *filled circles*, •, refer to t_{2g}

Table 2.1. List of Md and Bo values for alloying elements in bcc Fe

Element		Md (eV)	Bo
3d	Ti	2.497	2.325
	V	1.610	2.268
	Cr	1.059	2.231
	Mn	0.854	1.902
	Fe	0.825	1.761
	Co	0.755	1.668
	Ni	0.661	1.551
	Cu	0.637	1.361
4d	Zr	3.074	2.551
	Nb	2.335	2.523
	Mo	1.663	2.451
5d	Hf	3.159	2.577
	Ta	2.486	2.570
	W	1.836	2.512
	Re	1.294	2.094
Others	C	−0.230	0
	N	−0.400	0
	Si	1.034	0

Table 2.2. List of Md and Bo values for alloying elements in Ni

Element		Md(eV)	Bo
3d	Ti	2.271	1.098
	V	1.543	1.141
	Cr	1.142	1.278
	Mn	0.957	1.001
	Fe	0.858	0.857
	Co	0.777	0.697
	Ni	0.717	0.514
	Cu	0.615	0.272
4d	Zr	2.944	1.479
	Nb	2.117	1.594
	Mo	1.550	1.611
	Tc	1.191	1.535
	Ru	1.006	1.314
	Rh	0.898	1.068
	Pd	0.779	0.751
	Ag	0.659	0.391
5d	Hf	3.020	1.518
	Ta	2.224	1.670
	W	1.655	1.730
	Re	1.267	1.692
	Os	1.063	1.500
	Ir	0.907	1.256
	Pt	0.764	0.920
	Au	0.627	0.528
Others	Al	1.900	0.533
	Si	1.900	0.589

$$\overline{\mathrm{Bo}} = \sum \mathrm{X}_i \cdot (\mathrm{Bo})_i \ . \tag{2.4}$$

Here, X_i is the atomic fraction of the ith component in the alloy, $(\mathrm{Md})_i$ and $(\mathrm{Bo})_i$ are the respective values for the ith component. The summation extends over the components, $i = 1, 2, \ldots, n$. For simplicity, unless otherwise stated, the unit of the Md parameter, eV, is omitted in this paper.

2.4 Estimation of Alloy Properties Using Alloying Parameters

2.4.1 Nickel Alloys

Ni-based superalloys are heat-resistant alloys widely used for the blades of jet engines and industrial gas turbines.

New PHACOMP Method

It is well-known that the precipitation of topologically close-packed (TCP) phases (e.g., the σ phase) in the fcc (γ) matrix deteriorates the mechanical properties of a material remarkably. The composition of alloys that are free from such precipitates have been predicted by the phase computation (PHACOMP) method [20, 21]. The concept of "alloy design" was only solidified when PHACOMP was developed in 1964. In this method, the precipitation of the TCP phases in the alloy is predicted using the average electron vacancy number, Nv. Here, the electron vacancy number, Nv, is the number of electron vacancies or holes existing above the Fermi energy level in the d-band. It is expressed approximately as Nv $= 10.66-(e/a)$, where e/a is the electrons-per-atom ratio. For example, the Nv value for the 4A group of elements, Cr, Mo, and W, is 6.66. The compositional average of Nv is denoted as $\overline{\mathrm{Nv}}$, and when the $\overline{\mathrm{Nv}}$ value exceeds a certain value, the TCP phase is expected to be formed in the γ matrix. This Nv-PHACOMP method has been employed widely for the design and the quality control of nickel-based superalloys. However, there are many difficulties and contradictions in this method [2].

As explained earlier, the Md parameter is concerned with the electronegativity and the atomic radius of elements, both of which are classical parameters to be used as measures of the solid solubility in alloy. So, instead of the Nv parameter, this Md parameter is applied to the prediction of the occurrence of the TCP phases in nickel alloys [1, 2, 22]. Two typical phase diagrams of Ni$-$Co$-$Cr and Ni$-$Cr$-$Mo are shown in Fig. 2.6. In each phase diagram, the iso-$\overline{\mathrm{Md}}$ line of 0.925 (eV) traces closely the $\gamma/\gamma+\sigma$ phase boundary, whereas the iso-$\overline{\mathrm{Nv}}$ line is far away from the boundary. This Md method is called the new PHACOMP and has been applied successfully to practical alloys with multiple components [23, 24]. It has also been applied to the prediction of solidification microstructures in the weld nickel alloys [25, 26].

In Table 2.2, recently calculated values of Md are given for several elements of interest (e.g., Ru, Ir, Y). The third-generation single crystal superalloys containing some addition of Ru [27] and Ir [28] seem to show a reduced tendency to form undesirable TCP phases [29]. The Md values of these elements are relatively low.

Alloying Vector

The concept of alloying vectors is also important for practical alloy design. As shown in Fig. 2.7, the vector starts from the position of pure Ni and ends at the position of Ni-10 mol%M binary alloy in the $\overline{\mathrm{Bo}}-\overline{\mathrm{Md}}$ diagram, where M is an alloying element in Ni. The directions and the magnitudes of these vectors vary with alloying elements [30]. It is noted that the vector directions are similar among the same group elements in the periodic table – for example, among Ti, Zr, and Hf (4A group elements); V, Nb, and Ta (5A elements); and Cr, Mo, and W (6A elements). This is attributable to the fact that both

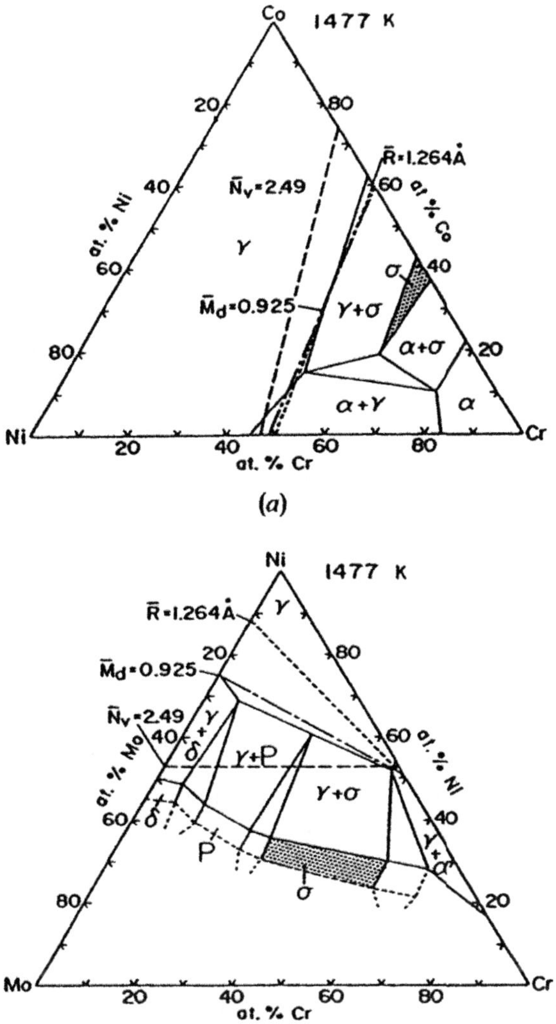

Fig. 2.6. Phase diagrams of (**a**) Ni–Co–Cr and (**b**) Ni–Cr–Mo. Here, $\overline{R}$ is the compositional average of atomic radii of elements in the alloy

the Bo and the Md parameters change according to the order of elements in the periodic table, as explained before.

Any superalloy consists of the γ' (Ni_3Al) phase and the γ (fcc) matrix phase; and the γ' phase is a precipitation hardening phase in the alloy. The γ' stabilizing elements such as Al, Ti, V, Nb, and Ta have relatively lower θ angles than the γ stabilizing elements such as Cr, Mo, W, and Re, as shown in the figure [31]. Thus, the θ angle is an indication of the stability of the γ'

phase with respect to the γ phase, and hence, it is related closely to the γ' volume fraction of the alloy [31].

Target Region for Alloy Design

Nineteen conventional cast superalloys are plotted in the $\overline{\mathrm{Bo}}-\overline{\mathrm{Md}}$ diagram as shown in Fig. 2.7 [30]. The contour lines showing the 0.2% yield strength level at 1255 K are indicated in the figure. The 0.2% yield strength shows a maximum $\overline{\mathrm{Md}}$ value of about 0.98 and a $\overline{\mathrm{Bo}}$ value of about 0.67. The creep strength also shows a maximum around this position. Furthermore, all of the single-crystal superalloys are located near the maximum position, as indicated by the shaded area in Fig. 2.7. Every alloy lying on such a maximum position contains a large volume fraction (about 60%) of the γ' phase without any TCP phases precipitating in it, resulting in the high strength of the alloy. Thus, a target region for alloy design can be specified concretely on the $\overline{\mathrm{Bo}}-\overline{\mathrm{Md}}$ diagram.

The $\overline{\mathrm{Md}}$ value in the target region, 0.98, is calculated using the alloy composition, so it is higher than the value for the γ phase, 0.925, shown in Fig. 2.6. However, assuming that the $\overline{\mathrm{Md}}$ value is 0.925 for the γ phase and 1.02 for the γ' phase, and also that the volume fraction of the γ' phase is 60%, then the $\overline{\mathrm{Md}}$ value for the whole alloy becomes about 0.98, in agreement with the $\overline{\mathrm{Md}}$ value of the target region.

Alloy Modification Using the $\overline{\mathrm{Bo}}-\overline{\mathrm{Md}}$ Diagram

The concept of alloying vectors is, for example, useful for alloy modification. In order to keep the alloy position on the $\overline{\mathrm{Bo}}-\overline{\mathrm{Md}}$ diagram nearly unchanged

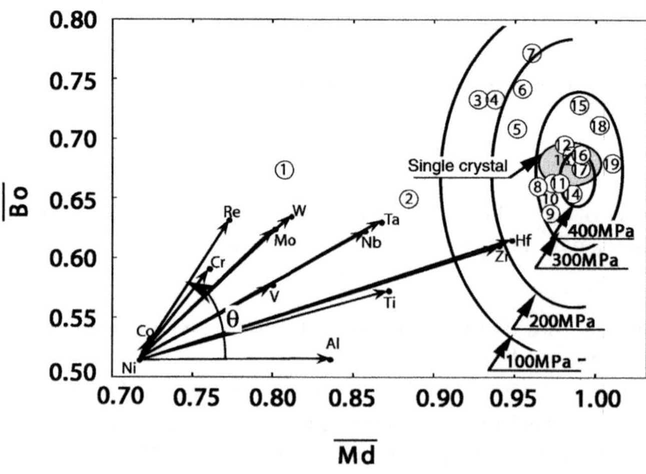

Alloy	0.2% Yield stress, 1255K
① INCONEL 600	28Mpa
② INCONEL 702	48
③ HASTELLOYR235	124
④ INCONEL700	138
⑤ GMR 235D	–
⑥ UDIMET 500	228
⑦ UNITEMP AF1753	–
⑧ ALLOY 713LC	283
⑨ M-22	359
⑩ ALLOY 713C	303
⑪ B-1900	414
⑫ NICROTUNG	359
⑬ MAR-M200	469
⑭ PDRL 162	448
⑮ UDIMET 700	303
⑯ MAR-M246	379
⑰ TRW 1900	–
⑱ NIMONIC115	241
⑲ IN 100	372

Fig. 2.7. The locations of commercial conventional cast Ni alloys in a $\overline{\mathrm{Bo}}-\overline{\mathrm{Md}}$ diagram

with the modification, alloy compositions may be adjusted among the elements having a similar vector direction. For example, the Re addition may require the reduction of the Cr content in the alloy, since their vectors are pointed in the similar direction. Such a compositional change is actually seen in the modification from PWA 1480 (the first-generation superalloy) to PWA 1484 (the second-generation superalloy). Namely, the Cr content of PWA 1484 decreases by 6 mol%, whereas the Re content increases by 1 mol% compared to the respective values of PWA 1480. Further decrease of the Cr content and corresponding increase of the Re content is seen in René N6 (the third-generation superalloy) [32].

2.4.2 High-Cr Ferritic Steels

Recently, there have been increasing demand for operating power-generating plants under ultra-super-critical conditions [33]. For this reason, there have been strong requests for the development of advanced ferritic steels that tolerate the severe operating conditions found in power plants.

Alloying Vector

In Fig. 2.8, characteristics of each alloying element are represented by an alloying vector starting from the position of pure Fe and ending at the position of the Fe-1 mol%M binary alloy in the $\overline{Bo}-\overline{Md}$ diagram for Fe-based alloys [9]. These alloying vectors are drawn using the Md and Bo values listed in Table 2.1. All of the ferrite stabilizing elements (e.g., Cr, and Mo) are located in the upper-right region (except for Mn), whereas all of the austenite stabilizing elements (e.g., Ni and Co) are located in the lower-left region in this diagram.

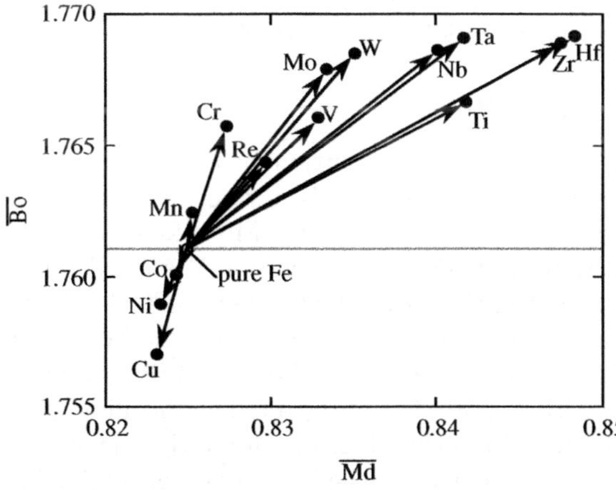

Fig. 2.8. Alloying vectors of elements in bcc Fe

It is known empirically that the elements having high Bo and low Md are principal alloying elements in most structural alloys. The high Bo element strengthen alloys by forming strong chemical bonds with the neighboring atoms, and the low Md element increase the phase stability without forming any undesirable phases in the alloy. This is also true in the ferritic steels. In other words, principal alloying elements have a high $\overline{\mathrm{Bo}}/\overline{\mathrm{Md}}$ ratio; namely, a large slope of alloying vectors shown in Fig. 2.8. The ratio changes in the following order: Cr > Mo > W > Re > V > Nb > Ta > Zr > Hf > Ti. Consistent with this order, Cr, Mo, W, V, and Nb are indeed principal alloying elements in most ferritic steels. The only exception lies in Re. But recently, the addition of Re has been found to improve creep strength significantly [9]. On the other hand, for austenite stabilizing elements, the alloying vector has the opposite direction, and the $\overline{\mathrm{Bo}}/\overline{\mathrm{Md}}$ ratio changes in the following order: Ni > Co > Cu.

δ Ferrite Formation

It is desirable to suppress the formation of the δ ferrite in ferritic steels because of the detrimental effect on their creep resistance and fracture toughness. The measured volume fraction of the δ ferrite existing in the steels normalized at 1323 K, are plotted against the $\overline{\mathrm{Md}}$ parameter as shown in Fig. 2.9. For Ni-free steels, the δ ferrite starts forming as the $\overline{\mathrm{Md}}$ value exceeds 0.852, and its volume fraction increases with increasing $\overline{\mathrm{Md}}$ value. Also, as is evident from this figure, the existence of Ni in the steel increases the critical $\overline{\mathrm{Md}}$ value for the δ ferrite formation. This is also the case of other austenite stabilizing elements, Co and Cu. Thus, the addition of Ni, Co, or Cu into the steel gives room to increase the amount of ferrite stabilizing elements such as Mo and W without δ ferrite formation, resulting in an increase in creep properties at high temperatures. In fact, any advanced steels contain such austenite stabilizing elements (e.g., 3 mass% Co). In addition, it is stressed here that δ ferrite formation is predicted more accurately by the $\overline{\mathrm{Md}}$ parameter than the Cr equivalent, a parameter widely used in steels [34].

Evolution of Ferritic Steels

Masuyama [35] has shown the evolution process of ferritic steels for boiler applications. For example, 9% Cr steels have been developed in the sequence, T9 → (F9) → T91 → NF616. Here, T91 (Mod. 9Cr−1Mo) is modified from T9 (9Cr−1Mo) by optimizing the contents of V and Nb, both of which form carbonitrides and strengthen the steel. Then NF616 is obtained from T91 by decreasing the Mo content, while increasing the W content. As shown in Fig. 2.10, this evolution is traceable in the $\overline{\mathrm{Bo}}-\overline{\mathrm{Md}}$ diagram [9]. The evolution of T9 → (F9) → T91 → NF616 is interpreted as the alloy modification toward higher $\overline{\mathrm{Md}}$ and higher $\overline{\mathrm{Bo}}$ in the diagram. Similarly, the evolution of the steels for turbine rotors can be traced in the $\overline{\mathrm{Bo}}-\overline{\mathrm{Md}}$ diagram [9]. Thus, this $\overline{\mathrm{Bo}}-\overline{\mathrm{Md}}$ diagram provides us a clue to alloy design.

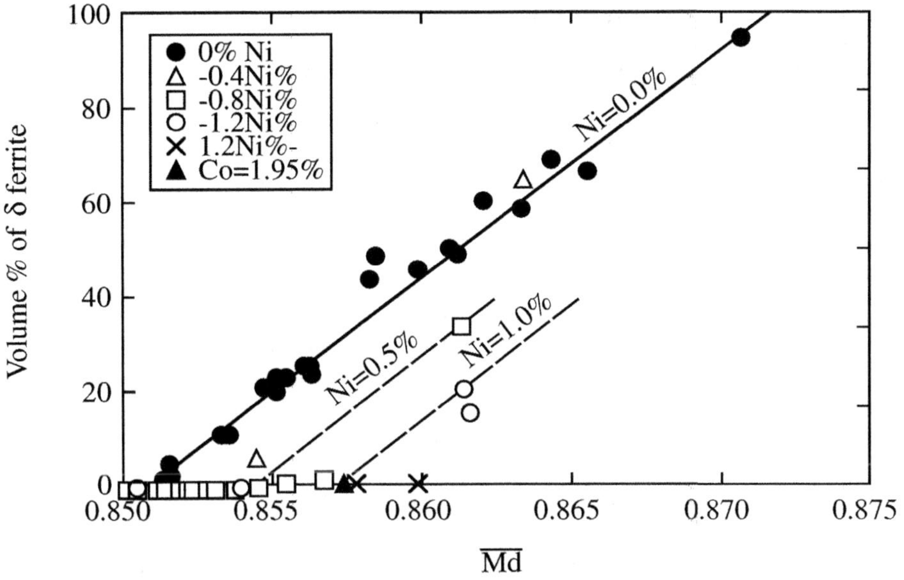

Fig. 2.9. Correlation of the volume fraction of δ ferrite with the $\overline{\mathrm{Md}}$ parameter

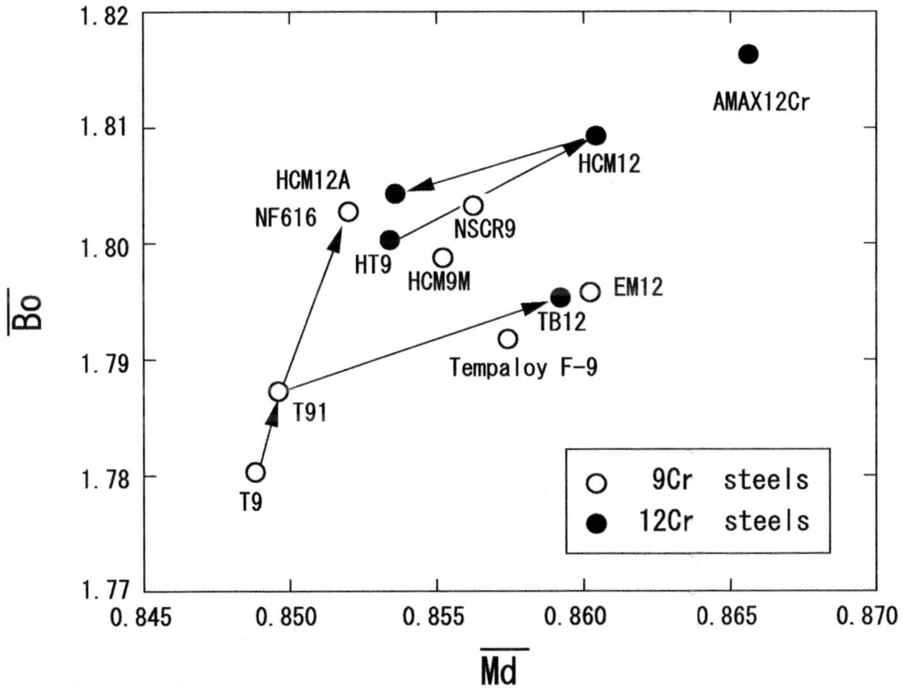

Fig. 2.10. Location of 9–12%Cr steels in the $\overline{\mathrm{Bo}}-\overline{\mathrm{Md}}$ diagram

2.5 Design of Structural Alloys

2.5.1 Nickel-Based Single-Crystal Superalloys

A single-crystal superalloy, TUT92, has been designed using the new PHACOMP method [1] while taking into account fabrication conditions [6]. The chemical composition of TUT92 is Ni−10%Cr−12%Al−1.5%Ti−2.05%Ta−2.30%W−0.77%Mo−0.25%Re (in mol%). It shows a better combination of creep resistance and hot corrosion resistance than PWA 1484 and CMSX-4. Also, it is stressed here that TUT92 and other second-generation single-crystal superalloys such as PWA 1484 and CMSX-4 are located in a very limited region around an $\overline{\mathrm{Md}}$ value of 0.985 and a $\overline{\mathrm{Bo}}$ value of 0.665 in the diagram, as shown in Fig. 2.11 [30, 31]. René N6 (the third-generation superalloy) is, however, located at the slightly lower $\overline{\mathrm{Md}}$ position so as to avoid solute segregation [32] which is induced by the existence of a large amount of Re (5.4 mass% Re) in it.

Recently, in order to reduce the cost of materials, high-Cr but Re-free alloys have been designed for the use in blades in industrial gas turbines [36]. The creep rupture strength is measured at a stress of 196 MPa and a temperature of 1193 K, close to the service conditions of a gas turbine. The measured life of the designed alloy is found to be comparable to that of the second-generation superalloy containing 3 mass% Re.

2.5.2 High-Cr Ferritic Steels

A high-strength high-Cr ferritic steel for steam turbine rotors has been designed using this approach [9]. A ten-ton turbine rotor has been made as shown

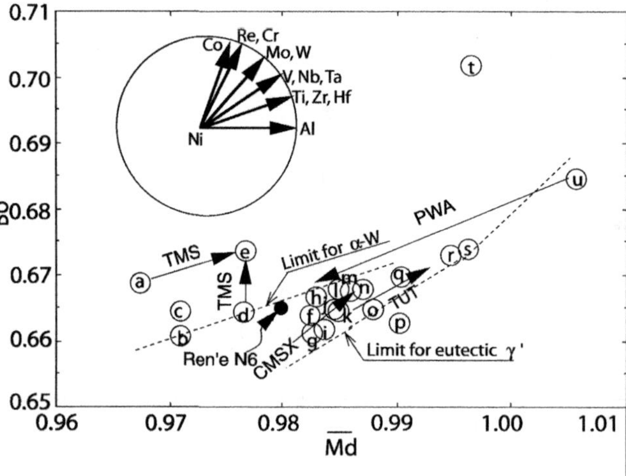

Alloy	Creep* Life, Ms	H.C.I.,** kg/m²
(a) TMS 12-1	10.4	0.52
(b) SC 53A	8.6	0.55
(c) SC 83	8.7	0.50
(d) TMS 12	6.5	0.56
(e) TMS 26	7.2	0.50
(f) SRR 99	1.7	0.17
(g) CMSX-2	1.6	0.23
(h) PWA 1484	2.7	0
(i) CMSX-3	—	—
(j) MXON	2.3	—
(k) TUT 92	4.3	0
(l) TUT 103	2.2	—
(m) CMSX-4	3.7	0
(n) CMSX-4G	3.1	0
(o) NASAIR100	3.6	0.17
(p) CMSX-6	—	—
(q) TUT 31D	7.2	0
(r) TUT 95	8.2	0
(s) RR 2000	—	—
(t) RR 2060	—	—
(u) PWA 1480	0.8	0.16

*Creep test at 1313K, 137MPa
**Hot corrosion index

Fig. 2.11. Locations of single-crystal superalloys in the $\overline{\mathrm{Bo}}$−$\overline{\mathrm{Md}}$ diagram

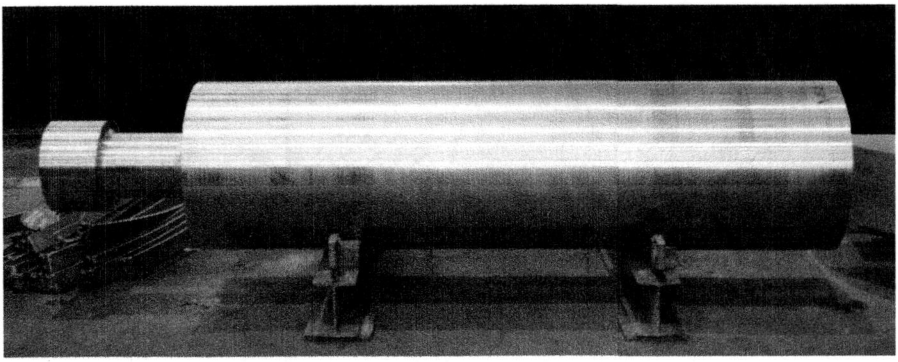

Fig. 2.12. Ten-ton turbine rotor

in Fig. 2.12. The chemical composition of the steel is Fe−0.09%C−10%Cr−4%W−3.5%Co−0.2%Ni−0.1%Mo−0.15%V−0.06%Nb−0.08%Mn−0.08%Si−0.007%B−0.018%N in mass%. The creep rupture life of this alloy is longer than 4000 hours (14.4 Ms) in the creep test conditions of a temperature of 923 K and an applied stress of 157 MPa, which is the longest lifetime among the turbine rotor steels so far developed in the world. It is also found experimentally that the addition of 0.2%Re increases the creep rupture life considerably.

2.6 Crystal Structure Maps for Intermetallic Compounds

Intermetallic compounds (e.g., TiAl, $LaNi_5$) are promising materials expected to be used for various applications in future. The Bo−Md diagram is useful for predicting crystal structure of intermetallic compounds such as aluminides and silicides containing transition metals, M [5]. One example is shown in Fig. 2.13 for MAl compounds. The crystal structures are clearly separated in this diagram. As explained earlier, both the Bo and the Md parameters change according to the position of elements in the periodic table. Recalling that the crystal structures of any compounds also change according to the periodic table, we may say that these Bo and Md values are indeed suitable parameters for constructing structure maps. Bo−Md structure maps have been made for a variety of compounds. The physical properties of the compounds are so sensitive to the crystal structure that these structure maps will provide us a convenient tool for the design of intermetallic compounds [37].

2.7 Hydrogen Storage Alloys

Hydrogen storage alloys are key materials for the development of future clean hydrogen energy systems. A large amount of hydrogen is stored in the alloy through metal hydrides. Many hydrogen storage alloy have been developed,

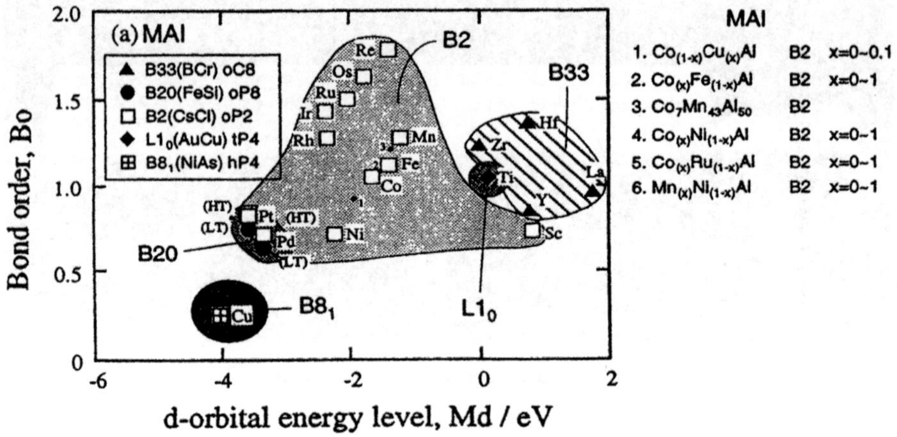

Fig. 2.13. Bo−Md structure maps for MAl, where M's are transition metals

but the total amount of hydrogen stored in any alloys is still low in view of practical applications in automobiles. In order to get a clue regarding the development of new alloys, a series of calculations has been performed by employing appropriate cluster models. For example, a cluster model used for the $LaNi_5$ system is shown in Fig. 2.14. This model is constructed based on the crystal structure of $LaNi_5H_6$ [38].

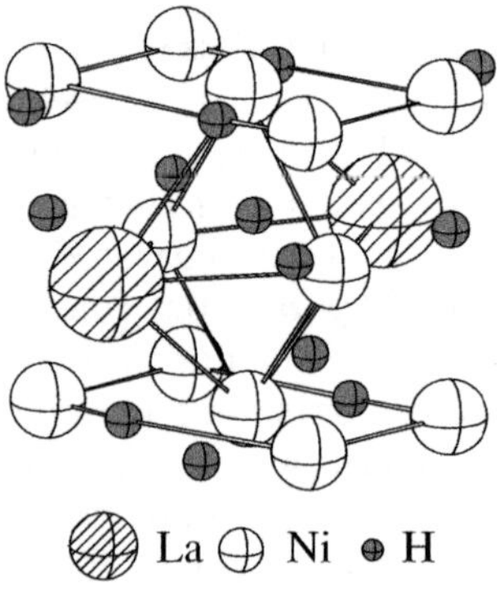

Fig. 2.14. Cluster model used in the calculation for the $LaNi_5$ system

2.7.1 Metal-hydrogen Interaction

Most hydrogen storage alloys such as $LaNi_5$ (AB_5-type), $ZrMn_2$ (AB_2-type), TiFe (AB-type), and Mg_2Ni (A_2B-type) consist of hydride forming elements, denoted by A, and non-forming elements, denoted by B. Needless to say, the A element has a larger affinity for hydrogen than the B element in the binary metal-hydrogen system. Therefore, the role of each element in the alloy has been understood in the following way: the hydride-forming element (e.g., La) may make a strong chemical bond with hydrogen, whereas the hydride non-forming element (e.g., Ni) may work to reduce such a strong A–H bond, so that hydrogen is released readily from the A atom in the hydrogen desorption process.

However, this naive understanding is not true according to our calculations [39–47]. For example, a contour map of the electron densities is shown in Fig. 2.15 for the $LaNi_5$ hydride on the atomic plane containing La, Ni, and H atoms. As is evident from this figure, a relatively high electron-density region near hydrogen extends to the Ni atom site, but not to the La atom site. This result clearly indicates that hydrogen interacts more strongly with Ni atoms than La atoms in the $LaNi_5$ system [39,40], despite the larger affinity of La atoms for hydrogen than Ni atoms in the binary metal-hydrogen system. Similar results are also obtained for the Mg_2Ni system [41], $ZrMn_2$ system [42], and TiFe system [43,44]. In all cases, hydrogen interacts more strongly with B atoms (Ni, Mn or Fe) than A atoms (La, Mg, Zr or Ti).

However, it is noted that the B–H (e.g., Ni–H) interaction in the alloy is still weaker compared to the A–H (e.g., La–H) interaction in pure metal hydride, AH_2, (e.g., LaH_2), judging from the values of the heat of formation of the hydride. It is $-209.2\,\mathrm{kJ/mol}\ H_2$ for LaH_2 and $-30.1\,\mathrm{kJ/mol}\ H_2$ for $LaNi_5H_6$. The hydrogen desorption process can be activated readily because of such a weak B–H interaction operating in every hydrogen storage alloys [45]. On the other hand, LaH_2 is too stable to be dehydrided at moderate temperatures. Therefore, we may say that the existence of weak hydrogen-metal interactions is characteristic of hydrogen storage alloys [37].

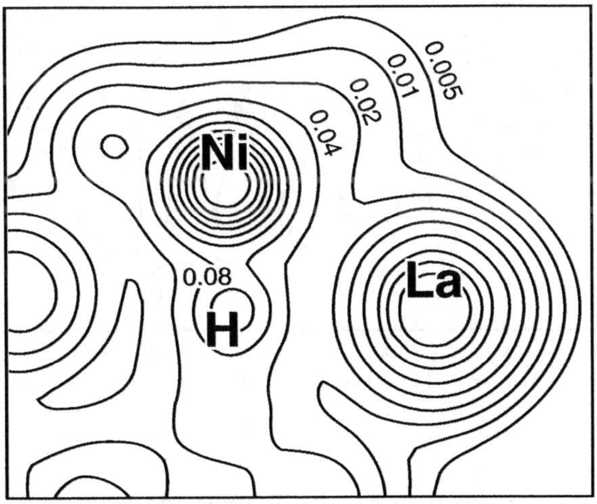

Fig. 2.15. Contour map of the electron densities in the $LaNi_5$ system. The numbers indicate the numbers of electrons per a.u.^{3} (1 a.u. = 0.529 nm)

2.7.2 Roles of Hydride-Forming and Non-Forming Elements

The hydride-forming element, A, does play an important role in the formation of the B–H chemical bond [48, 49]. In order to make clear the roles of hydride forming and non-forming elements in hydrogen storage alloys, a series of calculations has been performed using a Ni-based octahedral cluster, M_2Ni_4H (inset in Fig. 2.16), where M's are various elements presented on the horizontal axis.

From this figure, it is apparent that the Ni–H bond order is large when the M–H bond order is small. In particular, the Ni–H bond order is enhanced when M's are hydride-forming elements such as Na, Ca, and La. This means that hydrogen interacts strongly with Ni atoms if the hydride-forming elements exist in the neighborhood. On the other hand, when M's are the hydride non-forming elements such as Fe, Co, and Ni, the Ni–H bond order is very small, and hydrogen may not be absorbed in such an octahedral cluster. Thus, it is evident from Fig. 2.16 that the magnitude of the Ni–H bond order correlates well with the heat of hydride formation of the M metal or the heat of hydrogen dissolution in the M metal, ΔH.

Similar results are also obtained for the Ti-based and the Mg-based octahedral clusters [48–51]. Thus, there is a general trend that the B–H bond

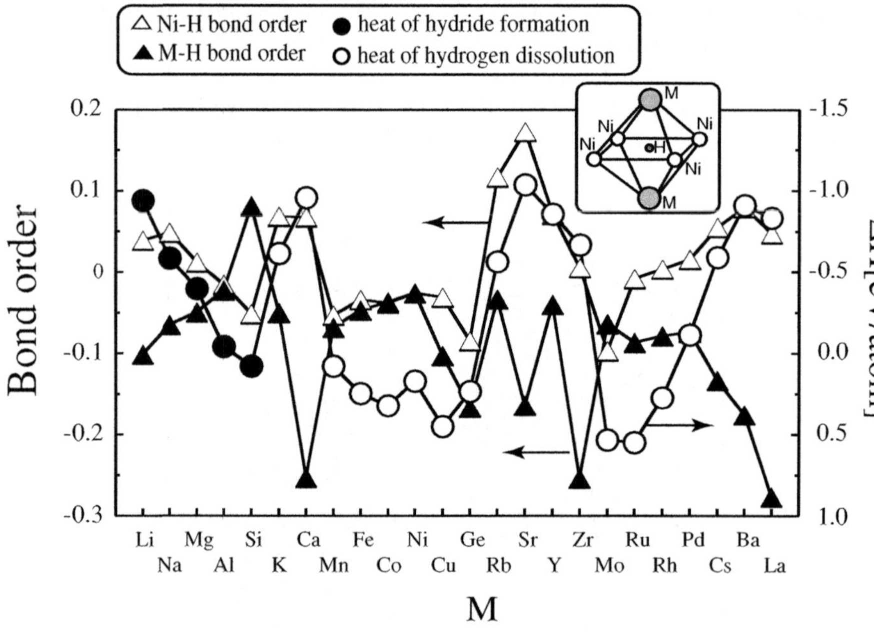

Fig. 2.16. Bond orders between Ni and H atoms and between M and H atoms in the M_2Ni_4H cluster, and comparison with the heat of hydride formation of M metal or the heat of hydrogen dissolution in M metal, ΔH

becomes strong only when the hydride forming element, A, exists adjacent to B and H atoms. Otherwise, it remains weak or even antibonding, such that it is difficult for hydrogen to be absorbed in the alloy. Thus, both A and B are indeed essential elements in hydrogen storage alloys.

The trend mentioned above is probably due to the fact that the B−H interatomic distance is shorter than the A−H interatomic distance, because of the smaller atomic size of the B element than the A element in most cases. Since hydrogen occupies an interstitial site at the center of a distorted octahedron or tetrahedron with unequal edge lengths, the B−H interaction is inevitably strengthened, while the A−H interaction is weakened. Thus, a polyhedron made of A and B elements provides a unique interstitial space for hydrogen in the A−B alloy, which is very different from the situation in pure A metal. Furthermore, the H-1*s* energy levels appear in the energy range much closer to the valence band of the B element than to that of the A element. This phenomenon also causes a stronger B−H interaction, compared to the A−H interaction. For example, in the case of the $LaNi_5$ system, the H-1*s* level appears below the Ni-3*d* band, and the Ni−H bond is formed preferentially in it.

2.7.3 Criteria for Alloy Design

Alloy Cluster Suitable for Hydrogen Storage

As explained above, the B−H bond is formed when atoms of the A element exists in the neighborhood. Therefore, the effective number of the B−H bonds is supposed to be large when the concentration of A element is high in the alloy. In other words, the total amount of hydrogen absorbed in the alloy will increase with increasing A/B compositional ratio. However, a disproportionation reaction tends to take place if the concentration of A element is too high in the alloy.

This phenomenon can be explained assuming that there are three types of tetragonal clusters in the A−B binary alloy as shown in Fig. 2.17(a)–(c) [50, 51]. If the A/B compositional ratio is high, or when the A−A bond is stronger than the A−B or the B−B bond, the clusters shown in Fig. 2.17(a) are dominant in the alloy. In such a case, hydrogen will interact mainly with A elements when hydrogen is introduced into the alloy, resulting in the formation of strong A−H bonds. The disproportionation reaction, $AB + H \rightarrow AH + B$, will thus take place easily. In this case, AH is so stable that the onset of this reaction is undesirable in view of the activation of the hydrogen desorption process.

By contrast, when the A/B compositional ratio is low, or when the B−B bond is stronger than the A−A or the A−B bond, the clusters shown in Fig. 2.17(c) are dominant in the alloy. In this case, it is very hard for hydrogen to be absorbed in such a B-atom-abundant cluster.

Therefore, in order to absorb and desorb hydrogen smoothly without the onset of a disproportionation reaction, the A/B compositional ratio should

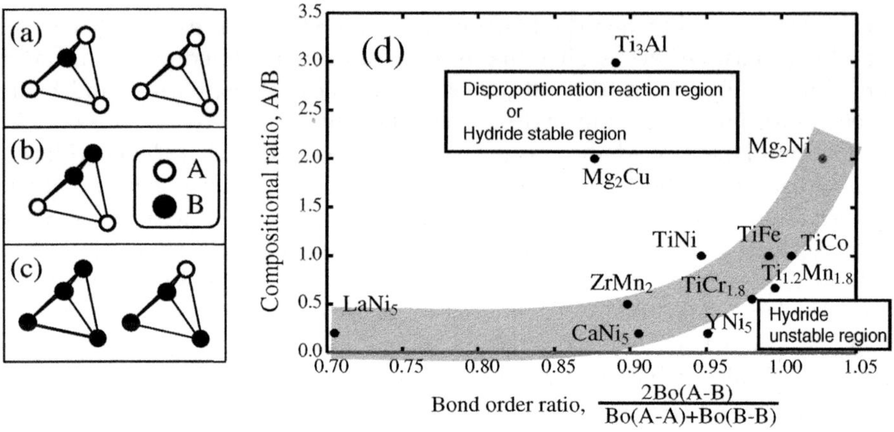

Fig. 2.17. (**a**)–(**c**) Three types of tetrahedral clusters. (**d**) Correlation between the A/B compositional ratio and the bond order ratio

be controlled in a proper manner, depending on the chemical bond strength between the A and B elements. In other words, optimum clusters shown in Fig. 2.17(b) may be made in the alloy only when the bond order ratio, 2Bo(A–B)/[Bo(A–A) + Bo(B–B)], takes a suitable value for a given A/B compositional ratio.

Alloy Compositions

A series of calculations is performed to obtain the bond order ratio, 2Bo (A–B)/[Bo(A–A) + Bo(B–B)], for various alloy systems using the tetrahedral cluster, and the results are shown in Fig. 2.17(d) [47,48]. Here, this bond order ratio has a physical meaning similar to the ordering energy parameter, $\Omega_{AB}(= V_{AB} - 1/2(V_{AA} + V_{BB}))$, where V_{AB}, V_{AA}, and V_{BB} are the respective bond strengths between the atoms given in the subscript. Needless to say, the atomic arrangement shown in Fig. 2.17(a) or (c) is present in the clustering-type alloy system, but the one shown in Fig. 2.17(b) is present in the ordering-type alloy system. The hydrogen storage alloys except for bcc V alloys [52,53], have an ordered atomic arrangement, so that their characteristics can be represented using an ordering energy parameter, Ω_{AB}. But instead of Ω_{AB}, the bond order ratio is used here, because it is a parameter that can be easily estimated from the calculations, but still holds a similar physical meaning to Ω_{AB}.

As shown in Fig. 2.17(d), there exists a strong correlation between the A/B compositional ratio and the bond order ratio. For example, all the typical hydrogen storage alloys, $LaNi_5$, $ZrMn_2$, TiFe, and Mg_2Ni, are located on a narrow band illustrated in the figure. Here, for Mg_2Ni, the bond order ratio is high, because the Mg–Ni bond order is much larger than the Mg–Mg bond order. Also, for $LaNi_5$, the bond order ratio is low, because the La–Ni bond order is smaller than the La–La bond order. All the alloys located in the upper region above this narrow band tend to decompose during hydrogenation and/or tend to form a stable hydride. One example is the Mg_2Cu system, in which a disproportionation reaction, $Mg_2Cu + 2H_2 \rightarrow 2MgH_2 + Cu$, takes place, and a stable hydride, MgH_2, is formed during hydrogenation. Therefore, the region lying above the narrow band is called the disproportionation reaction region or hydride stable region. On the other hand, the alloys located in the region below this narrow band, tend to form unstable hydrides; YNi_5 is an example of this. Therefore, this region is called the hydride unstable region.

Similar results are also obtained for the octahedral clusters as shown in Fig. 2.18. Thus, the present result is independent of the clusters used for the calculation. Once A and B elements are chosen, the A/B compositional ratio is determinable from the relationship with the bond order ratio shown in Figs. 2.17(d) and 2.18.

Mg-Based Alloys

This method is applied to the Mg-based alloys. The results obtained from the calculation using octahedral clusters are shown in Fig. 2.19. In this figure, the decomposition temperatures at a constant hydrogen pressure of 0.1 MPa are also denoted for some hydrides. The location of the narrow band is extrapolated from Fig. 2.18 and shown here as a dotted band.

The hydriding properties of the Mg-based alloys correlate well with the bond order ratio as shown in Fig. 2.19. For example, in the case when the A/B

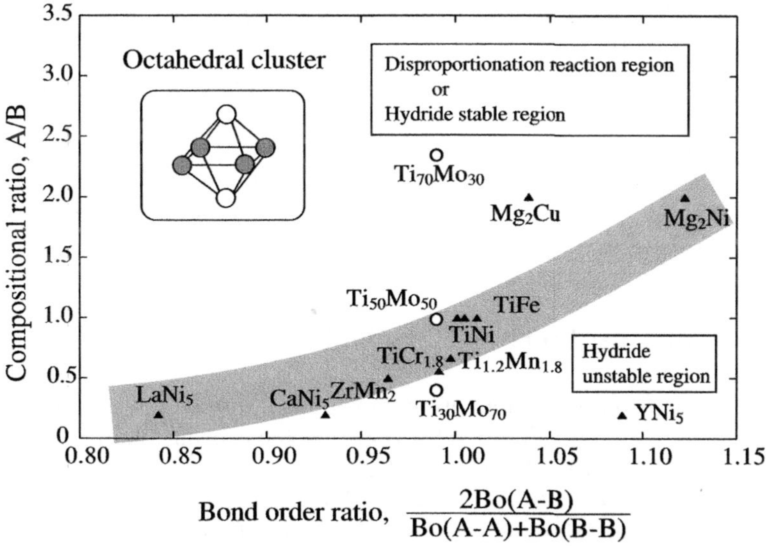

Fig. 2.18. Correlation between the A/B compositional ratio and the bond order ratio for an octahedral cluster

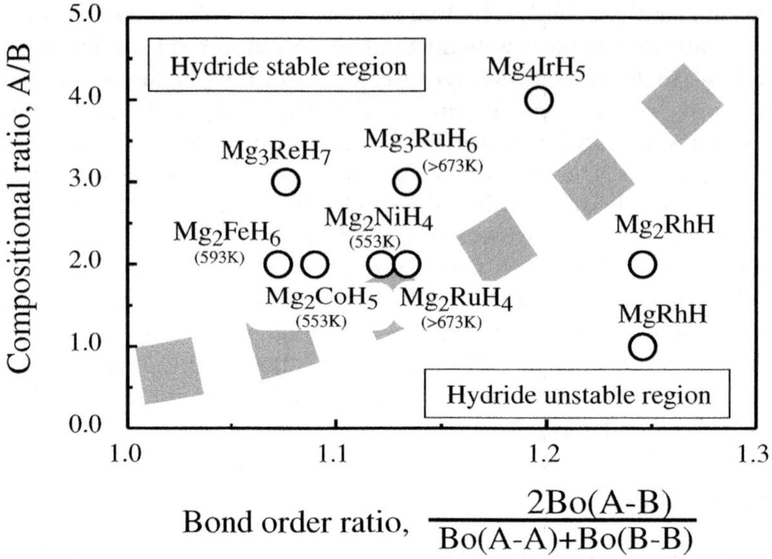

Fig. 2.19. Correlation between the A/B compositional ratio and the bond order ratio for Mg-based alloys

compositional ratio is fixed at 2, the total amount of hydrogen stored in the hydrides changes in the order $Mg_2FeH_6 > Mg_2CoH_5 > Mg_2NiH_4 = Mg_2RuH_4 >$ Mg_2RhH, so that it decreases with increasing bond order ratio. This means that the hydrogenation proceeds further in an alloy that has a lower bond order ratio. This is reasonable, since an alloy with a lower bond order ratio is located in the hydride stable region, whereas an alloy with a higher bond order ratio is located in the hydride unstable region. Even for the other A/B compositional ratios, the alloys located in the region above the dotted band tend to form hydrides with a high H content (e.g., Mg_3ReH_7). On the other hand, the alloys located in the region below the band tend to form hydrides with the low H content (e.g., MgRhH). These results are consistent with the results shown in Figs. 2.17 and 2.18.

2.8 Proton-Conducting Perovskite-type Oxides

The molecular orbital method presented is also applicable to a proton conducting oxide, $SrTiO_3$. The level structures have been simulated, and the results are shown in Fig. 2.20(a)–(f) [54].

As shown in Fig. 2.20(a), in pure $SrTiO_3$, there is a band gap of about 3.5 eV between the O-2*p* valence band and the Ti-3*d* conduction band, in agreement with experiments. As shown in Fig. 2.20(b), when the Sc ion is doped into $SrTiO_3$, an acceptor level appears just above the valence band.

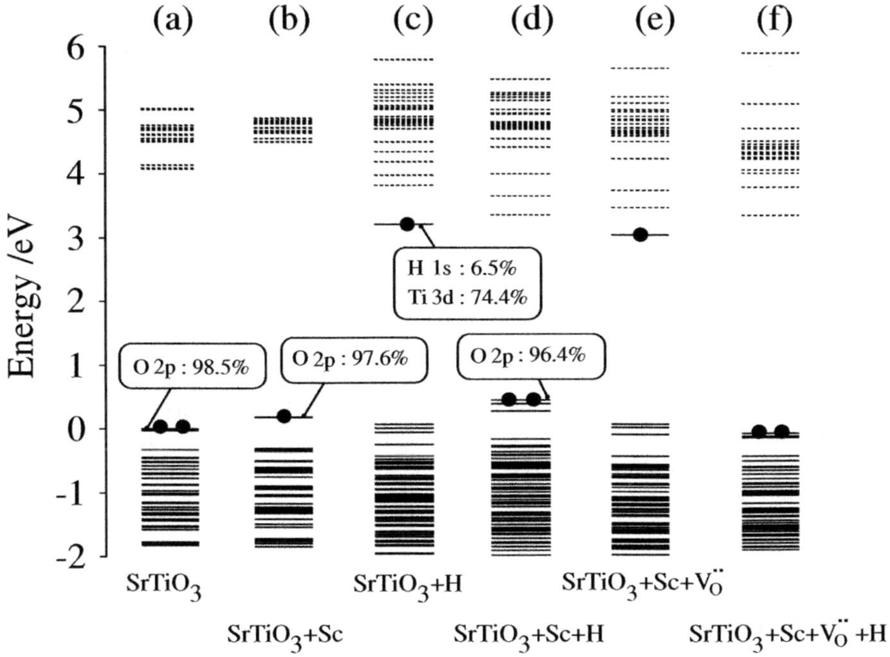

Fig. 2.20. Energy level structures near the band gap for (**a**) pure $SrTiO_3$, (**b**) Sc-doped $SrTiO_3$, (**c**) $SrTiO_3$ containing H, (**d**) Sc-doped $SrTiO_3$ containing H, (**e**) Sc-doped $SrTiO_3$ containing an oxygen ion vacancy, (**f**) Sc-doped $SrTiO_3$ containing both an oxygen ion vacancy and H

On the other hand, when hydrogen is introduced into $SrTiO_3$, a donor level appears below the conduction band as shown in Fig. 2.20(c). The molecular orbital of the donor level is composed mostly of the Ti-3*d* and O-2*p* electrons, but still there is a small occupancy of the H-1*s* electrons. When both the Sc ion and hydrogen coexist in $SrTiO_3$ as shown in Fig. 2.20(d), charge transfer takes place from the donor level to the acceptor level. As a result of this charge compensation, the effective ionicity of hydrogen becomes about +0.2. Also, the chemical bond strengths between constituent ions are modified largely by dopants. For example, the Sc doping tends to strengthen the chemical bond between hydrogen and oxygen ions, but weakens the chemical bond between the oxygen ion and the surrounding metal ions.

In addition, as shown in Fig. 2.20(e), when an oxygen ion vacancy exists in the Sc-doped $SrTiO_3$, the defect level appears below the conduction band and acts as a donor level. Therefore, when an Sc ion, oxygen ion vacancy, and hydrogen coexist in $SrTiO_3$, charge compensation takes place among their donor and acceptor levels, as is shown in Fig. 2.20(f).

Similar results are also obtained in $SrZrO_3$ [55, 56]. The local electronic structure around hydrogen is found to change largely with the acceptor dopants in $SrZrO_3$, which probably induces a large displacement of oxygen ions near hydrogen, and affects protonic conductivity of the doped oxide [57,58].

2.9 Conclusion

A molecular orbital approach is useful for the design and development of not only structural alloys but also for functional alloys. In the super-computer age, such a computer-aided approach is superior to the trial-and-error approach to save time and cost necessary for alloy development.

Acknowledgement. The authors acknowledge the staff of the Computer Center at the Institute for Molecular Science in the Okazaki National Institute for the use of the VPP5000, SGI 2800/3800 supercomputers. This research was supported by the Grant-in-Aid for Scientific Research from the Ministry of Education, Culture, Sports, Science and Technology of Japan, and from the Japan Society for the Promotion of Science.

References

1. M. Morinaga, N. Yukawa, H. Adachi, H. Ezaki, *Superalloys 1984*, ed. by M. Gell et al. (The Metallurgy Society of AIME, Warrendale, 1984), p. 523
2. M. Morinaga, N. Yukawa, H. Ezaki, H. Adachi, Phil. Mag. **51**(2), 223 and 247 (1985)
3. M. Morinaga, S. Kamado, Model. Simul. Mater. Sci. Eng. **1**, 151 (1993)
4. R. Ninomiya, H. Yukawa, M. Morinaga, K. Kubota, J. Alloys Compounds **215**, 315 (1994)
5. Y. Harada, M. Morinaga, J. Saito, Y. Takagi, J. Phys.: Condens. Matter **9**, 8011 (1997)
6. K. Matsugi, Y. Murata, M. Morinaga, N. Yukawa, Mat. Sci. Eng. **A172**, 101 (1993)
7. M. Morinaga, M. Kato, T. Kamimura, M. Fukumoto, I. Harada, K. Kubo, *Titanium '92, Science and Technology*, ed. by F.H. Froes, I. Caplan, vol. 1 (The Metallurgy Society of AIME, Warrendale, 1993), p. 217
8. M. Niinomi, D. Kuroda, M. Morinaga, Y. Kato, T. Yashiro, *Non-Aerospace Applications of Titanium*, ed. by F.H. Froes, P.G. Allen, M. Niinomi (The Metallurgy Society of AIME, Warrendale, 1998), p. 217
9. M.Morinaga. R. Hashizume, Y. Murata, *Materials for Advanced Power Engineering 1994*, ed. by D. Coutsouradis et al. (Kluwer, Dordrecht, 1994), p. 329
10. E.D. Ellis, G.S. Painter, Phys. Rev. **82**, 2887 (1970)
11. H. Adachi, M. Tsukada, C. Satoko, J. Phys. Soc. Japan **45**, 875 (1978)
12. J.C. Slater, *Quantum Theory of Molecules and Solids*, Vol. 4 (McGraw-Hill, New York, 1974)
13. For example, see *Advances in Quantum Chemistry*, Vol. 29 (Academic Press, Burlington, 1997), p. 1
14. M. Morinaga, N. Yukawa, H. Adachi, J. Phys. F: Met. Phys. **15**, 1071 (1985)

15. M. Morinaga, N. Yukawa, H. Adachi, J. Phys. Soc. Japan **53**, 653 (1984)
16. Y. Matsumoto, M. Morinaga, T. Nambu, T. Sakaki, J. Phys.: Condens. Matter, **8**, 3619 (1996)
17. H. Shibutani, M. Morinaga, K. Kikuchi, J. Atomic Energy Soc. Japan **40**(1), 70 (1998)
18. S. Inoue, J. Saito, M. Morinaga, S. Kano, J. Phys.: Condens. Matter **6**, 5081 (1994)
19. H. Yukawa, M. Maeda, M. Morinaga, H. Nakamatsu, Bull. Soc. Discret. Variation. Xα **13**, 75 (2000)
20. W.J. Boesch, J.S. Slaney, Met. Prog., **86**, 109 (1964)
21. L.R. Woodyatt, C.T. Sims, H.J. Beattie, Jr., Trans. AIME, **236**, 519 (1966)
22. M. Morinaga, Y. Murata, H. Yukawa, *Materials Design Approaches and Experiences*, ed. by J.-C. Zhao, M. Fahrmann, T.M. Pollock (The Metallurgy Society of AIME, Warrendale, 2001), p. 15
23. P. Caron, *Superalloys 2000*, ed. by T.M. Pollock et al. (The Metallurgy Society of AIME, Warrendale, 2000), p. 737
24. P. Caron, T. Khan, *Materials Design Approaches and Experiences*, ed. by J.-C. Zhao, M. Fahrmann, T.M. Pollock (The Metallurgy Society, 2001), p. 81
25. M.J. Cieslak, G.A. Knorovsky, T.J. Headley, A.D. Romig, Metall. Trans. **17A**, 2107 (1986)
26. J.S. Ogborn, D.L. Olson, M.J. Cieslak, Mater. Sci. Eng. **A203**, 134 (1995)
27. K.S. O'Hara, W.S. Walston, E.W. Ross, R. Oarolia, US Patent 5,482,789 (1996)
28. T. Kobayashi, Y. Koizumi, S. Nakazawa, T. Yamagata, H. Harada, *Advanced in Turbine Materials, Design and Manufacturing*, ed. by A. Strang et al. (Institute of Metals, Sendai 1997), p. 766
29. P. Caron, T. Khan, Aerosp. Sci. Technol. **3**, 513 (1999)
30. M. Matsugi, Y. Murata, M. Morinaga, N. Yukawa, *Superalloys 1992*, ed. by S.D. Antolovich et al. (The Metallurgy Society of AIME, Warrendale, 1992), p. 307
31. Y. Murata, S. Miyazaki, M. Morinaga, R. Hashizume, *Superalloys 1996*, ed. by R.D. Kissinger et al. (The Metallurgy Society of AIME, Warrendale, 1996), p. 61
32. W.S. Walston, K.S.O'Hara, E.W. Ross,T.M. Pollock, W.H. Murphy, *Superalloys 1996*, ed. by R.D. Kissinger et al. (The Metallurgy Society of AIME, 1996), p. 27
33. W. Schlachter, G.H. Gessinger, *High Temperature Materials for Power Engineering 1990*, ed. by E. Bachelet et al. (Kluver Academic Publishers, Pordrecht, 1990), p. 1
34. H. Ezaki, M. Morinaga, K. Kusunoki, Y. Tsuchida, Tetsu-to-Hagane, **78**(8), 1377 (1992)
35. F. Masuyama, *Proceedings of the 78th Annual Meeting of the Kyushu Branch of the Iron and Steel Institute of Japan*, 25 September 1992, p. 1
36. Y. Murata, R. Hashizume, A. Yoshinari, N. Aoki, M. Morinaga, Y. Fukui, *Superalloys 2000*, ed. by T.M. Pollock et al. (The Metallurgy Society of AIME, Warrendale, 2000), p. 285
37. M. Morinaga, H. Yukawa, Adv. Eng. Mater. **3**(6), 381 (2001)
38. J.L. Soubeyroux, L. Pontonnier, S. Miraglia, O. Isnard, D.Z. Fruchart, Phys. Chem. Bd. **179**, 187 (1993)
39. H. Yukawa, Y. Takahashi, M. Morinaga, Intermetallics **4**, S215 (1996)
40. H. Yukawa, T. Matsumura, M. Morinaga, J. Alloys Compounds **293-295**, 227 (1999)

41. Y. Takahashi, H. Yukawa, M. Morinaga, J. Alloys Compounds **242**, 98 (1996)
42. T. Matsumura, H. Yukawa, M. Morinaga, J. Alloys Compounds **279**, 192 (1998)
43. H. Yukawa, T. Takahashi, M. Morinaga, Comput. Mat. Sci. **14**, 291 (1999)
44. T. Nambu, H. Ezaki, H. Yukawa, M. Morinaga, J. Alloys Compounds **293-295**, 213 (1999)
45. H. Yukawa, M. Morinaga, Adv. Quantum Chem. **29**, 83 (1997)
46. T. Nambu, H. Ezaki, M. Takagi, H. Yukawa, M. Morinaga, J. Alloys Compounds **330**, 318 (2002)
47. M. Morinaga, H. Yukawa, Mater. Sci. Eng. **A329-331C**, 267 (2002)
48. K. Nakatsuka, M. Yoshino, H. Yukawa, M. Morinaga, J. Alloys Compounds **293-295**, 222 (1999)
49. M. Morinaga, H. Yukawa, K. Nakatsuka, M. Takagi, J. Alloys Compounds **330**, 20 (2002)
50. K. Nakatsuka, M. Takagi, M. Nakai, H. Yukawa, M. Morinaga, *Proceedings of the International Conference on Solid-Solid Phase Transformations '99(JIMIC-3)*, ed. by M. Koiwa et al. (The Japan Institute of Metals, Sendai, 1999), pp. 681–684
51. H. Yukawa, K. Nakatsuka, M. Morinaga, Solar Energy Mater. Solar Cells **62**(1–2), 75 (2000)
52. T. Matsumura, H. Yukawa, M. Morinaga, J. Alloys Compounds **284**, 82 (1999)
53. H. Yukawa, S. Ito, D. Yamashita, M. Morinaga, Adv. Quantum Chem. **42**, 263 (2003)
54. H. Yukawa, K. Nakatsuka, M. Morinaga, Solid State Ionics **116**, 89 (1989)
55. M. Yoshino, K. Nakatsuka, H. Yukawa, M. Morinaga, Solid State Ionics, **127**, 109 (2000)
56. M. Yoshino, K. Nakatsuka, H. Yukawa, M. Morinaga, *Proceedings of the 4th Pacific Rim International Conference on Advanced Materials and Processing (PRICM4)*, ed. by S. Hanada et al. (The Japan Institute of Metals, Sendai, 2001), p. 563
57. M. Yoshino, Y. Liu, K. Tatsumi, I. Tanaka, M. Morinaga, H. Adachi, Mater. Trans. **43**(7), 1444 (2002)
58. M. Yoshino, Y. Liu, K. Tatsumi, I. Tanaka, M. Morinaga, H. Adachi, Solid State Ionics 162–163, 127 (2003)

3

Chemical Bonding Around Lattice Imperfections in 3*d*-Transition Metal Compounds

Masataka Mizuno

3.1 Introduction

One of the central paradigms in modern materials science is that lattice imperfections dominate the macroscopic properties of materials. Each material has its own advantages, but most single materials are not appropriate for practical use. One of the most promising ways of modifying the properties of materials is to dope them with other elements. Interfaces play a key role in the joining of different types of materials. Great efforts to reveal the behavior and nature of lattice imperfections, such as point defects, dislocations, planar faults, and interfaces, have therefore been made in the past few decades. In order to discuss the influence of lattice imperfections, the change of the geometric and electronic structures induced by lattice imperfections has to be considered. Around lattice imperfections, there are lattice strains that arise from lattice mismatch between the matrix and the additional elements. The chemical bonding around lattice imperfections is expected to be different from that of perfect crystalline elements or compounds due to the change of the electronic structure. Therefore, theoretical approaches to study lattice imperfections can be classified into two categories: one is based on strain energy analyses, and the other is based on electronic structure analyses. Although the strain energy approach is more widely used to study imperfections in general, it does not mean that the electronic structure's contribution to the behavior of imperfections is minor. These two approaches are mutually dependent, since the strain originates from the spatial distribution of electron clouds around the imperfections. The major deficiency of electron theory for the study of imperfections is that it is difficult to obtain physical insight into electronic mechanisms behind the behavior and nature of imperfections. Electrons obey the Schrödinger equation whose solution is not easy to imagine without special training. In polyatomic systems, we need to perform numerical calculations to solve the Schrödinger equation accurately. In particular, when imperfections are included in the system, the calculation becomes large, and we need to work with very fast computers that were not available until recently. In the present work, we focus on the solute atoms in Fe_3C and the Fe/TiX interfaces as examples of lattice imperfections. These lattice imperfections play important roles in the production of practical steel products. The electronic structures of model clusters for these lattice imperfections are found by first-principle molecular orbital calculations using the DV-Xα cluster method. We discuss the chemical bonding of solutes and interfaces as lattice imperfections based on the mechanism of chemical bonding.

3.2 Computational Method

In this work, the DV-Xα cluster method [1,2] is used to calculate the electronic state of the model clusters from first-principles. The molecular orbital wave function is expressed as a linear-combination of atomic orbitals (LCAO) and is written as

$$\phi_l = \sum_i c_{il} \chi_i \,, \tag{3.1}$$

where c_{il}'s are coefficients. As the basis function χ_i, we use the numerical atomic orbitals obtained by solving the radial part of the Schrödinger equation for individual atoms in a given cluster. The matrix elements H_{ij} and S_{ij} in the secular equation $\mathbf{HC} = \mathbf{ESC}$ are evaluated as

$$H_{ij} = \sum_k \omega(r_k) \chi_i(r_k) \boldsymbol{h}(r_k) \chi_j(r_k) \,, \tag{3.2}$$

$$S_{ij} = \sum_k \omega(r_k) \chi_i(r_k) \chi_j(r_k) \,, \tag{3.3}$$

by the DV numerical integration scheme [3], where $\boldsymbol{h}$ is the one-electron Hamiltonian operator, r_k is one of the sample points, and $\omega(r_k)$ is the integration weight for the sample point.

In order to evaluate the magnitude of chemical bondings, the Mulliken population analysis [4] is employed. The orbital population, $Q_i(l)$, and the overlap population, $Q_{ij}(l)$, of the lth molecular orbital between χ_i and χ_j are defined by

$$Q_i(l) = \sum_l \sum_j f_l c_{il} c_{jl} \sum_k \omega(r_k) \chi_i(r_k) \chi_j(r_k) \,, \tag{3.4}$$

$$Q_{ij}(l) = \sum_l f_l c_{il} c_{jl} \sum_k \omega(r_k) \chi_i(r_k) \chi_j(r_k) \,, \tag{3.5}$$

where f_l is the occupation number of the lth molecular orbital. The orbital population is useful in determining the effective atomic charge. The bond overlap population between A and B atoms is given by

$$Q_{\mathrm{AB}} = \sum_{i \in \mathrm{A}} \sum_{j \in \mathrm{B}} \sum_{l} Q_{ij}(l) . \tag{3.6}$$

The overlap population can be used as a measure of strength of the covalent bonding.

3.3 Effect of Solute Atoms on the Chemical Bonding of Fe_3C

In steel technology, Fe_3C (cementite) plays a very important role in controlling or improving various properties of steel. Usually Fe_3C contains some alloying elements such as Cr and Mn when it is present in steel. The influence of the solute atoms on various properties of Fe_3C, i.e., the lattice parameter [5], magnetic behavior [6, 7], mechanical properties [8, 9], has been studied experimentally. On the theoretical side, with these structural features in mind, the chemical bonding of Fe_3C was discussed without detailed electronic structure calculations nearly fifty years ago. Petch [10] suggested that Fe_3C structure is more dominated by the Fe–Fe bonding than by the Fe–C bonding. On the other hand, Pauling [11] suggested that the Fe–C bonding was twice as strong as the Fe–Fe bonding. There have only been a few calculations performed on Fe_3C because of its complicated crystal structure. The band-structure of Fe_3C was calculated in two papers [12, 13]. Tight-binding calculations of a series of iron carbides was done by Cottrell [14]. As for solute atoms in Fe_3C, the influence of additional elements on the graphitization of Fe_3C was discussed using the idea of cohesive energy obtained by the extended Hückel molecular orbital method [15]. Yamamoto et al., however, suggested that the Fe–C bonding could be changed by additional elements. In this section, we discuss the effect of the solute atoms on Fe_3C from the viewpoint of chemical bonding.

3.3.1 Crystal Structure and Cluster Models for Fe_3C with Solute Atoms

The crystal structure of Fe_3C was first studied by Westgren and Phragmén [16, 17]. The position of atoms was first established by Lipson and Petch [18], and afterwards, a large number of studies have been made to determine the lattice parameters and the positions of atoms exactly [19–23]. In many works discussing Fe_3C, the lattice parameters are given in the *cab* setting using the extended symbol *Pmcn*, which is the same spacegroup as *Pnma*. If the *cab* setting is employed for Fe_3C, the lattice parameters are $a = 4.5248$ Å, $b = 5.0896$ Å, and $c = 6.7443$ Å. The primary slip plane [24], plane defects [25] and the orientation relationship with ferrite [26, 27] are usually discussed using the *cab* setting. In crystallographic databases, the standard setting, $a = 5.0896$ Å, $b = 6.7443$ Å, and $c = 4.5248$ Å, is employed. In this work, the *cab* setting is employed for consistency with previous works. The crystal structure of Fe_3C in the *cab* setting is shown in Fig. 3.1. It has an orthorhombic unit cell with four C atoms, four Fe_1 atoms, and eight Fe_2 atoms [23]. Fe_1 and Fe_2 denote Fe atoms at two different crystallographic sites: 8*d* and 4*c* sites in Wyckoff notation, respectively. Every C atom is surrounded by six Fe atoms at an average distance of 2.018 Å, forming a trigonal prism (Fig. 3.1(c)). In a trigonal prism of Fe_3C, two Fe_1 sites and four Fe_2 sites exist. The Fe_1 has 12 nearest-neighbor Fe atoms at an average distance of 2.630 Å, and two nearest-neighbor C atoms at an average distance of 1.978 Å. The Fe_2 has 11 nearest neighbor Fe atoms at an average distance of 2.591 Å, and two neighbor C atoms at an average distance of 2.038 Å.

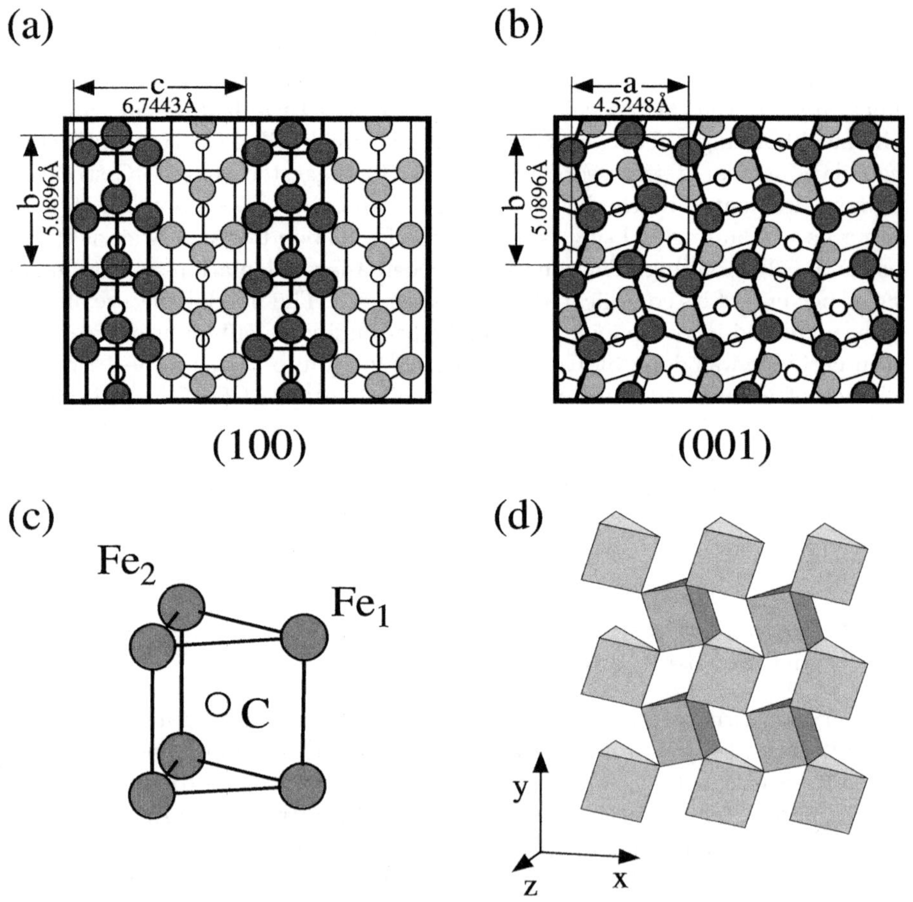

Fig. 3.1. The structure of Fe_3C (cementite). (**a**) The (100) projection. (**b**) The (001) projection. (**c**) Local structure around carbon atom, consisting of a *trigonal prism* composed of six Fe atoms and one carbon atom. (**d**) The arrangement of the *trigonal prisms* in a layer, as described in the text

The arrangement of Fe atoms is nearly closed-pack. The Fe−Fe distance in Fe_3C is only 2% (Fe_1) and 0.5% (Fe_2) greater than that in fcc-Fe (γ-Fe), i.e., 2.579 Å.

The prisms are joined together by sharing their corners and edges to make a layer (Fig. 3.1d). These layers are stacked along the c-axis with the sequence ABAB... The two types of layers, A and B, have the same trigonal prism arrangement and are mutually related by inversion symmetry. Between these layers in the unit cell, there are 28 Fe−Fe bonds whose average distance is 2.544 Å. There are eight Fe−C bonds with a distance of 2.372 Å, which is 18% longer than the nearest-neighbor Fe−C bond length. Therefore, the Fe−Fe bonding should be dominant for the adhesion of these layers. There is virtually no C−C direct bonding in Fe_3C because the nearest neighbor distance between C atoms is 3.09 Å, which is 100% greater that the C−C covalent bond length, 1.54 Å.

The model cluster, $Fe_{46}C_{10}-X_2$ (X = Ti, V, Cr, Mn, Fe, and Co), employed in this calculation for Fe_3C, is shown in Fig. 3.2. The $Fe_{46}C_{10}-X_2$ cluster has two layers composed of five trigonal prisms by sharing their edges and corners. According to the Mössbauer study by Schaaf et al. [7], Cr and Mn atoms prefer occupying the Fe_2 sites in Fe_3C. Therefore, we substituted one Fe_2 atom in each layer with $3d$ transition metal atoms, i.e. Ti, V, Cr, Mn, Co, and Ni. In order to maintain the inversion symmetry of the model cluster, two Fe atoms were substituted with the other metal atoms. Computation time is remarkably saved in this way. Two solute atoms, which are drawn as

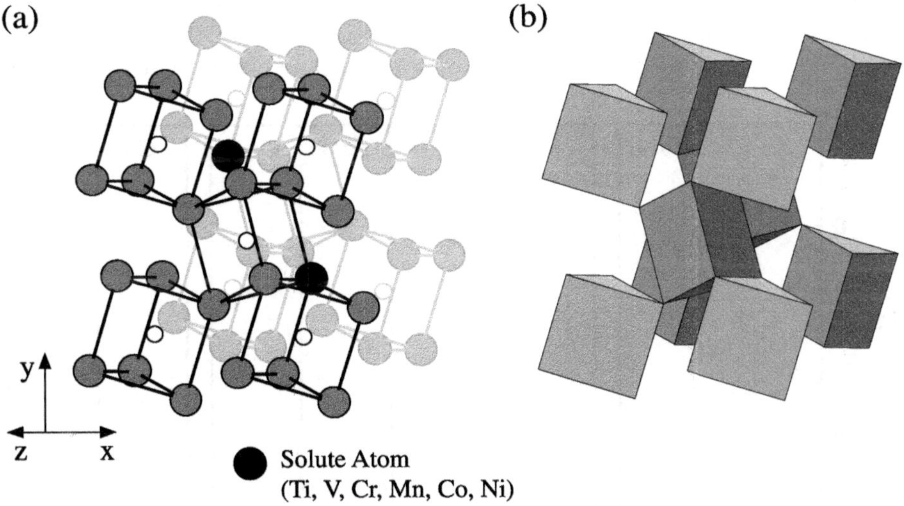

Fig. 3.2. The structure of the model cluster employed in this work. (**a**) The model cluster, $Fe_{46}C_{10}-X_2$. The *large gray circles* denote Fe atoms; the *large solid circles* denote solute atoms; the *small open circles* denote carbon atoms. (**b**) The arrangement of *trigonal prisms* in the model cluster

large solid circles in Figure 3.2, in the cluster are so separated that there is practically no interaction between them.

3.3.2 Pure Fe_3C

Figure 3.3 shows the total and partial density of states (DOS) curves for the valence states of the $Fe_{48}C_{10}$ cluster which is a model of pure Fe_3C. These DOS curves are made by broadening the discrete energy levels by a Gaussian function of 1.0 eV FWHM. They are shifted so as to set the Fermi level at zero. The large band around the Fermi level is mainly composed of the Fe-3*d* orbitals. The lower part of this band which shows two peaks at about −5 eV corresponds to the C-2*p* orbitals interacting with the Fe-3*d*, 4*s*, and 4*p* orbitals. The band located at about −12 eV is mainly composed of the C-2*s* orbital interacting with the Fe-3*d*, 4*s*, and 4*p* orbitals. The basic features of our DOS agree with those in the use of the LMTO (Linear Muffin Tin Orbital) calculation by Häglund et al. [13].

The metal–metal (M−M) and metal–C (M−C) overlap population diagrams for the $Fe_{48}C_{10}$ cluster are shown in Fig. 3.4. They are obtained by convoluting the overlap population at each molecular orbital with a Gaussian function of 1.0 eV FWHM. The bond overlap population can be obtained by summing the overlap population on each molecular orbital up to the Fermi level. In this figure, metal–metal refers to an average of overlap populations between the Fe_2 atom near the center of the cluster and its nearest neighbor Fe atoms. Metal–C refers to an average of overlap populations between the C atom in the prism located near the center of the cluster and its nearest neighbor Fe atoms. The values written in each panel are the bonding and the antibonding contributions to the bond overlap population, which were ob-

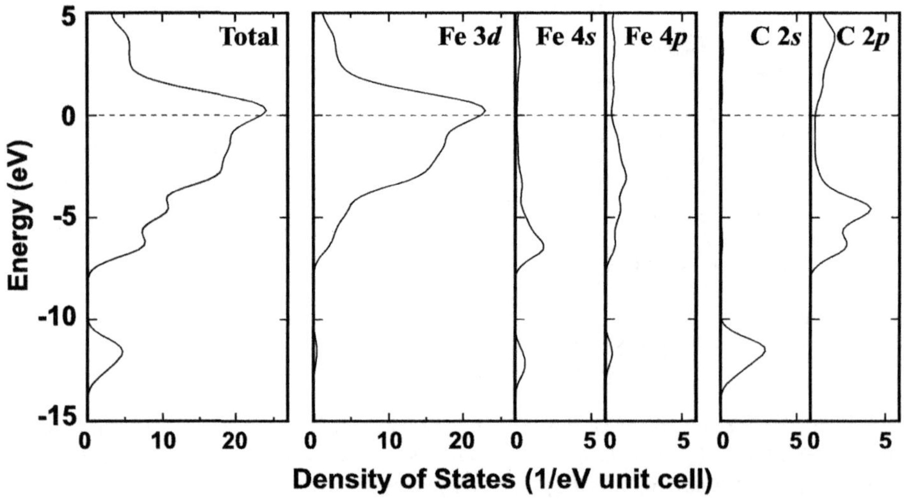

Fig. 3.3. Total and partial density of states for the $Fe_{48}C_{10}$ cluster

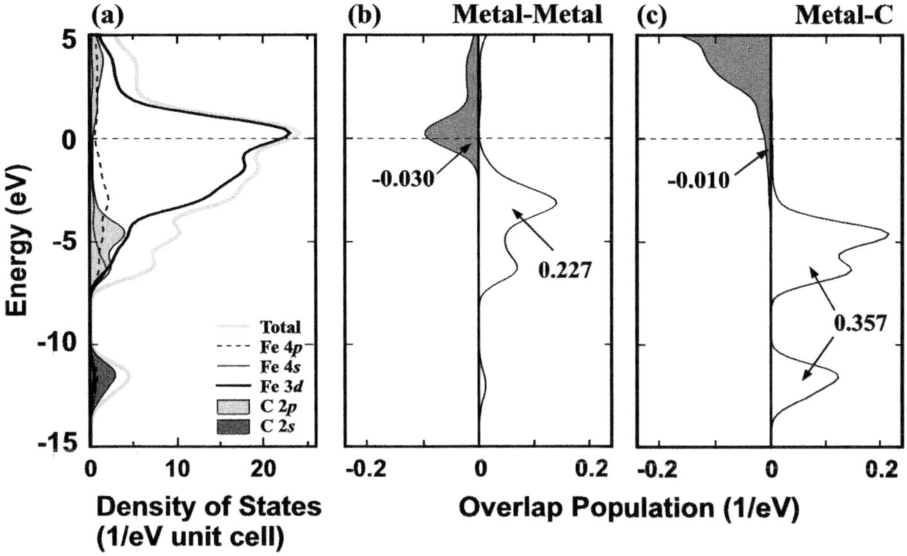

Fig. 3.4. Density of states and the overlap population diagrams for the $Fe_{48}C_{10}$ cluster. (**a**) Total and partial density of states. (**b**) Overlap population diagram for the M−M bond; (**c**) for the M−C bond

tained by summing the positive and negative values of the overlap population up to the Fermi level, respectively.

Regarding the M−M bonding, there is a large bonding band and a small antibonding band. The upper part of the bonding band is mainly composed of the Fe-3*d* orbitals and the lower part is composed of the C-2*p*–Fe-3*d*, 4*s*, and 4*p* hybridized orbitals. The antibonding band is mainly composed of the Fe-3*d* orbitals. The large bonding band is occupied by electrons completely. The Fermi level lies at the middle of the antibonding band. On the other hand, for the M−C bonding, there are two large bonding bands under the Fermi level. The upper bonding band is formed by the C-2*p*–Fe hybridization. The lower bonding band is formed by the C-2*s*–Fe hybridization. A large part of the antibonding band exists above the Fermi level. Therefore, around the Fermi level, the M−C bond is nearly non-bonding in character.

3.3.3 Fe_3C with Solute Atoms

The partial density of 3*d* and 4*sp* states of solute atoms are plotted for the $Fe_{46}C_{10}-Ti_2$, $Fe_{46}C_{10}-Cr_2$, $Fe_{48}C_{10}$, and $Fe_{46}C_{10}-Ni_2$ clusters in Fig. 3.5. The 4*sp* band becomes slightly larger as the atomic number of the solute atoms increases. However, the shape of the 4*sp* band does not change with the solute atoms. On the other hand, the 3*d* band is shifted downwards and more electrons occupy the 3*d* orbitals as the atomic number of the solute atoms increases.

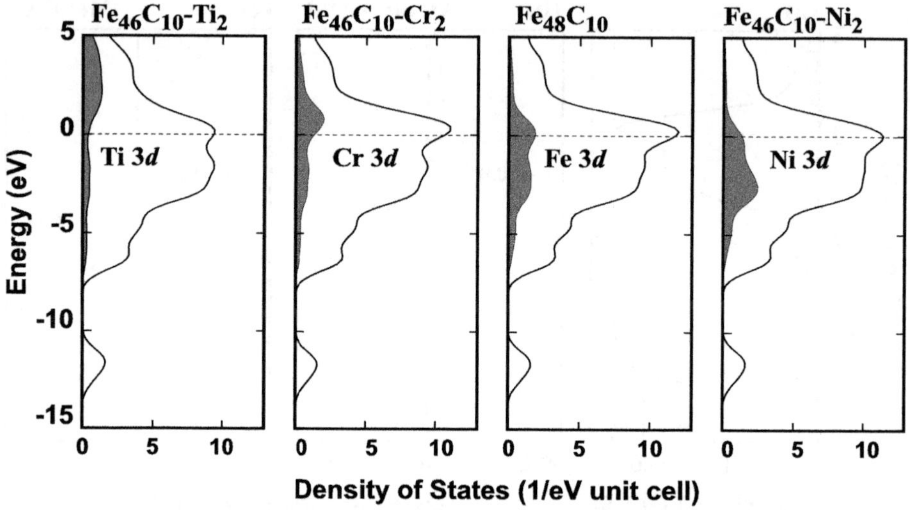

Fig. 3.5. Partial density of 3*d* and 4*sp* states of solute atoms

In order to estimate the influence of solute atoms on the M−M and M−C bondings in the $Fe_{46}C_{10}-X_2$ clusters, we calculated the bond overlap populations of the M−M and M−C bondings, and the results are shown in Fig. 3.6. In this figure, metal–metal refers to an average of the bond overlap populations between the solute atom and its nearest neighbor Fe atoms. Metal–C refers to an average of the bond overlap populations between the carbon atom and its nearest neighbor metal atoms in the trigonal prism containing the solute atom. The bond overlap population of M−C is about 0.34 irrespective of the solute atoms. On the other hand, the bond overlap population of M−M strongly depends on the atomic number of the solute atom. It shows a maximum value of 0.22 when the solute is Cr.

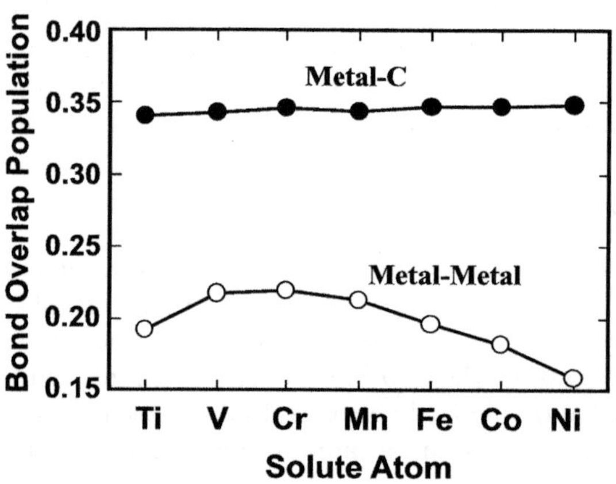

Fig. 3.6. Variation of bond overlap population with solute atom

The net charge of the solute atoms and C atoms are plotted in Fig. 3.7. In the undoped cluster, a small charge transfer from Fe to C is found. The ionicity of the C atom is -0.3, which is almost independent of the atomic number of the solute atom. The net charge of the solute atom decreases almost linearly with the atomic number of the solute atom. As a result, a solute atom that has a smaller atomic number than Fe, i.e., Ti, V, Cr, and Mn, becomes more ionic than Fe in Fe_3C.

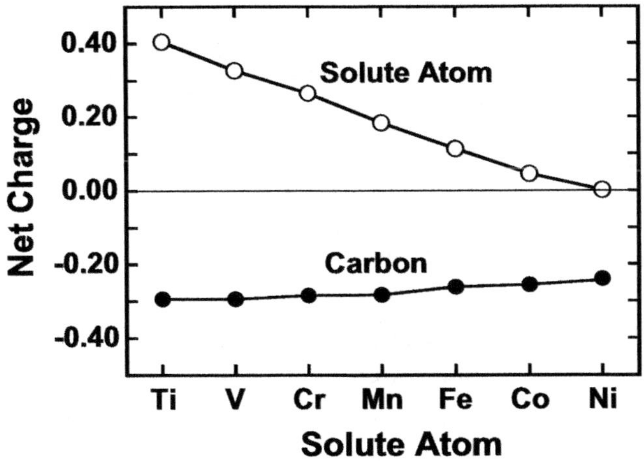

Fig. 3.7. Variation of net charge with solute atom

3.3.4 Effect of Solute Atoms on Fe_3C

As seen in Fig. 3.6, the solute atoms have a larger influence on the M–M bonding than the M–C bonding. In order to consider the solution effect in detail, the M–M overlap population diagram of the clusters with the solute atoms were examined. Figure 3.8 shows the results for the $Fe_{46}C_{10}-Ti_2$, $Fe_{46}C_{10}-Cr_2$, $Fe_{48}C_{10}$, and $Fe_{46}C_{10}-Ni_2$ clusters. In the $Fe_{48}C_{10}$ cluster, the Fermi level lies in the antibonding band of the M–M bond. When two Ni atoms are substituted for the Fe atoms, the antibonding band is shifted downwards. Consequently the Fermi level moves up, and more electrons occupy the antibonding orbitals. As a matter of fact, the bonding contribution decreases from 0.227 to 0.207 and the antibonding contribution increases from 0.030 to 0.048 when Ni is substituted for Fe as shown in Fig. 3.5–3.8. On the other hand, when Cr is substituted for Fe atom, the antibonding band is shifted upwards. Therefore, the Fermi level moves down, and the number of electrons occupying the antibonding orbitals is decreased. As a result, the bonding contribution increases from 0.227 to 0.233, and the antibonding contribution decreases from 0.030 to 0.013. When Ti is doped, the Fermi energy becomes much lower in energy; it crosses the bonding band. The bonding contribution

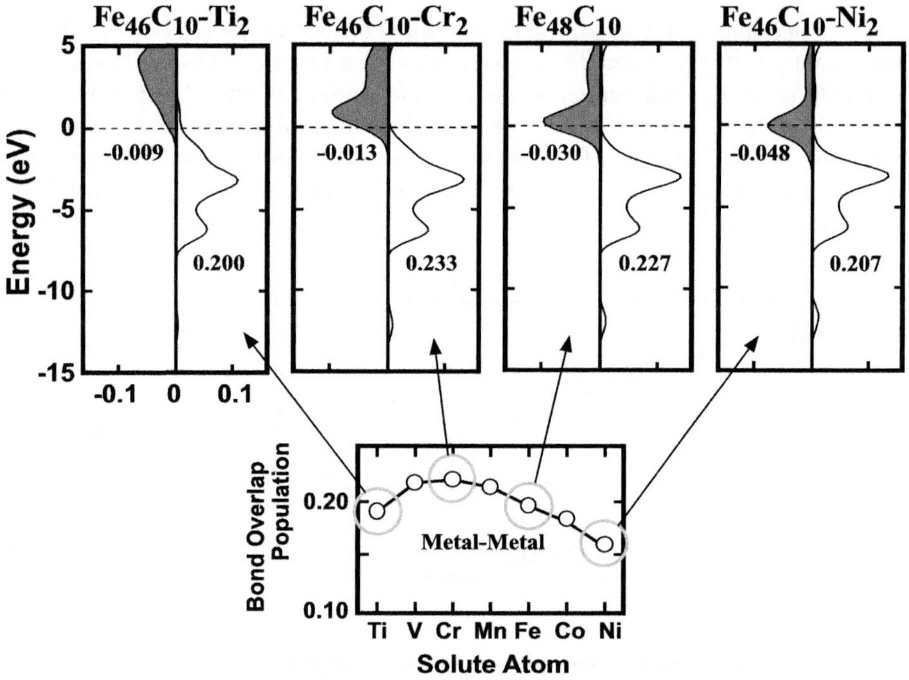

Fig. 3.8. Variation of overlap population of the M$-$M bond with solute atom

therefore decreases concurrently with a decrease in the antibonding contribution. The balance of the bonding and the antibonding contributions is found to determine the atomic number dependence of the bond overlap population shown in Fig. 3.8.

In order to elucidate the mechanism of the M$-$M bonding in more detail, the partial overlap populations between atomic orbitals are examined. Figure 3.9 shows the partial overlap population diagram for 3*d*–3*d*, 3*d*–4*sp*, and 4*sp*–4*sp* of the M$-$M bond. The component 3*d*–4*sp* is the sum of (solute) 3*d*–4*sp* (Fe) and 4*sp* (solute)-3*d* (Fe). Two things can be learned from Fig. 3.9. One is that the antiboding contribution of the 3*d*–3*d* and 3*d*–4*sp* increase notably with an increase in the atomic number of the solute atom. This is not the case for the 4*sp*–4*sp* antibonding. With an increase in the atomic number of the solute, the nuclear potential is deepened as the number of electrons increases: they change the shape and position of the 3*d* band considerably, as shown in Fig. 3.5. The effects are much smaller for the 4*sp* band. The other point we should emphasize in Fig. 3.9 is the bonding contribution of the 3*d*–4*sp* and the 4*sp*–4*sp*. The magnitude of the 4*sp*–4*sp* bonding increases with increasing atomic number of the solute. On the other hand, the bonding contribution of the 3*d*–3*d* simply decreases with increasing atomic number. This decrease is due to the contraction in the size of the 3*d* orbitals with an increase in the atomic number of the solute atoms.

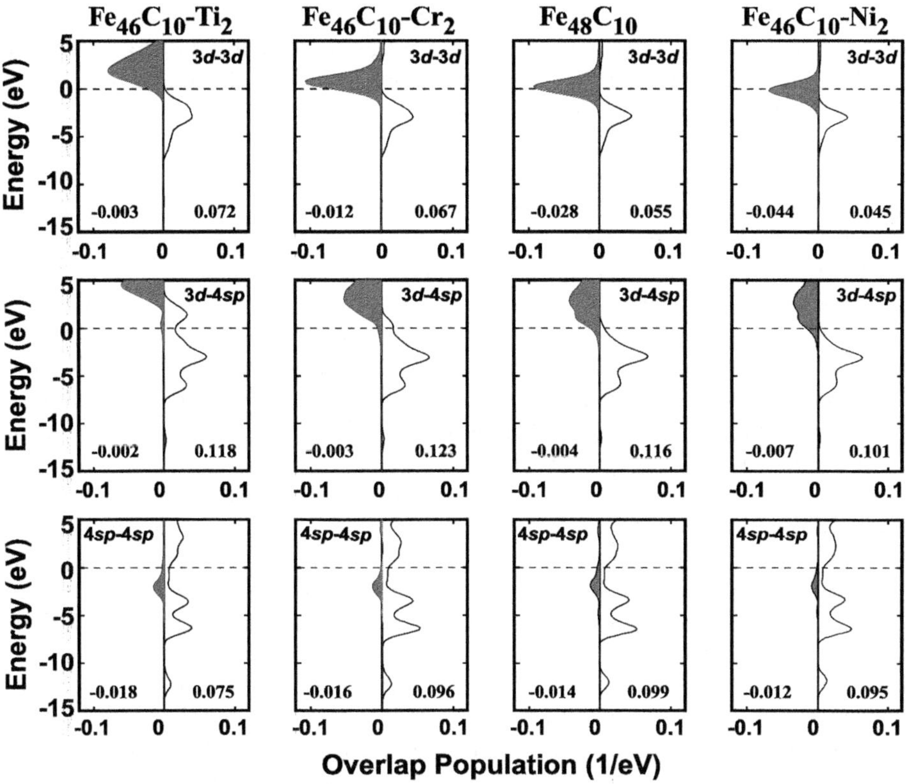

Fig. 3.9. Partial overlap population diagrams of 3*d*–3*d*, 3*d*–4*sp*, and 4*sp*–4*sp* for the M–M bond

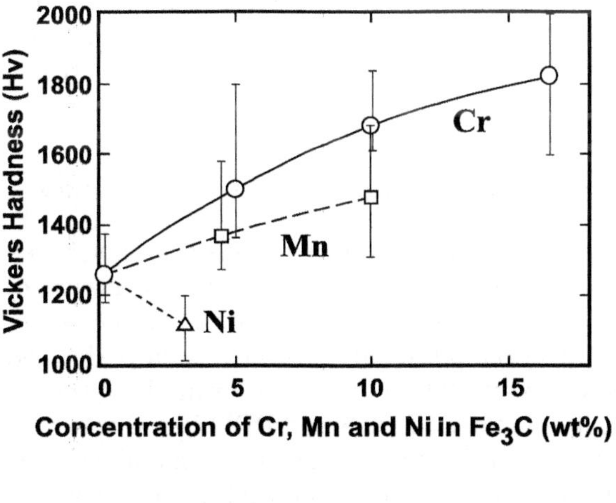

Fig. 3.10. Solute concentration dependence of Vickers hardness reported by Inoue et al. Ref. 8

In contrast to the M−M bond, the M−C bond exhibits little dependence on the atomic number. As shown in Fig. 3.4, the bonding band and the antibonding band of the M−C bond are largely separated, and molecular orbitals near the Fermi level are nearly of non-bonding character. Therefore, the changes of the number of electrons in both bonding and antibonding orbitals for different solute atoms are very small. This is the reason the magnitude of the M−C bond is almost independent of the atomic number. As a consequence, the M−M bond is expected to play a more important role on the effect of the solute atoms in Fe_3C than the M−C bond.

The Vickers hardness of Fe_3C-based solid solutions as measured by Inoue et al. [8] is shown in Fig. 3.10. Vickers hardness decreases when there is Ni doping. On the other hand, Vickers hardness increases with Cr and Mn doping. The concentration dependence is greater for Cr than for Mn. The dependence has been explained by the weakening of the M−C bond associated with an increase in the atomic number from Ti to Ni [8], since bulk TiC has the highest hardness and melting point among 3*d* transition-metal carbides. Later transition elements such as Fe, Co, and Ni do not form stable carbides in equilibrium. It may be natural to suspect that the M−C bond is weakened with an increase in the atomic number of M even when 3d transition metals are present in Fe_3C. However, we found that the M−C bond strength in Fe_3C does not depend on the atomic number of the solute atom. It is the M−M bond–not the M−C bond–that shows the great solution effects. The experimental results shown in Fig. 3.10 can be well-explained assuming that the Vickers hardness is determined by the magnitude of covalent bond strength around the solute atoms, because the M−M bond overlap population decreases with the order of Cr, Mn, Fe, Ni according to the model being presented.

The primary slip plane of Fe_3C is (001) [24]. The (001) plane is parallel to the layers composed of trigonal prisms (Fig. 3.1a). As described previously, the dominant chemical bonds between these layers are the M−M bonds. The strength of the M−M bond–rather than that of the M−C bond–may determine the critical stress for dislocation motion in Fe_3C. The result of geometric consideration is therefore consistent with our conclusion obtained by electronic calculations.

3.3.5 Effect of Solute Atoms in Other Transition Metal Carbides

Let us now consider the effect of solute atoms in other transition metal carbides. Before we come to that, we should discuss the electronic structure of pure transition-metal carbides without solute atoms. The electronic structure calculations for other 3*d* transition-metal carbides have been performed using model clusters composed of about 100 atoms. As an example of the results, the total and partial density of states (TDOS and PDOS) and the overlap population diagrams for the metal–C (M−C) and metal–metal (M−M) bonds of model clusters for Cr_7C_3, Mn_3C, Co_3C, and Ni_3C are shown in Figs. 3.11–3.14. Regarding the M−M bonding, E_F in Cr_7C_3 lies between the

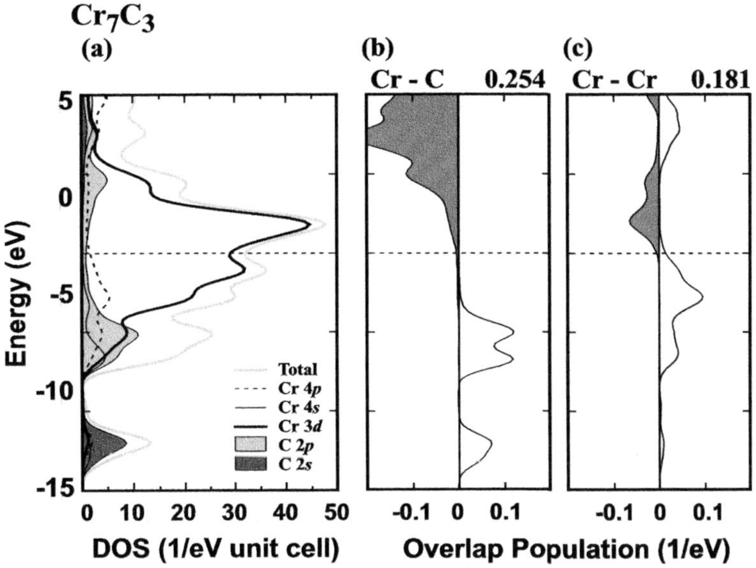

Fig. 3.11. Density of states and the overlap population diagrams for Cr_7C_3. (**a**) Total and partial density of states. (**b**) Overlap population diagram for the M–M bond; (**c**) for the M–C bond

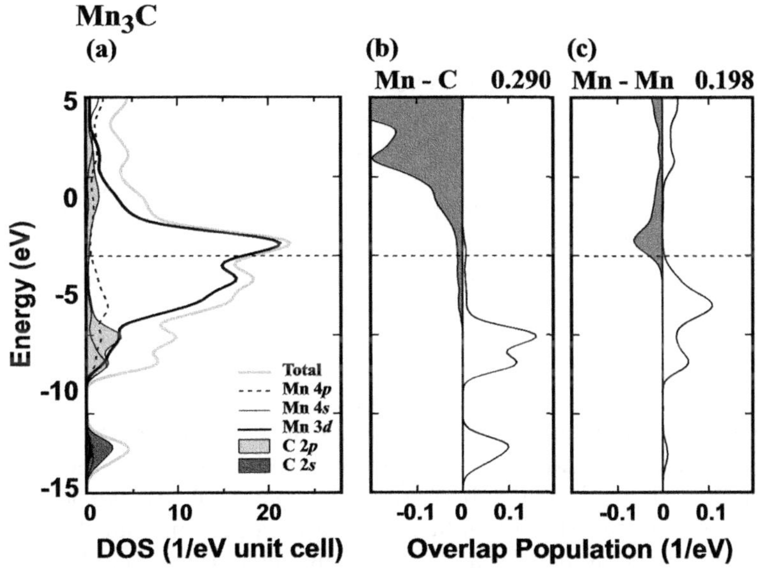

Fig. 3.12. Density of states and the overlap population diagrams for Mn_3C. (**a**) Total and partial density of states. (**b**) Overlap population diagram for the M–M bond; (**c**) for the M–C bond

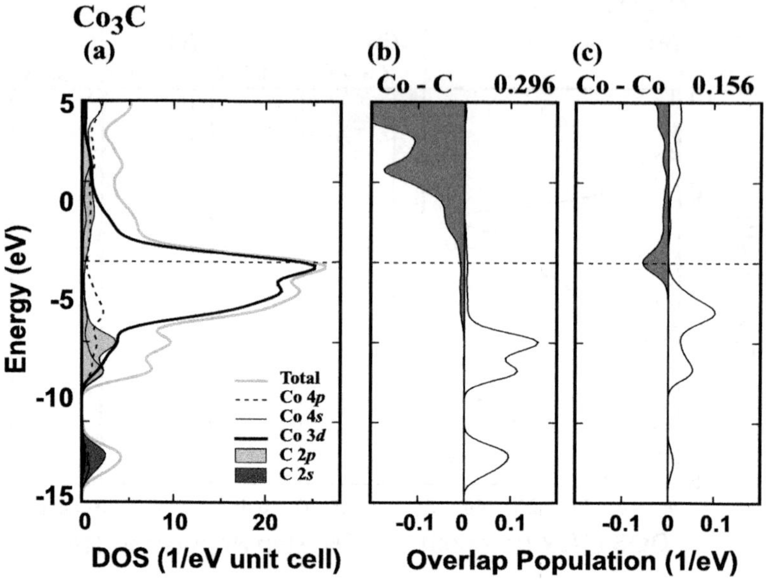

Fig. 3.13. Density of states and the overlap population diagrams for Co_3C. (**a**) Total and partial density of states. (**b**) Overlap population diagram for the M–M bond; (**c**) for the M–C bond

bonding and antibonding bands, and the antibonding band contributes nearly nothing to the Cr–Cr bonding. When going from Cr_7C_3 to Ni_3C, E_F moves upwards, and more electrons occupy the antibonding orbitals. From Mn_3C to Ni_3C, the strength of the M–M bonding decreases with increasing atomic number of the metal atoms. However, the bond overlap population of the Cr–Cr bond in Cr_7C_3 is smaller than that of the Mn–Mn bonding in Mn_3C in spite of a smaller antibonding contribution in Cr_7C_3. The reason is that the carbon content of Cr_7C_3 is larger than that of other carbides. Because the formation of the M–C bond decreases the strength of the M–M bond, an increase of the carbon content leads to a decrease of the bond overlap population of the M–M bond. Another feature is that the $3d$-band gets narrower with an increase in the atomic number of the metal atoms. This trend is similar to pure metals. As one proceeds leftward on the periodic table, the nuclear potential gets larger and contracts the size of the $3d$ orbitals. The interaction between $3d$ orbitals decreases because of the contraction of the $3d$ orbitals.

In contrast to the M–M bonding, the antibonding component of the M–C bonding is located above E_F, and the bonding character around E_F is nearly non-bonding in all carbides considered here. The bond overlap populations of the M–C bond are nearly same values in Mn_3C, Fe_3C, and Co_3C, which crystallize into the same structure. However, because the crystal structures of Cr_7C_3 and Ni_3C are different from the other carbides, the bond overlap population of the single M–C bond is not applicable to the comparison of the

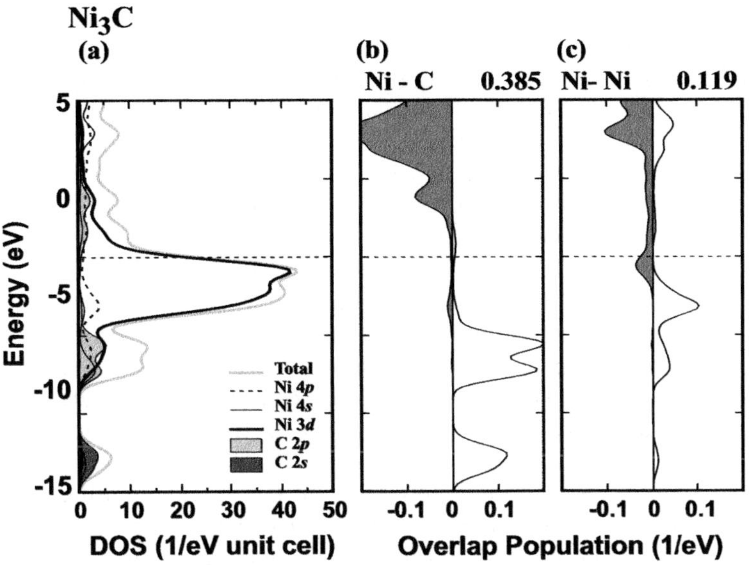

Fig. 3.14. Density of states and the overlap population diagrams for Ni_3C. (**a**) Total and partial density of states. (**b**) Overlap population diagram for the M–M bond; (**c**) for the M–C bond

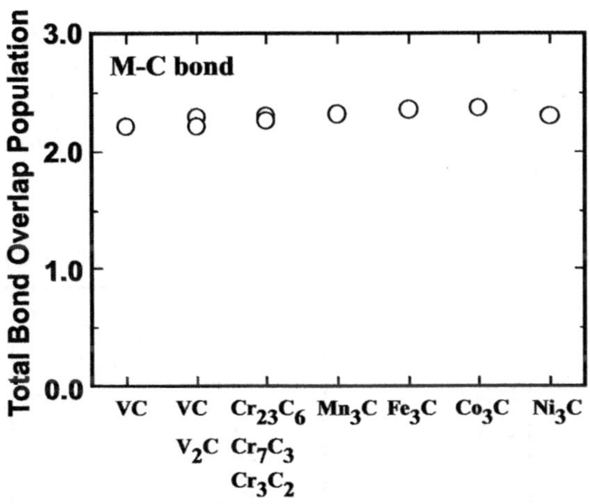

Fig. 3.15. Total bond overlap populations on C sites for M–C bonding

strength of the M–C bond. In all carbides in question, the carbon atom is located in a trigonal prism or an octahedron composed of six metals. Every carbon atom in Cr_7C_3, Mn_3C, and Ni_3C has three, two, and zero the second or third nearest neighbor atoms, respectively. In order to take the difference of the local structure into account, the total bond overlap population was evaluated by summing up the bond overlap populations for the M–C bond around the carbon atoms and plotted in Fig. 3.15. Irrespective of the sort of metal atoms or the carbon content, the total bond overlap populations are almost constant. This result indicates that the covalent bondings formed around the carbon atoms in 3*d* transition-metal carbides is almost independent of the kind of metal atoms or the carbon content. One of the reasons is that throughout the 3*d* transition-metal carbides from TiC to Ni_3C, the bonding character around E_F is non-bonding, and there is no antibonding contribution to the M–C bonding. These features about the M–C and M–M bonding in 3*d* transition-metal carbides are illustrated in Fig. 3.16.

Based on the effect of the solute atoms on Fe_3C, we can figure out the effect of the solute atoms on 3rd transition-metal carbides. As illustrated in Fig. 3.17, if the 3*d* transition-metal carbides containing the solute atoms have a large atomic number, E_F is shifted upwards. In the case of solute atoms with a smaller atomic number, E_F moves downwards. From the overlap population diagrams for the M–M bond, the change of the M–M bonding induced by the solute atoms can be estimated. In the case of Fe_3C, the solute atoms with a larger atomic number weaken the Fe–Fe bonding because of the increase of the antibonding contribution. When the solute atoms contained in Fe_3C have a smaller atomic number, the E_F moves downwards, and the strength of the Fe–Fe bond increases because of the decrease of the antibonding contribution. In the case of chromium carbides, any solute atoms in 3*d* transition metals decrease the strength of the M–M bond, because in pure chromium carbides, E_F is located between the bonding and antibonding band, and the shift of E_F always induces a loss of bonding components. Another factor is the change in the magnitude of the bonding and antibonding components because of the change of the M–M interaction. The 3*d*–3*d* interaction, which is the dominant component in M–M bonding, decreases with increasing atomic

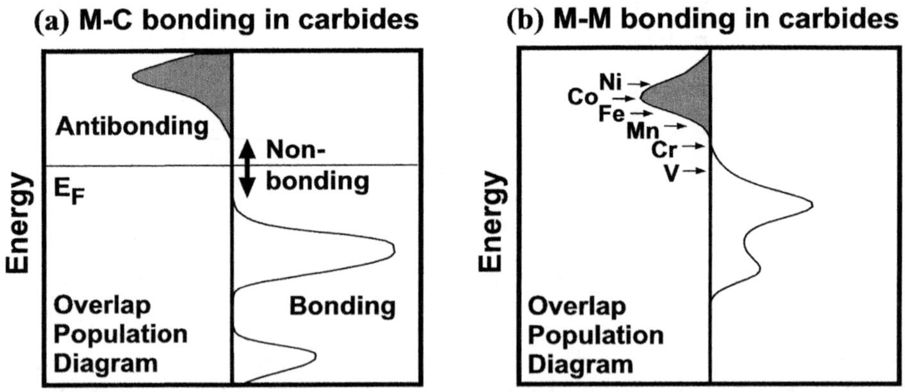

Fig. 3.16. Schematic illustrations of overlap population diagrams for 3*d* transition-metal carbides. (**a**) For the M–C bond. (**b**) For the M–M bond. *Arrows* denote the position of the Fermi level

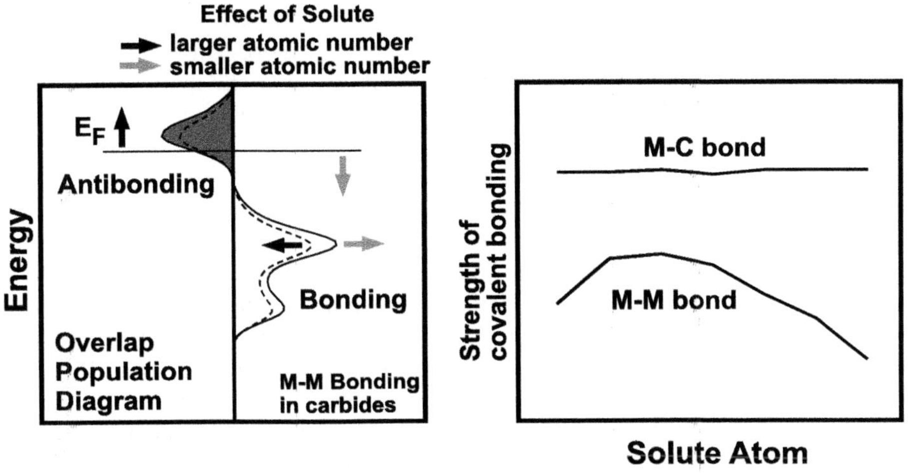

Fig. 3.17. Schematic illustrations for the effect of solute atoms in $3d$ transition-metal carbides. (**a**) Effect of solute atoms on the M−M bonding. *Black* and *gray arrows* denote the effect of the solute atoms which have larger and smaller atomic number, respectively. (**b**) The change in the strength of the M−C and M−M bonding induced by solute atoms

number of the metal atoms. This decrease is due to the contraction in the size of the $3d$ orbitals with an increase of the atomic number of the metal atoms. However, a trend of the interaction among $4sp$ orbitals is opposite to the $3d$–$3d$ interaction. Consequently, the M−M interaction has a maximum around Cr. In comparison with the effect of the shift of E_F, the change of the magnitude of the bonding and antibonding components induced by solute atoms has a similar effect on the strength of the M−M bonding. Therefore, the M−M bonding strongly depends on the atomic number of the solute atoms. On the other hand, the solute atoms have a smaller influence on the M−C bonding than on the M−M bonding, because the M−C bonding is nearly non-bonding in character around E_F. As mentioned above, the M−M bonding is more dominant in carbides with a smaller C content. Therefore, the influence of the solute atoms on the M−M bonding is more significant in carbides with a smaller C content.

3.4 Chemical Bonding at the Fe/TiX (X = C, N, or O) Interfaces

Interfaces play a key role in the control of material properties and in the joining of different types of materials. Some compounds that have good coherence with an Fe matrix have been used for precipitation hardening or as an intra-granular ferrite nucleation site in the steel industry [28–30]. One group of the most popular compounds for this application is composed of the compounds with a rock-salt-type structure, such as TiC or VC. In order to select the compounds for these applications, the lattice-misfit strain at the metal/compound interface have often been taken into consideration. However, there have been no discussions about the chemical bonding at the interface of these systems.

Metal/oxide ceramic interfaces play key roles in a number of structural and electronic applications. In recent years, in order to understand the interfaces, several electronic structure calculations have been performed for some model systems, such as the metal/MgO(001) [31–38] and metal/Al_2O_3(0001) interfaces [39–46]. In these works, the equilibrium interfacial distance or the stable configuration of atoms at the interfaces were evaluated by minimization of total energy using supercell models or image-interaction models. Regarding the chemical bondings at the interfaces, the basic feature of the interaction between metal and oxygen atoms was shown by Johnson and Pepper [39]. They made the calculation of (Fe, Ni, Cu, Ag)/Al_2O_3 interfaces using a self-consistent scattered-wave method with small cluster models composed of seven atoms and found that the strength of covalent bondings at the interface was decreased in the series Fe, Ni, Cu, and Ag because of an increase in the occupation of antibonding orbitals established by the metal–oxygen interaction. Regarding the metal–oxygen bonding, a similar trend was pointed out by Schönberger et al. who made a more realistic first-principle calculation of (Ti, Ag)/MgO interfaces [32]. However, strength of covalent bondings is determined by the balance of the bonding and antibonding contributions. It is important to discuss not only the occupation of antibonding orbitals but also the variation of bonding contributions due to the presence of metal atoms. In these previous works, little attention has been given to the metal–metal bond at metal/ceramic interfaces except for Heifets et al. [35], who reported that the Ag–Mg bond at the Ag/MgO interface was negligibly small using an ab initio Hartree–Fock calculation with a supercell model. However, it does not mean that the metal–metal bond is negligible in general. In this section, we discuss the chemical bonding at the Fe/TiC, Fe/TiN, and Fe/TiO interfaces with Baker–Nutting orientation relationship. These interfaces have been used for precipitation hardening or as intragranular ferrite nucleation sites in the steel industry. The electronic structures of these interfaces were calculated in order to investigate the mechanism of the interfacial bond with interests of the difference among Fe/TiC, Fe/TiN, and Fe/TiO interfaces.

3.4.1 Model Clusters for the Fe/TiX Interfaces

The orientation relationship often observed between bcc Fe and rock-salt-type carbides or nitrides is (001)Fe//(001)MX and [110]Fe//[100]MX, which is called the Baker–Nutting orientation relationship [47, 48], as shown in Fig. 3.18. In accord with this relationship, we employed two model clusters shown in Fig. 3.19 for the calculations of the Fe/TiX interfaces. The

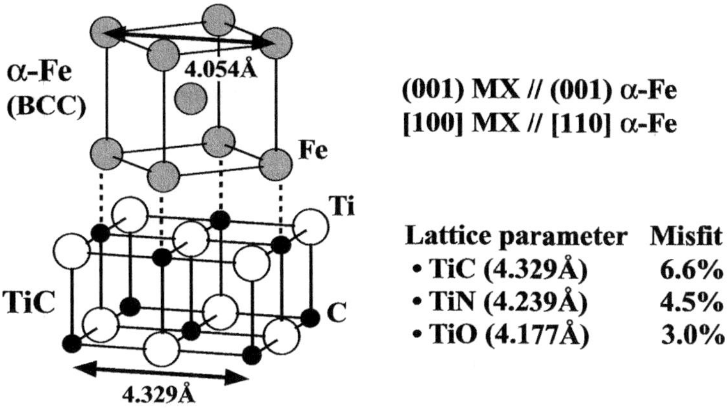

Fig. 3.18. Baker–Nutting orientation relationship between rock-salt-type MX compounds and α-Fe (bcc)

model clusters are composed of two bcc Fe metal layers and two TiX compound layers. The left model cluster in Fig. 3.19 is the Fe-on-X model where Fe atoms are present on the top of X atoms. The middle model cluster is the Fe-on-Ti model where Fe atoms are located on the top of Ti atoms. The table in Fig. 3.18 shows lattice parameters of bulk crystals as measured by experiment and the lattice misfit between bcc Fe and TiX with the Baker–Nutting orientation relationship. In our calculation, the lattice parameter of the Fe matrix parallel to the interface was expanded so as to fit it with the lattice parameters of TiX for simplicity. Therefore,

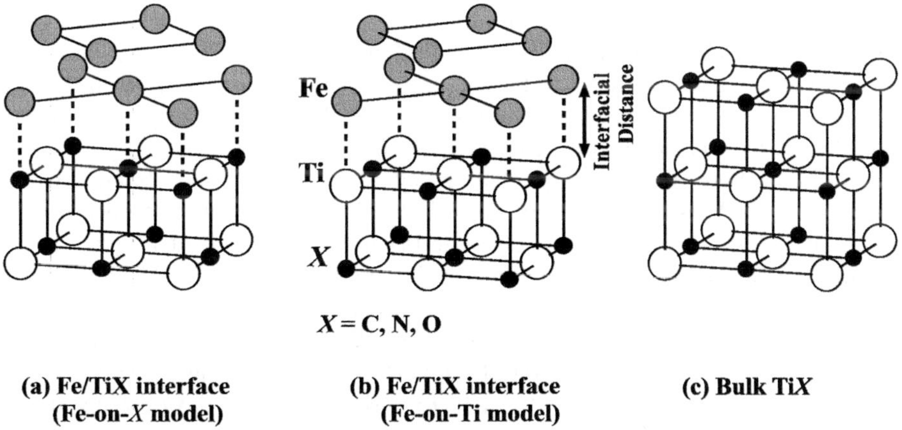

Fig. 3.19. The structure of model clusters employed in this work (X = C, N, or O). (**a**) $Fe_9-Ti_9X_9$ cluster for the Fe/TiX interfaces where the Fe atoms are on X atoms. (**b**) $Fe_9-Ti_9X_9$ cluster for the Fe/TiX interfaces where the Fe atoms are on Ti atoms. (**c**) $Ti_{14}X_{13}$ cluster for bulk TiX

all Fe atoms at the interface are positioned on the top of either Ti or X atoms in these clusters. In order to estimate the influence of the interfacial distance, the calculation of these models with various interfacial distances between Fe and TiX layers in contact were performed. For the calculations of bulk TiX, model clusters composed of 27 atoms were employed. In the case of TiO, the model cluster was embedded in the Madelung potential generated by approximately 2500 point charges. Spin polarization was taken into account in the calculations of the interfaces containing Fe atoms.

3.4.2 1 Bulk TiC, TiN, and TiO

Initially, the chemical bondings in bulk TiC, TiN, and TiO were investigated. Several authors have already reported a series of calculations for the rock-salt-type structure compounds, including TiC, TiN, and TiO [49–56]; TiC and TiN [57–62]; TiC and TiO [63,64]; and TiO [65–73]. In addition, many calculations for TiC have been performed using the band-structure method [74–78] and the cluster method [79–87]. The band-structure method has been generally used to calculate the electronic structure of perfect crystalline materials, whereas a cluster calculation provides a good approximation for the bulk electronic structure when valence orbitals are localized to a range of a few ionic shells. The cluster calculation is more flexible and useful applied to imperfections such as interfaces.

The PDOS of the valence band and the overlap population diagrams for the Ti−Ti and Ti−X bonds of these model clusters are shown in Fig. 3.20. The PDOS curves are made by broadening the discrete energy levels by a Gaussian function of 1.0 eV FWHM. The overlap population diagrams are obtained by convoluting the overlap population at each molecular orbital with a Gaussian function of 1.0 eV FWHM. These curves of the DOS and overlap population are shifted so as to set the Fermi level (E_F) at zero. The value written in each panel is the bond overlap population obtained by summing up the overlap population at each molecular orbital up to E_F.

In TiC, there are two large bands below E_F. The upper band located from −5 to 0 eV is mainly composed of the C-2*p* levels interacting strongly with the Ti-3*d*, 4*s*, and 4*p* orbitals. The lower band located at about −10 eV is mainly composed of the C-2*s* levels. The component of the Ti-3*d*, 4*s*, and 4*p* orbitals in the C-2*s* band is smaller than that in the C-2*p* band. As can be seen in the overlap population diagram, these two bands contribute greatly to the Ti−C bonding. The Ti-3*d* band above E_F is antibonding for the Ti−C.

In TiN, the basic feature of the PDOS and overlap population diagrams is similar to those of TiC. The separation between the Ti-3*d* and N-2*sp* bands in TiN is larger than that between the Ti-3*d* and C-2*sp* in TiC by 1.8 eV. The Ti orbitals show slightly less interaction with the N-2*sp* orbitals than in TiC. Consequently the magnitude of overlap population for the N-2*s* and N-2*p* bands are smaller than in the case of TiC.

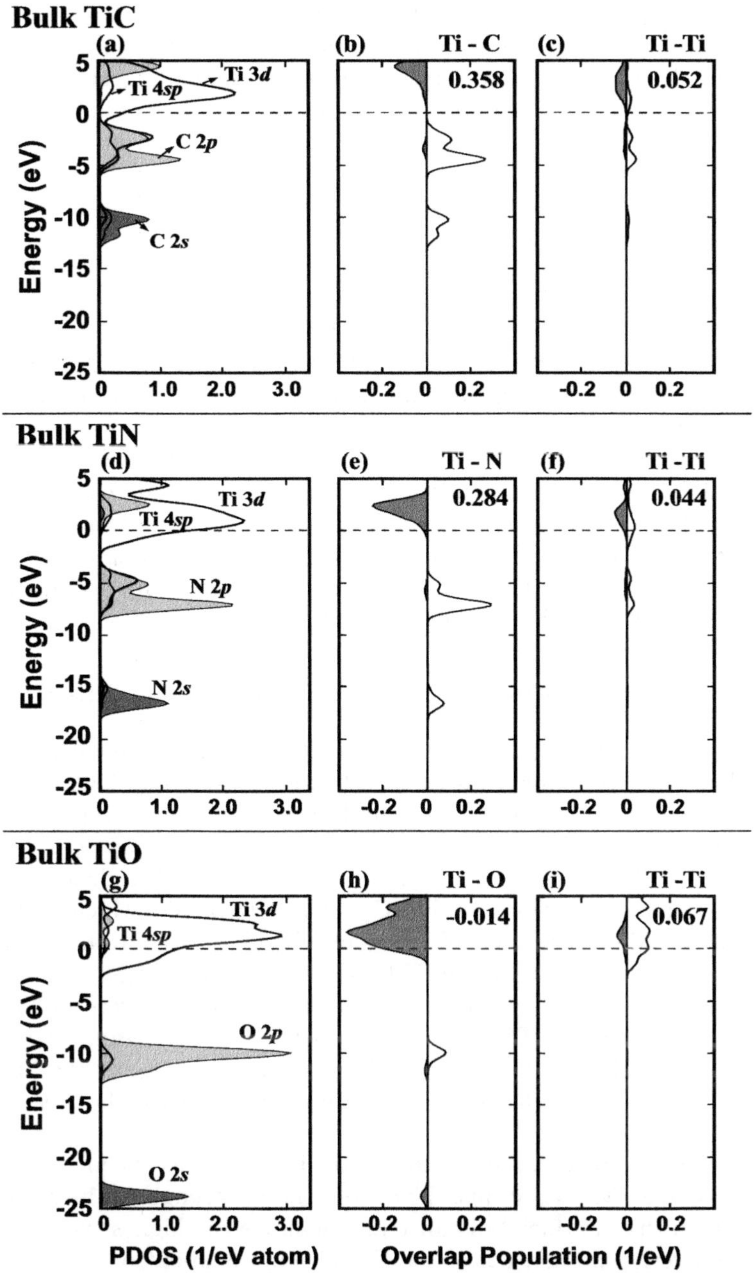

Fig. 3.20. Partial density of states and the overlap population diagrams for bulk TiX (X = C, N, or O). *Left panels*, (**a**), (**d**), and (**g**), show the density of states. *Middle panels*, (**b**), (**e**), and (**h**), show overlap population diagrams for the Ti–X bond. *Right panels*, (**c**), (**f**), and (**i**), show overlap population diagrams for the Ti–Ti bond

In TiO, the O-2*p* band is widely separated from the Ti-3*d* band. The component of the Ti-3*d*, 4*s*, and 4*p* orbitals in the O-2*p* band is very small. Therefore, the covalent interaction for the Ti−O bonding is much smaller than that for the Ti−C bonding in TiC or the Ti−N bonding in TiN. The O-2*s* band located at −23 eV has a small antibonding contribution to the Ti−O bonding because of the repulsive interaction between the O-2*p* and Ti-3*p* core. The Ti-3*d* band located above E_F reduces the bonding nature of the Ti−X bond and is considered to be an antibonding band. The antibonding band is empty in TiC and nearly empty in TiN. However, it is partially occupied in the case of TiO. This is the reason the bond overlap population of Ti−O is much smaller than that of Ti−X in TiC and TiN, and shows rather negative value. The feature of the PDOS for TiC, TiN, and TiO in this chapter agree well with previous band-structure calculations as referred to previously.

The Ti−Ti bond is much weaker than the Ti−X bond in TiC and TiN. In TiC and TiN, bonding components for the Ti−Ti bond are composed of the C-2*p* or N-2*p* band and the Ti-3*d* band, though these bonding contributions are very small. On the other hand, in TiO, the Ti−Ti bonding contribution is essentially zero in the O-2*p* band, but in the Ti-3*d* band, its magnitude is somewhat larger than that of TiC and TiN.

3.4.3 Preferred Position of Fe Atoms at the Fe/TiC Interface

We have performed the calculations for the Fe/TiC interface with various interfacial distances ranging from 1.6 to 2.4 Å. Figures 3.21 and 3.22 show bond overlap populations for the Fe-on-C and Fe-on-Ti models, respectively. The labels Fe−C and Fe−Ti respectively refer to the bond overlap populations for Fe−C and Fe−Ti bonds at the interface. The label Fe−TiC denotes the sum of bond overlap populations for Fe−C and Fe−Ti bonds, which corresponds to the magnitude of the total of interfacial bond. The coordination number is taken into account for the Fe−TiC bond, as seen in Figs. 3.21 and 3.22, the Fe atom has one first nearest-neighbor (NN) and four second NN atoms of TiX at the interface. The bond length between the Fe atom and the first NN atom is equal to the interfacial distance. In the range of the interfacial distances, 1.6 to 2.4 Å, the second NN are about 1.5 times farther than the first nearest neighbors.

In the Fe-on-C model, the Fe atom at the interface has one first NN C atom and four second NN Ti atoms. The Fe−C bond strength is increased with a decrease in the interfacial distance and achieves the highest value at 1.8 Å. The bond overlap population of the Fe−Ti bond is about one fourth of that of the Fe−C bond. However, taking account of the coordination number, the Fe−Ti bond has as large contribution to the Fe−TiC bond as the Fe−C bond.

In the Fe-on-Ti model, the Fe atom at the interface has one first NN Ti atom and four second NN C atoms. In comparison with the Fe-on-C model, both first and second NN bonds are weaker than those in the Fe-on-Ti model.

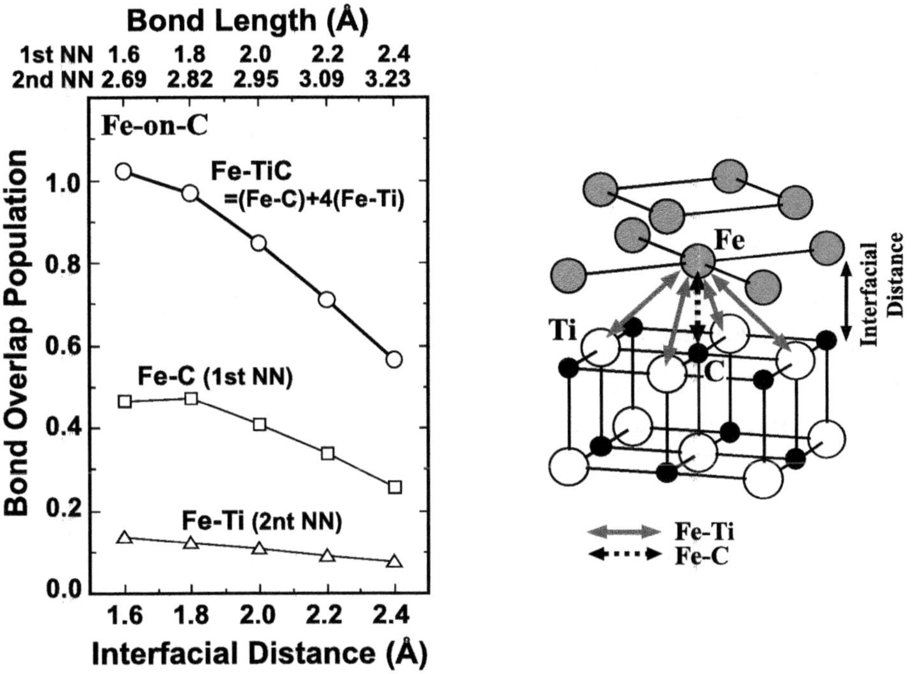

Fig. 3.21. Bond overlap populations of the Fe-on-C model for the Fe/TiC interface as a function of the interfacial distance

Therefore, the strength of the Fe–TiC bond in the Fe-on-C model is about twice that in the Fe-on-Ti model at the same interfacial distance. From metallic radii [88], the Fe–Ti bond length is considered to be longer than the bond length of Fe–C. In iron carbide, Fe_3C, the Fe–C and the Fe–Fe bond lengths are 2.018 Å, and 2.611 Å, respectively. At a suitable interfacial distance for the Fe–Ti bond in the Fe-on-Ti model, the Fe–C bond length is too long for formation of the Fe–C bond. On the other hand, when the interfacial distance is suitable for the Fe–C bond in the Fe-on-C model, the Fe–Ti bond-length is still within the range of the bonding interaction. Since the number of second NN bonds is four times greater than the first NN bonds, the Fe-on-X geometry is expected to be more preferable from the viewpoint of interfacial bond strength. However, the accurate determination of the interfacial distance is rather difficult. The Fe–TiC bond strength calculated in this work does not display a maximum value. In order to estimate the equilibrium interfacial distance, it is necessary to include the bonding interactions near the interface that are ignored in above discussion. They are the Fe–Fe bond in Fe layers and the Ti–C bond in TiC. However, the model clusters employed in this work are not large enough to account for and to quantitatively discuss the strength of these bonds. In the following section, we employ Fe-on-X models with a fixed interfacial distance of 2.0 Å for simplicity and examine the origin

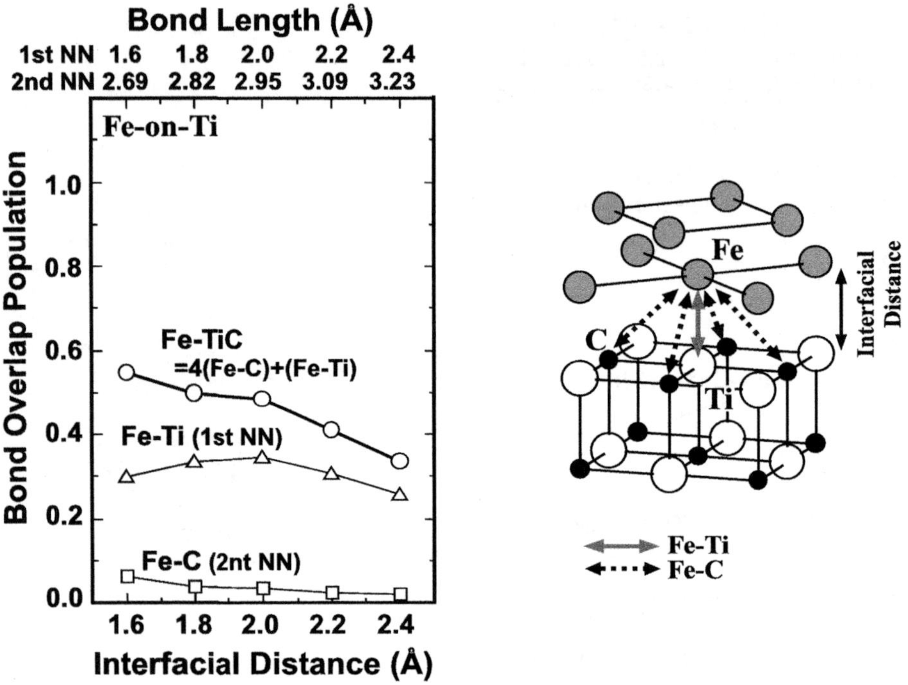

Fig. 3.22. Bond overlap populations of the Fe-on-Ti model for the Fe/TiC interface as a function of the interfacial distance

of the chemical bonding at these interfaces. The Fe−X distances are 2.018 Å, 1.899 Å and 2.163 Å, in Fe_3C, Fe_4N, and FeO, respectively. The value 2.0 Å, for the interfacial distance is determined as an approximate average of Fe−X distances in bulk Fe−X compounds.

TiO layers of the Fe/TiO cluster was embedded in an ionic TiO lattice in the same way as the bulk TiO cluster. In order to estimate the influence of the Madelung potential on the chemical bonding, we have also performed the calculation of the Fe/TiO cluster without the Madelung potential and compared the calculated results with or without the Madelung potential.

3.4.4 Analysis of the Chemical Bonding at the Fe/TiC, Fe/TiN, and Fe/TiO Interfaces

Figure 3.23 shows the bond overlap populations of the Fe/TiC, Fe/TiN, and Fe/TiO clusters. At these interfaces, one Fe atom has one Fe−X bond and four Fe−Ti bonds. The marks labeled "4 (Fe−Ti)" in Fig. 3.23 denote the bond overlap population of the Fe−Ti bond multiplied by a factor of four. The Fe−Ti bond strength is increased in the sequence, Fe/TiC, Fe/TiN, Fe/TiO. Regarding the Fe−X bond, the Fe−C and Fe−N bonds display strong covalent bonding. On the other hand, the bond overlap population of the Fe−O bond

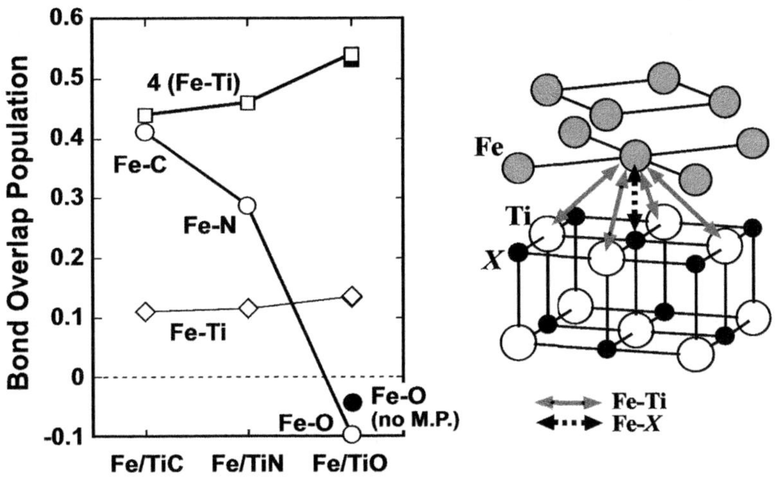

Fig. 3.23. Bond overlap populations at the Fe/TiX interfaces (X = C, N, or O). *Solid marks* denote bond overlap populations of the Fe/TiO cluster without the Madelung potential

is −0.1, which means the Fe−O bond is slightly antibonding in character. Summing up the bond overlap populations of the Fe−X and Fe−Ti bonds, which correspond to the total of interfacial bond strength, the value is decreased in the order of Fe/TiC, Fe/TiN, Fe/TiO. In our calculation, we take into account the effect of the Madelung potential only in the case of Fe/TiO. However, when the Madelung potential in the Fe/TiO cluster is not taken into account, the bond overlap-populations do not change remarkably. Therefore, the difference of the chemical bonding between the Fe−O and the Fe−C or Fe−N bond cannot be ascribed to the inclusion of the Madelung potential. In what follows, we only discuss the chemical bonding of the Fe/TiO cluster with the Madelung potential for the Fe/TiO interface.

The net charges of atoms in the clusters are plotted in Fig. 3.24. The net charges of Ti atoms at the interfaces are almost the same as those in bulk TiX. On the other hand, the net charges of X atoms at the interfaces are slightly smaller than those in bulk TiX. The Fe atoms at the interfaces are charged negatively, though the values of the net charges are very small for all of the interfaces. Furthermore, these results show that no significant charge transfer takes place between Fe and TiX layers at the interface. Therefore, the ionic interaction between Fe and TiX seems to be small even in the case of TiO, which is more ionic than TiC and TiN.

The PDOS of Fe, Ti (3d + 4s + 4p), and X (X = C, N, or O) atoms in the valence band at the interfaces and the overlap population diagrams for the Fe−X and Fe−Ti bonds are shown in Fig. 3.25. In all of the interfaces, the Fe component in the up-spin band appears below E_F and peaks at −1.8 eV, while in the down-spin band, the Fe component lies around E_F and peaks at

1.5 eV. The shapes of the PDOS of the Ti and X atoms at the interfaces are similar to those in bulk TiX.

At the Fe/TiC interfaces, the Fe band strongly interacts with the C-2p orbital. This causes a large bonding component in the valence band from −7 to 0 eV and a small antibonding component around E_F with respect to the Fe–C bond. From −13 to −10 eV a small bonding contribution exists mainly composed of the C-2s interacting with Fe orbitals.

The PDOS and overlap population diagrams at the Fe/TiN interface have similar features to those at the Fe/TiC interface. However, the bonding interaction is slightly smaller, and the antibonding component is larger, than at the Fe/TiC interface. Therefore, the bond overlap population of the Fe–N bond is smaller than that of the Fe–C bond.

At the Fe/TiO interface, the O-2p band is located at around −10 eV, which is just below the bottom of the Fe band. Regarding Fe–O bonding, the O-2p band is completely bonding, but the bonding contribution is much smaller than the other two interfaces. In contrast, the Fe band is completely antibonding. The antibonding band below E_F is larger than the bonding one. This makes the bond overlap population of Fe–O a negative value, −0.098.

As seen before, the position of the X-2s and 2p bands become lower in the order of C, N, O. As a result, the bonding interaction becomes smaller while the number of electrons occupying the antibonding state increases. This is the reason the bond overlap population of the Fe–X bond is smaller in the order of C, N, O.

As for the Fe–Ti bond, at all of the interfaces, the Fe band below E_F displays a small bonding tendency. At the Fe/TiO interface, the Fe–O bond

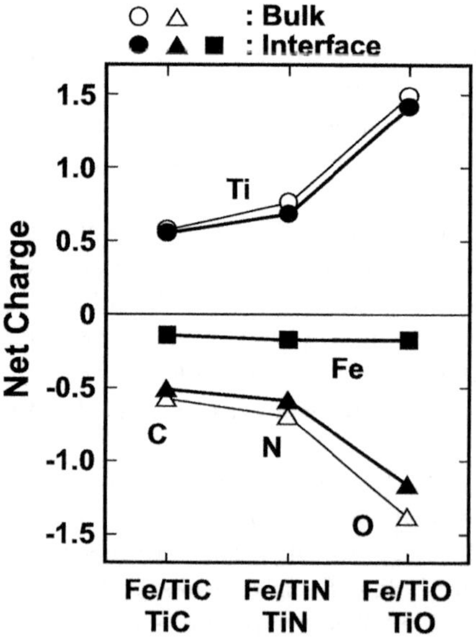

Fig. 3.24. Net charges at the Fe/TiX interfaces and in bulk TiX (X = C, N, or O). *Open marks* denote the net charge of Ti and X in bulk TiX, and *filled marks* denote the net charge of Fe, Ti, and X at the Fe/TiX interfaces

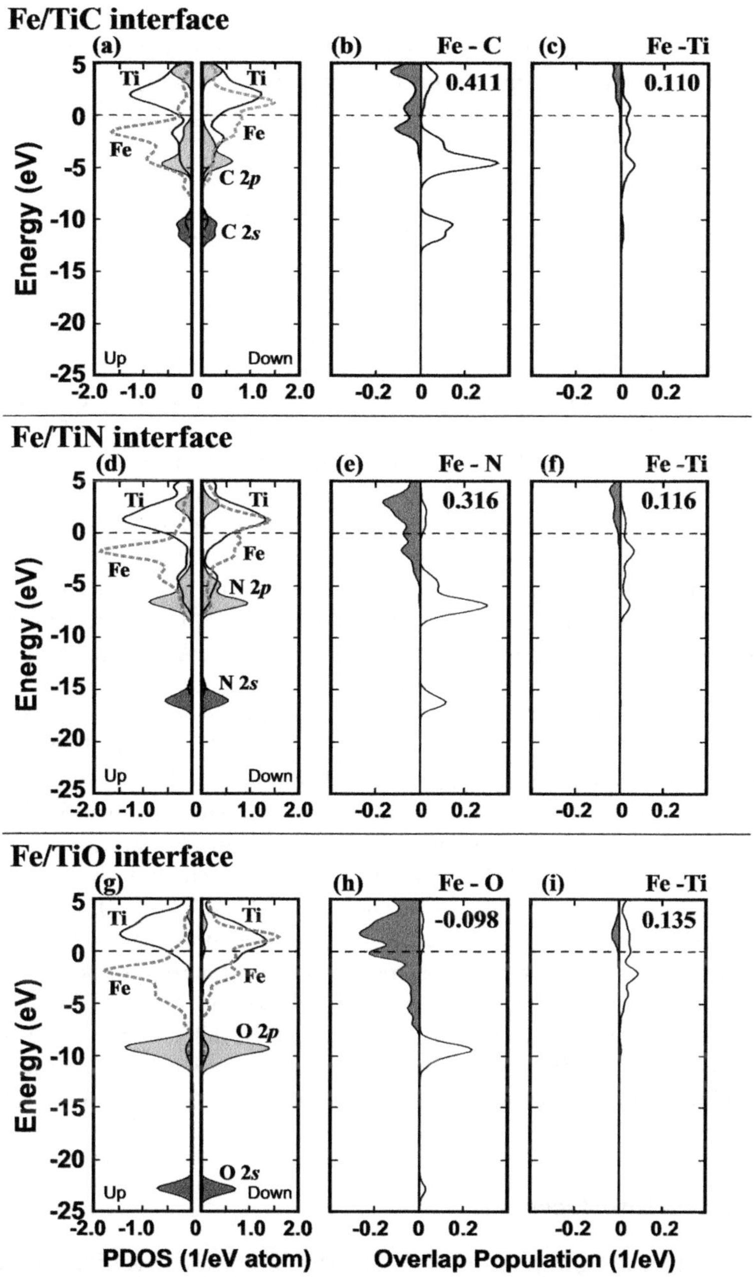

Fig. 3.25. Partial density of states and the overlap population diagrams for the Fe/TiX interfaces (X = C, N, or O). The *left panels*, (**a**, **d**), and (**g**), show density of states. The *middle panels*, (**b**, **e**), and (**h**), show overlap population diagrams for the Fe–X bond. The *right panels*, (**c**, **f**), and (**i**), show overlap population diagrams for the Fe–Ti bond

is antibonding; therefore, the dominant chemical bond is the Fe–Ti bond. As mentioned above, previous theoretical calculations pay little attention to the metal–metal bonding at metal/ceramic interfaces. In fact, in the case of the Ag/MgO interface, it was reported that the Ag–Mg bond at the interface was negligibly small [35]. In the other calculation for the Ag/MgO interface, it was shown that the Ag band was located at the almost same energy range as the O-2p band, but there was no apparent overlap between the Ag and Mg bands [32]. In the present work, at the Fe/TiO interface, the Fe band interacts with the Ti band around E_F to contribute to the Fe–Ti bonding. In recent work for the V/MgO interface [89] through the use of the DV-Xα molecular orbital calculation, it was suggested that the V–Mg bond displays covalent bonding, and the strength of the V–Mg bond is much larger than that of the V–O bond, in contrast to the Ag/MgO interface. Therefore, it seems to be important to consider the metal–metal bond formed at the interface between transition metals with unfilled d-bands and some ceramics.

The TiX compounds act as preferential nucleation sites of ferrite (bcc Fe) when steels are cooled from the austenite (fcc Fe) region. However, experimental data on the nucleation potency are not well-established. Yamamoto et al. concluded that TiN was a good ferrite nucleant because of its small lattice mismatch with ferrite [30]. Gregg and Bhadeshia, however, reported that TiO acted as a good nucleant, while TiN did not [90]. Although the mismatch is smaller by 1.5% in Fe/TiO than in Fe/TiN, they were reluctant to accept the lattice-matching model. Recently, Zhang et al. investigated the ferrite nucleation effectiveness of pure polycrystal ceramics (TiC, TiN, TiO, VN, AlN, and Al_2O_3) [91]. Their results contradicted the data by Gregg and Bhadeshia. They reported that VN, which has the smallest mismatch with ferrite, is the most effective, while TiO, which has the second smallest mismatch, is the least effective. Regarding TiC, TiN, and TiO, they reported that the ferrite nucleation effectiveness was decreased in the order of TiC, TiN, TiO. This order is opposite to that expected from the lattice mismatch with ferrite. In other words, good lattice coherence could not be the major requirement for a good ferrite nucleant. According to the present calculation, the Fe/TiX interfacial bond strength decreases in the order of TiC, TiN, TiO, which is the same as that in the report by Zhang et al. It suggests that the chemical bonding would be more important than the lattice mismatch for the ferrite nucleation in the Fe/TiX system.

3.4.5 Other Interfaces

To gain systematic physical insight into the interfacial bonding mechanism, calculations of the Cr/TiX, Ni/TiX, TM/MgO (TM = Cr, Fe, or Ni) interfaces have been made using the same type of model clusters. The interfacial distances in these interfaces were set to the same value, 2.0 Å, as in Fe/TiX interfaces. The PDOS of the atoms at the interfaces and the overlap population diagrams for the TM–M and TM–X bonds are shown in Figs. 3.26–3.28. The

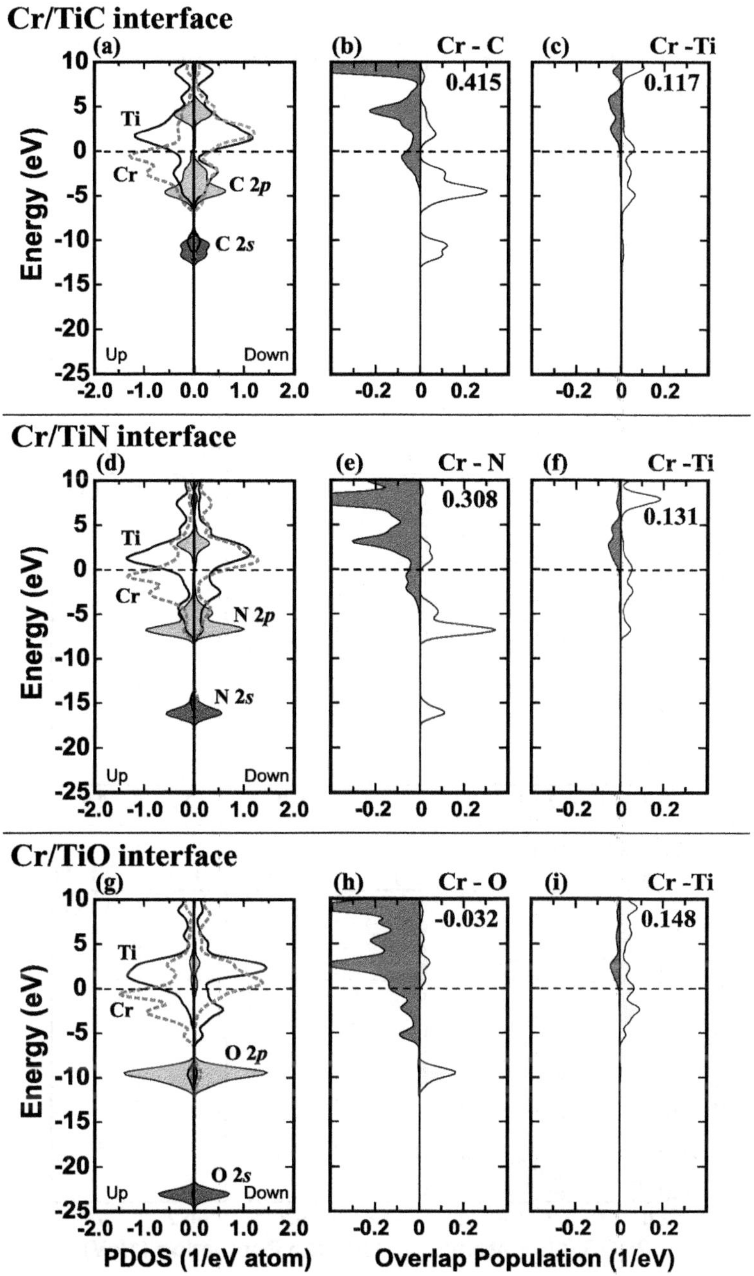

Fig. 3.26. Partial density of states and the overlap population diagrams for the Cr/TiX interfaces (X = C, N, or O). The *left panels*, (**a**), (**d**), and (**g**), show the density of states. The *middle panels*, (**b**), (**e**), and (**h**), show overlap population diagrams for the Cr–X bond. The *right panels*, (**c**), (**f**), and (**i**), show overlap population diagrams for the Cr–Ti bond

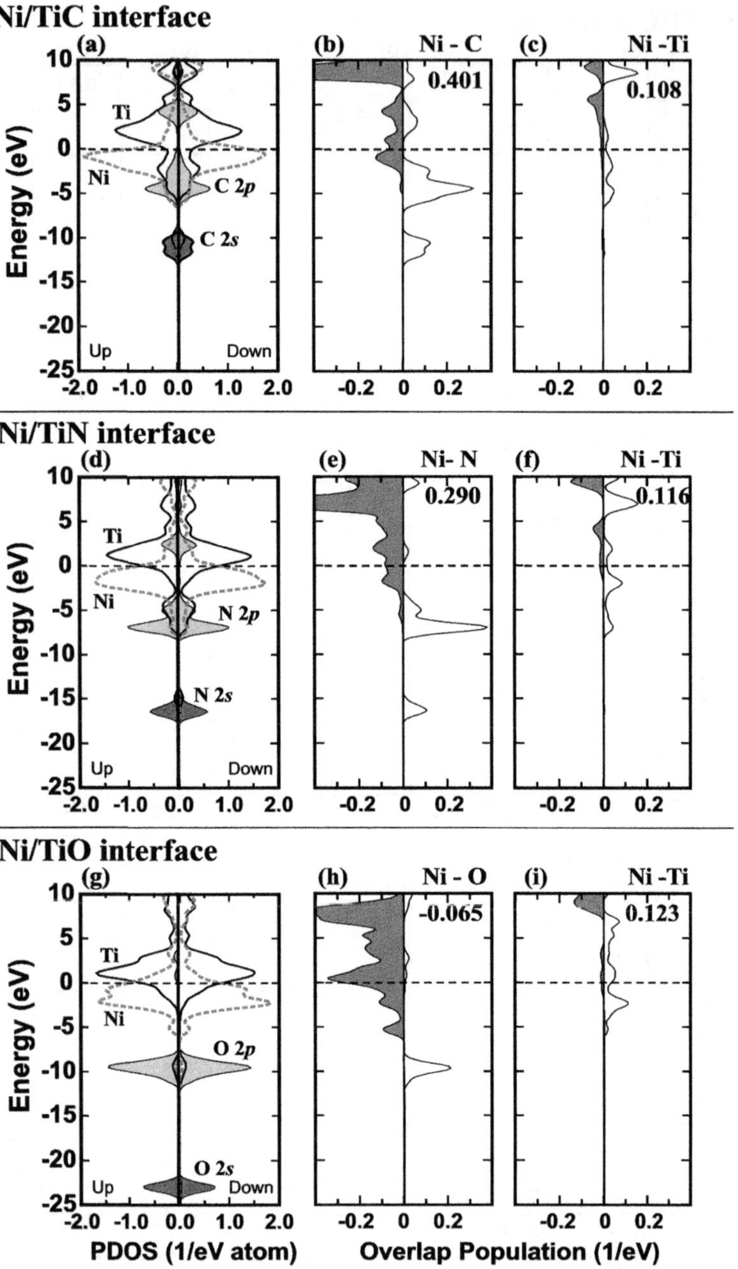

Fig. 3.27. Partial density of states and the overlap population diagrams for the Ni/TiX interfaces (X = C, N, or O). The *left panels*, (**a**), (**d**), and (**g**), show the density of states. The *middle panels*, (**b**), (**e**), and (**h**), show overlap population diagrams for the Ni–X bond. The *right panels*, (**c**), (**f**), and (**i**), show overlap population diagrams for the Ni–Ti bond

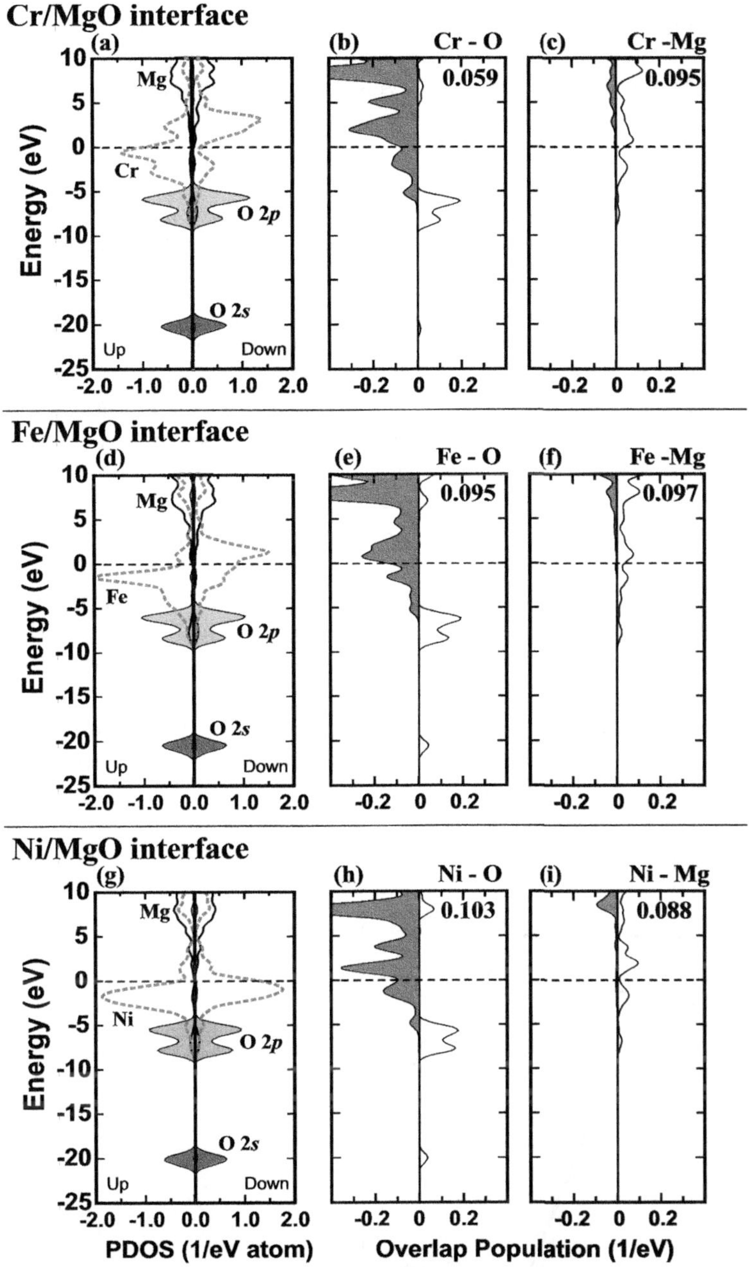

Fig. 3.28. Partial density of states and the overlap population diagrams for the TM/MgO interfaces (TM = Cr, Fe, or Ni). The *left panels*, (**a**), (**d**), and (**g**), show the density of states. The *middle panels*, (**b**), (**e**), and (**h**), show overlap population diagrams for the TM–O bond. The *right panels*, (**c**), (**f**), and (**i**), show overlap population diagrams for the TM–Mg bond

overlap population diagrams for the Cr/TiX and Ni/TiX interfaces show the same features observed in the Fe/TiX interface. One of the clear differences is that the PDOS of the TM moves downwards, and the separation between the Ti and TM bands increases in the sequence Cr, Fe, Ni. This effect leads to the decrease of the TM–Ti bond strength. Regarding the TM–X bonding, both increases of the magnitude of the bonding band and the occupation of the antibonding band occur as the TM band moves downwards. Because these two changes have opposite contributions to the TM–X bonding, the TM–X bonding does not change remarkably with the kind of TM used. In the case of the TM/MgO interfaces, the bond overlap populations of the TM–O bond have positive values in contrast to the TM/TiO interfaces. In comparison with the TM/TiO interfaces, the O-2s and 2p bands in the TM/MgO interfaces are located close to the Fermi level, and the occupation of the antibonding band is small. Consequently, the antibonding contributions to the TM–O bonding are smaller, which leads to an increase of the TM–O bond strength.

Figure 3.29 shows the bond overlap populations of all the interfaces appearing in this study. The TM–M bond strength decreases with an increase in the atomic number of TM due to the downward shift of the TM band. This is a common tendency of the M–M bonding in 3d transition-metal carbides discussed previously. However, the change in the TM–X bond strength is more significant because the change in the position of the X-2s and 2p bands is larger than the TM bands. At all the interfaces, importance of the TM–M bonding compared with the TM–X bonding should be a general feature, whereas the difference of the bonding character at the interfaces mainly arises from the TM–X bonding.

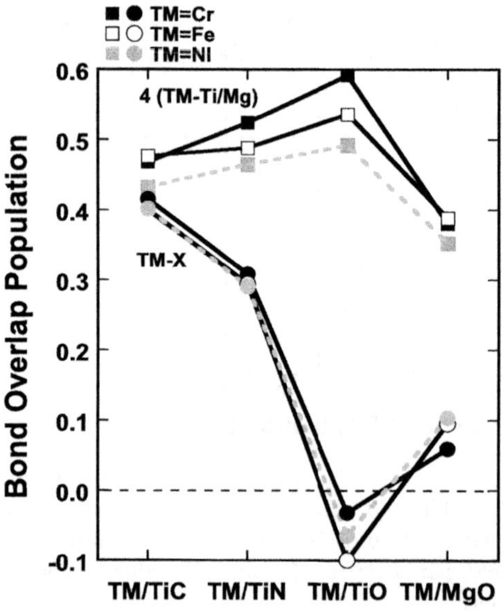

Fig. 3.29. Bond overlap populations at the TM/TiX and TM/MgO interfaces (TM = Cr, Ni, or Fe; X = C, N, or O)

3.5 Conclusions

First-principle calculations using model clusters have been made in order to study the effect of solute atoms on Fe_3C and the Fe/TiX (X = C, N or O) interfaces as examples of lattice imperfections. The change in the chemical bonding between metal atoms plays an important role in Fe_3C-containing solute atoms. On the other hand, the Fe–X bonding at the Fe/TiX interfaces characterize the chemical bonding at the Fe/TiX interfaces.

The strength of the M–C bond in Fe_3C shows little change with a change in solute atoms, but the M–M bond strength strongly depends on the atomic number of the solute atoms. The results can be well-explained using overlap population diagrams for the M–M and M–C bonds. The Fermi level lies at the middle of the antibonding band of the M–M bond in pure Fe_3C. When the solute atoms are present, the position of the M–M antibonding band changes. The number of electrons occupying the antibonding orbitals is therefore altered; the net overlap population of the M–M bond decreases with increasing atomic number of the solute. On the other hand, the M–C bond is nearly non-bonding in character around the Fermi level. The solute atoms have a smaller influence on the M–C bonding than on the M–M bonding.

We found a good correlation between the magnitude of the M–M bond strength and the experimental Vickers hardness of the Fe_3C solid solutions. Since the M–M bond is dominant between (001) planes in Fe_3C, it is natural that the M–M bond strength shows strong effects on the hardness, which is mainly determined by the (001) slip. The present result is contrary to the previous views: it was believed that the M–C bond strength rather than the M–M bond determines the hardness of the Fe_3C solid solutions.

At the Fe/TiX interfaces, no significant charge transfer occurs between the Fe atoms and the TiX. Ionic interaction between Fe atoms and TiX is small, and covalent bonding is predominant at the interface. At these interfaces, the Fe-on-X geometry is more preferable than Fe-on-Ti because of the long-range interaction of the Fe–Ti bond. The reason is that the Fe–Ti bond is found to be effective even at the second nearest neighbor (NN) sites and the number of second NN bonds is four times greater than the first NN bonds. The strength of the Fe–Ti bond is increased in the order of Fe/TiC, Fe/TiN, Fe/TiO. On the other hand, the strength of the Fe–X bond is decreased in this order. In particular, at the Fe/TiO interface, the Fe–O bond is slightly antibonding in character. The change of the strength of the Fe–X bond is much larger than that of the Fe–Ti bond. Therefore, the interfacial bond-strength, which is the sum of the strength of the Fe–X and Fe–Ti bonds, decreases in the order of Fe/TiC, Fe/TiN, Fe/TiO. There is a similar tendency for the chemical bonding at the Cr/TiX and Ni/TiX interfaces.

Several compounds, including TiC, TiN, and TiO, have been applied as intragranular ferrite nucleation sites in steels. However, experimental data on the nucleation potency are not well-established. The interfacial bond-strength as obtained by our calculation agrees well with the ferrite nucleation potency most recently found by experiments – the result which is not predicted by the lattice mismatch with ferrite. This suggests that the chemical bonding contribution seems to be larger than the mismatch strain contribution for the ferrite nucleation in these systems.

References

1. H. Adachi, M. Tsukada, C. Satoko, J. Phys. Soc. Japan **45**, 875 (1978)
2. D.E. Ellis, G.S. Painter, Phys. Rev. B **2**, 2887 (1970)
3. F.W. Averill, D.E. Ellis, J. Chem. Phys. **59**, 6412 (1973)
4. R.S. Mulliken, J. Chem. Phys. **23**, 1833 (1955)
5. M.J. Duggin, D. Cox, L. Zwell, Trans. Metall. AIME **236**, 1342 (1966)
6. G.P. Huffman, P.R. Errington, R.M. Fisher, Phys. Stat. Sol. **22**, 473 (1967)
7. P. Schaaf, S. Wiesen, U. Gonser, Acta. Metall. Mater. **40**, 373 (1992)
8. A. Inoue, T. Ogura, T. Masumoto, Bull. Japan Inst. Met. **13**, 653 (1974)
9. M. Umemoto, Z.G. Liu, K. Masuyama, K. Tsuchiya, Scripta Mater. **45**, 391 (2001)
10. N.J. Petch, J. Iron. Steel Inst. **149**, 143 (1944)
11. L. Pauling, J. Amer. Chem. Soc. **69**, 542 (1947)
12. E.J. Garba, R.L. Jacobs, J. Phys. Chem. Solids **50**, 101 (1989)
13. J. Häglund, G. Grimvall, Phys. Rev. B **44**, 2914 (1991)
14. A.H. Cottrell, Mater. Sci. Technol. **9**, 277 (1993)
15. S. Yamamoto, Y. Kobayashi, N. Inoyama, A. Kirihata, *Proceedings of the 4th International Symposium on the Physical Metallurgy of Cast Iron*, ed. by G. Ohira, T. Kusakawa, E. Niyama (Materials Research Sociery, Pittsburgh, 1990), pp. 103
16. A. Westgren, G. Phragmén, J. Iron Steel Inst. **105**, 241 (1922)
17. A. Westgren, G. Phragmén, J. Iron Steel Inst. **109**, 159 (1924)
18. H. Lipson, N.J. Petch, J. Iron Steel Inst. **142**, 95 (1940)
19. D. Meinhardt, Arch. Eisenhütt. **30**, 51 (1959)
20. K. Löhberg, Arch. Eisenhütt. **32**, 409 (1961)
21. D. Meinhardt, O. Krisement, Arch. Eisenhütt. **33**, 493 (1962)
22. F.H. Herbstein, J. Smuts, Acta Cryst. **17**, 1331 (1964)
23. E.J. Fasiska, G.A. Jeffrey, Acta Cryst. **19**, 463 (1965)
24. A.S. Keh, Acta Met. **11**, 1101 (1963)
25. Z. Nishiyama, A. Kore'eda, S. Katagiri, Trans. JIM **5**, 115 (1964)
26. K.W. Andrews, Acta. Metall. **11**, 939 (1963)
27. D.S. Zhuo, G.J. Shiflet, Metall. Trans. A **23A**, 1259 (1992)
28. Y. Ito, M. Nakanishi, J. Japan Weld. Soc. **44**, 815 (1975)
29. N. Mori, H. Honma, S. Okita, M. Wakabayashi, J. Japan Weld. Soc. **50**, 174 (1981)
30. K. Yamamoto, T. Hasegawa, J. Takamura, Tetsu-to-Hagené **74**, 465 (1988)
31. C. Li, A.J. Freeman, Phys. Rev. B **43**, 780 (1991)
32. U. Shönberger, O.K. Andersen, M. Methfessel, Acta. Metall. Mater. Suppl. **40**, S1 (1992)
33. D.M. Duffy, J.H. Harding, A.M. Stoneham, Acta. Metall. Mater. Suppl. **40**, S11 (1992)
34. C. Li, R. Wu, A.J. Freeman, Phys. Rev. B **48**, 8317 (1993)
35. E. Heifets, E.A. Kotomin, R. Orlando, J. Phys. Condens. Matter **8**, 6577 (1996)

36. J. Cho, K.S. Kim, C.T. Chan,, Z. Zhang, Phys. Rev. B **63** 113408 (2001)
37. J. Goniakowski, C. Noguera, Phys. Rev. B **66**, 085417 (2002)
38. L. Giordano, J. Goniakowski,, G. Pacchioni, Phys. Rev. B **67**, 045410 (2003)
39. K.H. Johson, S.V. Pepper, J. Appl. Phys. **53**, 6634 (1982)
40. K. Nath, A.B. Anderson, Phys. Rev. B **39**, 1013 (1989)
41. M. Kohyama, S. Kobe, M. Kinoshita, R. Yamamoto, J. Phys. Chem. Solids **53**, 345 (1992)
42. W. Zhang, J.R. Smith, Phys. Rev. B **61**, 16883 (2000)
43. I.G. Batyrev, A. Alavi, M.W. Finnis, Phys. Rev. B **62**, 4698 (2000)
44. I.G. Batyrev, L. Kleinman, Phys. Rev. B **64**, 033410 (2001)
45. D.J. Siegel, L.G. Hector, Jr., J.B. Adams, Phys. Rev. B **65**, 085415 (2002)
46. X. Wang, J.R. Smith, M. Scheffler, Phys. Rev. B **66**, 073411 (2002)
47. G. Baker, J. Nutting, Special Report Number 64, (Iron and Steel Institute, London, 1959), p. 1
48. M. Tanino, T. Nishida, J. Japan Inst. Met. **29**, 794 (1965)
49. V, Ern, A.C. Switendick, Phys. Rev. **137**, A1927 (1965)
50. V.A. Gubanov, E.Z. Kurmaev, G.P. Shveikin, J. Phys. Chem. Solids **38**, 201 (1977)
51. A. Neckel, P. Rastl, R. Eibler, P. Weinberger, K. Schwarz, J. Phys. C **9**, 579 (1976)
52. A.L. Ivanovsky, V.A. Gubanov, G.P. Shveikin, E.Z. Kurmaev, J. Less-Common Met. **78**, 1 (1981)
53. L.M. Sheludchenko, Y.N. Kucherenko, V.G. Aleshin, J. Phys. Chem. Solids **42**, 733 (1981)
54. A.L. Ivanovsky, V.I. Anisimov, D.L. Novikov, A.I. Lichtenstein, V.A. Gubanov, J. Phys. Chem. Solids **49**, 465 (1988)
55. V.P. Zhukov, V.A. Gubanov, O. Jepsen, N.E. Christensen, O.K. Andersen, J. Phys. Chem. Solids **49**, 841 (1988)
56. S. Kim, R.S. Williams, J. Phys. Chem. Solids **49**, 1307 (1988)
57. L.I. Johansson, P.M. Stefan, M.L. Shek, A. Nörlubd Christensen, Phys. Rev. B **22**, 1032 (1980)
58. H. Obara, Y. Endoh, Y. Ishikawa, T. Komatsubara, J. Phys. Soc. Japan **49**, 928 (1980)
59. P. Marksteiner, P. Weinberger, Phys. Rev. B **33**, 812 (1986)
60. A.L. Ivanovsky, V.I. Anisimov, A.I. Lichtenstein, V.A. Gubanov, J. Phys. Chem. Solids **49**, 479 (1988)
61. V.A. Pai, A.P. Sathe, V.R. Marathe, J. Phys., Condens. Matter **2**, 9363 (1990)
62. A.G. Nargizyan, S.N. Rashkeev, J. Phys. B, Condens. Matter **82**, 217 (1991)
63. L.M. Huisman, A.E. Carlsson, C.D. Gelatt, Jr., H. Ehrenreich, Phys. Rev. B **22**, 991 (1980)
64. V.A. Gubanov, A.L. Ivanovsky, G.P. Shveikin, D.E. Ellis, J. Phys. Chem. Solids **45**, 719 (1984)
65. L.M. Mattheiss, Phys. Rev. B **5**, 290 (1972)
66. J.H. Binks, J.A. Duffy, J. Solid State Chem. **87**, 195 (1990)
67. J. Klima, G. Schadler, P. Weinberger, A. Neckel, J. Phys. F, Met. Phys. **15**, 1307 (1985)
68. G. Hörmandinger, J. Redinger, P. Weinberger, Solid State Commun. **68**, 467 (1988)
69. F. Schlapansky, P. Herzig, R. Eibler, G. Hobiger, A. Neckel, Z. Phys. B, Condens. Matter **75**, 187 (1989)

70. P.K. Khowash, D.E. Ellis, J. Appl. Phys. **65**, 4815 (1989)
71. P.K. Khowash, D.E. Ellis, Phys. Rev. B **39**, 1908 (1989)
72. G. Hobiger, P. Herzig, R. Eibler, F. Schlapansky, A. Neckel, J. Phys., Condens. Matter **2**, 4595 (1990)
73. S.R. Barman, D.D. Sarma, Phys. Rev. B **49**, 16141 (1993)
74. H. Ihara, Y. Kumashiro, A. Itoh, Phys. Rev. B **12**, 5465 (1975)
75. H.R. Trebin, H. Bross, J. Phys. F, Met. Phys. **14**, 1453 (1984)
76. J. Redinger, R. Eibler, P. Herzig, A. Neckel, R. Podloucky, E. Wimmer, J. Phys. Chem. Solids **46**, 383 (1985)
77. J. Redinger, R. Eibler, P. Herzig, A. Neckel, R. Podloucky, E. Wimmer, J. Phys. Chem. Solids **47**, 387 (1986)
78. V.P. Zhukov, V.A. Gubanov, J. Phys. Chem. Solids **48**, 187 (1987)
79. T. Hori, H. Adachi, S. Imoto, Trans. JIM **18**, 31 (1977)
80. V.A. Gubanov, A.L. Ivanovsky, G.P. Shveikin, J. Weber, Solid State Commun. **29**, 743 (1979)
81. A.L. Ivanovsky, V.A. Gubanov, V.P. Zhukov, G.P. Shveikin, Phys. Stat. Sol. B **98**, 79 (1980)
82. V.A. Gubanov, E.Z. Kurmaev, D.E. Ellis, J. Phys. C **14**, 5567 (1981)
83. J.E. Lowther, J. Less-Common Met. **99**, 291 (1984)
84. J.E. Lowther, A. Andriotis, J. Phys. Chem. Solids **48**, 713 (1987)
85. A.L. Ivanovsky, D.L. Novikov, V.I. Anisimov, V.A. Gubanov, J. Phys. Chem. Solids **49**, 487 (1988)
86. A.H. Cottrell, Mater. Sci. Technol. **10**, 22 (1994)
87. A.H. Cottrell, Mater. Sci. Technol. **10**, 788 (1994)
88. L. Pauling, J. Am. Chem. Soc. **69**, 542 (1947)
89. Y. Ikuhara, Y. Sugawara, I. Tanaka, P. Pirouz, Interface Sci. **5**, 5 (1997)
90. J.M. Gregg, H. K.D.H. Bhadeshia, Acta Metall. Mater. **42**, 3321 (1994)
91. S. Zhang, N. Hattori, M. Enomoto, T. Tarui, ISIJ (The Iron and Steel Institute of Japan) Int. **36**, 1301 (1996)

4

Ceramics

T. Kamiya, N. Ohashi, and J. Tanaka

4.1 General Introduction

The development of ceramics science and technology has made huge contributions to recent progress in engineering, including electronics, mechanics, biotechnology, etc. For example, the miniaturization of electronic devices (e.g., mobile phones, video cassette recorders, and so on), has been supported by amazingly small circuit elements achieved by the development of ceramic technology. It is also known that ceramic technology has opened up a number of frontiers. For example, aerospace technology, (e.g., space shuttles), cannot keep developing unless the ceramics for extreme environments are realized. Moreover, an aging society, which is a serious social matter in developed countries, has large needs for reliable and non-toxic artificial bones and teeth. Indeed, advances in engineering and solutions for technological and social problems are addressed by ceramic science and technology.

A number of new ceramic materials were discovered and synthesized in the past century, e.g. high-T_C superconductors (HTSC), relaxor-ferroelectrics, p–n diodes, etc. The synthesis has mostly been done by trial-and-error procedures, empirical predictions and intuition. The HTSC hunters in the 1980s studied almost all elementary oxide reagents and a number of mortars in their lab. Some researchers have sought to design materials non-empirically, but have not achieved the same success. Needless to say, material design and characterization must be implemented together to realize advancement in ceramics engineering. The science and technology of material characterization has been developed, but the interpretation of the results in most cases has been based on a researcher's experience.

It should be noted that a ceramic system is complicated as shown in Fig. 4.1. The system is inhomogeneous and thermodynamically metastable, and further, the materials contain a lot of defects and interfaces, i.e., non-ideal materials. Moreover, elements in useful ceramics are heavy. For example, the high-performance piezoelectric compound of $Pb(Ti, Zr)O_3$ contains lead (atomic number 82). Computer simulation of electronic states in a solid with

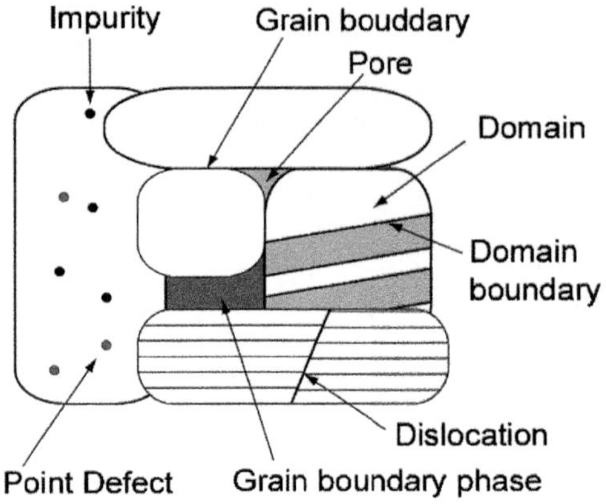

Fig. 4.1. Schematic diagram of defects and interfaces in ceramics

a huge electron number requires huge amount of memory, high-speed processors and a highly stable operating system. Thus, the complete simulation of ceramic material is dependent upon the frontiers of computational material science. However, recent developments in both computer hardware and software technology have enabled us to apply ab initio calculations to ceramics research. It is a great advancement for ceramic science and technology.

In this section, we introduce several examples of the use of the DV-Xα code for ceramics research. In the following section, the DV-Xα method is applied for the peak assignments of some spectroscopic characterizations. The transition state calculation is appropriate for the simulation of electron spectra, and thus, the spectral profiles were calculated taking into consideration the electron relaxation. Auger electron spectra calculated by the DV-Xα method will be described in 4-2-a. An application of the DV-Xα calculation to a magnetic resonance study is also introduced in 4-2-b. For this case, the partial density of states was used for the estimation of the super-hyperfine interaction tensor, and the simulated values were compared with those found from the observed parameters. Another important feature of the DV-Xα calculation is the evaluation of the bond overlap population (BOP) as the result of charge analyses. The magnitude of the BOP corresponds to the bond strength (covalency) of the chemical bonds, and it is useful for the qualitative prediction of the structural stability of the compounds. In 4-2-c, estimation of the most stable surface structure was done using the DV-Xα method and was used for the interpretation of ion-scattering spectroscopy.

In Sect. 4.4, several examples of the design and the property prediction of materials using the DV-Xα method are introduced. Most ceramic materials have the characteristic feature of ionic crystals. This is why stable atomic configurations in ionic crystals can be predicted by classical molecular dynamics (MD) calculations. The improvement in accuracy of the classical MD calculation assisted by the DV-Xα calculation is introduced in 4-3-a. Recent progress in oxide fabrication techniques, e.g., molecular beam epitaxy (MBE) methods, etc., enables us to artificially synthesize modulated superlattice structures. However, it takes a large amount of time for optimization of the growth conditions, and hence, a trial-and-error procedure is not appropriate for material exploration using MBE. The prediction of the structure–property relationship is thus very important for this kind of material synthesis. In 4-3-b, use of the DV-Xα method to obtain tight-binding band calculation parameters are introduced, and the result of the band calculation for the $SrZrO_3$–$SrTiO_3$ super-lattice system will be described as an example. Another example of the property prediction is introduced in 4-3-c, where the dielectric properties of perovskite oxide materials are estimated from the results of the DV-Xα calculation.

4.2 Characterization of Ceramics with the Assistance of DV-Xα Calculations

The advancement of the fine-powder preparation technique for ceramic sources and integration technology in ceramic fabrication processes are exemplified by the miniaturization of electric circuit elements. One of the typical examples is multi-layer ceramic capacitors (MLCCs). MLCCs have a layered structure, composed of alternating stacks of oxide dielectric layers and metal electrode layers. In particular, in the most highly integrated MLCCs, the thickness of the oxide dielectric layer is on the order of microns, and one layer is composed of a few grains of the oxide crystal. It is said that the next target of ceramic fabrication technology is the mounting/growth of ceramic chips directly onto silicon-based integrated circuits. The decrease in the thickness and the increase in complexity mean that an increase in the interfacial region per unit volume of the devices. Consequently, controlling the surface and interface is getting more important. Moreover, the artificially integrated or fabricated stacking structures are in non-equilibrium states, and they sometimes involve nonstoichiometric composition and structures. This means that the behavior of nonequilibrium defects plays important roles in these materials; for example, aging behavior is mostly caused by such kinds of defects.

One of the most useful probes for the analysis of the micro- or nanoregion is the focused electron beam, found in transmission electron microscopes (TEMs), secondary electron microscopes (SEMs), electron spectrometers in, for example, electron energy-loss spectroscopy (EELS) and Auger electron spectroscopy (AES). These instruments are designed for the characterization of chemical and electronic states in microregions. However, the sensitivity of these techniques is rather low. Indeed, an impurity or defect concentration of at least a few percent is necessary for quantitative analysis. Thus, other sensitive methods for the detection of defects are consequently necessary for complete characterization and understanding of the minor impurities and defects in materials. In this article, electron spin resonance (ESR) is introduced as a method for the detection of minor impurities and defects.

The interpretations of the obtained spectra are usually given empirically or semi-empirically. For an example of a semi-empirical assignment of AES spectra, it is well-known that the so-called Auger parameter, evaluated from the difference between the kinetic energy of the Auger electron peak and the binding energy of the peak of the X-ray photoelectron spectra, is a useful parameter to characterize the chemical state of the specimen. Comparing the Auger parameter of unknown species with that of the well-known species, we can discuss the chemical state empirically. In order to investigate the chemical or electronic states in detail, an electronic state calculation is one of the most powerful approaches. This is also the case for the assignment of ESR spectra; that is, the defect structure can be obtained empirically from the observed spectra, but the use of an ab initio calculation enables us to describe the defect structure more precisely. In the following subsections, the application of the DV-Xα calculations for spectral assignments and the interpretation of the spectra of ceramic materials will be introduced.

4.2.1 Assignments for Electron Spectroscopy

Introduction

Figure 4.2 is a schematic diagram of AES and X-ray photoelectron spectroscopy (XPS). In this figure, the $KL_{23}L_{23}$ Auger transition of oxygen and an O-1s core-level XPS in zinc oxide (ZnO) are illustrated. XPS is a relatively simple process in which an incident energy injection, by an electromagnetic wave, causes the ionization of molecular orbital (MO) Φ_0 – namely, the O-1s orbital. On the other hand, the emission process of AES is a three-electron process. After the ionization of MO Φ_0, one of the other electrons in the outer core, e.g., an electron at Φ_5, transfers to the MO Φ_0, and the extra energy corresponds to the energy difference between Φ_5 and Φ_0, which causes a second ionization, e.g., ionization of MO Φ_8.

On the basis of the linear combination of atomic orbitals (LCAO), the MOs (Φ_n's) in ZnO are described as follows:

$$\Phi_n = \sum_{j,k} c_{njk}\phi_{jk} + \sum_{l,m} c_{nlm}\phi_{lm} \ . \tag{4.1}$$

Here, ϕ_{jk} is the kth AO of the jth O, ϕ_{lm} is the mth AO of the lth Zn, and the value of Φ_n is normalized as $|\Phi_n|2 \equiv 1$ for spin-polarized AOs and $|\Phi_n|2 \equiv 2$ for spin-unpolarized AOs.

The intensity of an AES transition for the jth O can be estimated from the coefficients of the MO [1], as shown in (4.2).

$$I_{nm} \propto p_n^4 p_m^4 \ , \tag{4.2a}$$

$$p_n \equiv c_{njk} \ , \tag{4.2b}$$

$$p_m \equiv c_{mjk} \ , \tag{4.2c}$$

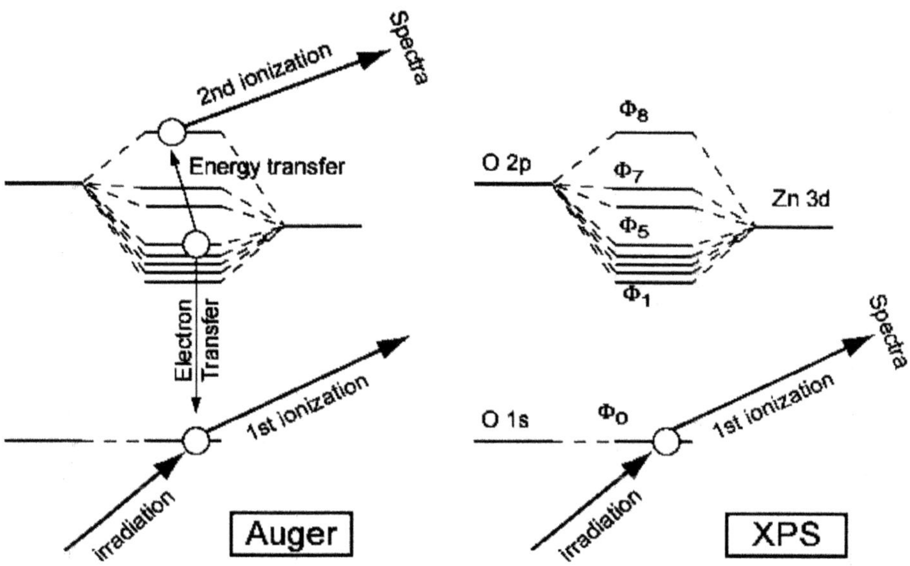

Fig. 4.2. Schematic diagram of Auger and photoelectron emission processes

Here, we assume that the AO of O-1s is completely localized, and the O-1s core-level (Φ_0) is the AO of O-1s. The result of a population analysis using the DV-Xα code is therefore appropriate for the simulation of AES transition intensity. In order to simplify the material in this chapter, the simplest cluster model, $[Zn_4O]^{6+}$, is employed for the estimation of AES transition intensity.

The observed AES spectrum of ZnO ceramic [2] doped with Bi is shown in Fig. 4.3. It seems that the oxygen $KL_{23}L_{23}$ spectrum is composed of three peaks, at 508, 514, and 518 eV. This indicates that the L shell of the oxygen ion in ZnO consists of two major states – the bonding (B) and anti-bonding (aB) states – which give way to three possible combinations: B-B, B-aB, and aB-aB. Two AES profiles are plotted in Fig. 4.3: one is obtained from the grain boundary doped with Bi and the other from a ZnO grain. The difference in the two spectral profiles is caused by the modification of the chemical bond between Zn and O ions due to Bi doping. The DV-Xα calculations are useful for spectral assignments of this kind of variation in electron spectra.

Another important issue for describing AES transitions is the estimation of the degree of electronic relaxation. The process of Auger electron emission is a three-electron process, as mentioned previously. This means that electronic relaxation takes place in a complicated way during the transition process. In other words, the relaxation process makes interpretations of the AES profile difficult. The DV-Xα code enables us to calculate the electronic transition state by using its transitional calculation subroutine. Ionization potential can be estimated with this method. The transitional state in the DV-Xα calculation introduces half-filled MOs – namely, an intermediate electronic config-

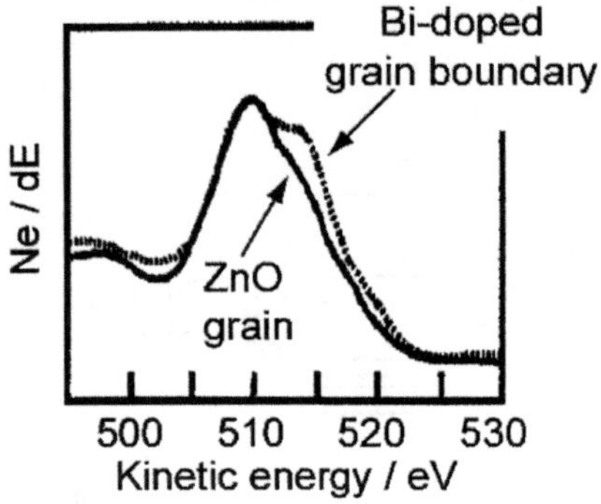

Fig. 4.3. $KL_{23}L_{23}$ spectra of ZnO ceramic

uration between the initial and final states – as illustrated in Fig. 4.4. For a transition state of a photoelectron emission (PES) process, only the target MO is set to a half-filled MO for the transition state calculation of the DV-Xα method, since the transition state (Fig. 4.4b) is defined as the intermediate state between the initial ground state (fully occupied state, Fig. 4.4a) and the final ionized state with one hole (Fig. 4.4c). On the other hand, for the simulation of the AES process, three half-filled holes are necessary, since the AES transition is a three-electron process. For the AES process, the initial state is set to the core-hole state (Fig. 4.4d) after the first ionization by irradiation and the final state to the two-hole state (Fig. 4.4f) as a result of Auger electron emission. Thus, three half-filled MOs appear in the transition state (Fig. 4.4e), which is an intermediate state between initial and final states.

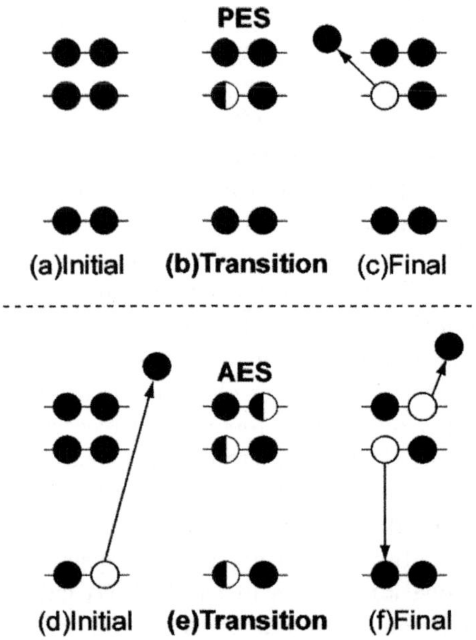

Fig. 4.4. Initial and final states of Auger and photoelectron emission processes and corresponding the "transitional state" for the DV-Xα calculation

Calculation

The cluster model for the $KL_{23}L_{23}$ AES simulation examined in [2, 3] was [Zn_4O], as illustrated in Fig. 4.5. Three half-filled MOs are introduced into the cluster model for the transitional state calculation. Since Auger electron emission is an intra-atomic process, all of the half-filled MOs should be mostly localized at the target ion/atom. If we use the cluster model [Zn_4O], the half-filled MOs are always localized at the O(1) site of this cluster and, it is easy to calculate the AES transition. On the other hand, in the [Zn_4O_4] cluster, the hole generated at the O(1) site sometimes transfers from the O(1) site to the O(2) site during an iteration to obtain a state of minimum energy. Thus, the use of the simplest cluster model is the easiest and simplest way to calculate the AES profiles.

The compounds used in ceramics, including ZnO, are mostly ionic crystals. Thus, the Madelung potential is an important issue for the description of ionic crystals. As illustrated in Fig. 4.6, the cluster model [Zn_4O] is surrounded by dummy atoms, which gives an electrostatic potential to the cluster.

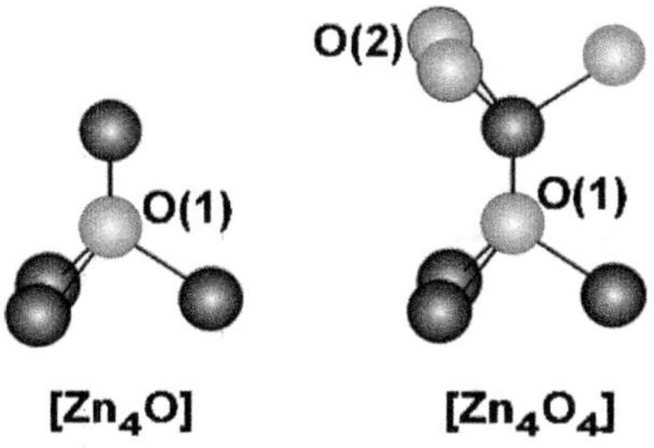

Fig. 4.5. Cluster models [Zn_4O] and [Zn_4O_4]

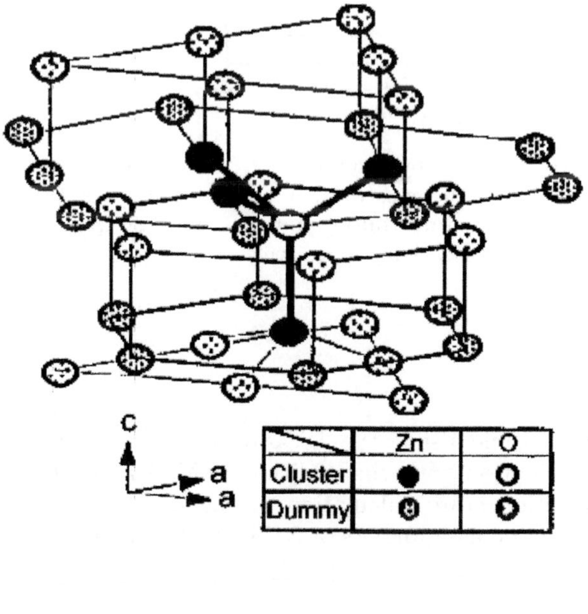

Fig. 4.6. [Zn_4O] cluster surrounded by dummy ions yielding a Madelung potential

Results

Figure 4.7 shows three AES spectra simulated by the DV-Xα method. Here, three kinds of results are compared. The spectrum obtained from the ground state calculation is considerably different from the other two simulated spectra. In particular, the center peak is the most intense for the ground state calculation, while the low-energy peak is the highest in the other spectra. This indicates that an electron relaxation after the initial ionization of the core level is essential for the profile of AES. Figure 4.7 also indicates that the calculated kinetic energy for the core-hole state is different from that of the transition state, though their peak shape is almost the same. Since the transition to a spin-triplet final state (^{3}P) is not allowed in KLL Auger transitions, the correlation energy between the resultant holes is expected to be low. This means that the difference in kinetic energy between the core-hole state (Fig. 4.4d) and transition state (Fig. 4.4e) is ascribed to the difference in relaxation energy for the core-hole.

The variation of the AES profile, i.e., Fig. 4.3, was ascribed to a change in ionicity of the Zn–O bond by doping. The DV-Xα calculation for the [$BiZn_3O$] cluster indicates that the ionic charge of the oxygen ion changes by Bi doping in ZnO and that the change in ionicity results in a variation of the AES profile.

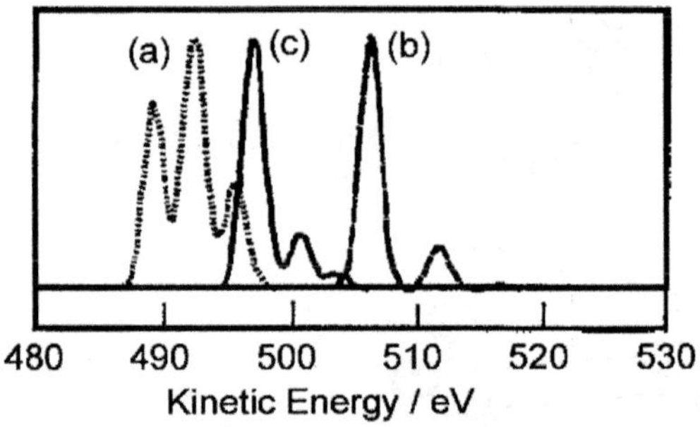

Fig. 4.7. Auger electron spectra simulated by assuming 4.2. The coefficients in 4.2 were evaluated by DV-Xα calculation. The spectra **a**, **b**, and **c** corresponds to the ground state calculation, the core-hole state (initial state of the AES transition) calculation, and the transition-state calculation, respectively

Remarks

The simulation of the AES profile using the DV-Xα method was introduced. This method, in combination with the addition of Madelung potential (Fig. 4.6) and the transition state calculation of the DV-Xα method (Fig. 4.5), is actually useful for the assignments of AES profiles. The authors calculated AES profiles of a nitride system, e.g., boron nitride. The simulated AES profiles for nitrides qualitatively agreed well with the observed profiles. The systematic change in nitrogen $KL_{23}L_{23}$ spectra was ascribed to the ionicity and covalency of nitrides. The focused electron beam techniques are getting more important, since the nano-sized grains and structures are attractive topics in ceramic technology. Thus, AES analyses, especially those excited by finely focused electron beams, will be a key technology for the development of the next generation of ceramic science and technology.

4.2.2 Prediction of Atomic Arrangements

Introduction

The absence of periodicity is the most important issue in the study of solid materials. Point defects are the zero-dimensional examples of a break in periodicity, dislocation is a one-dimensional example, and an interface is a two-dimensional example. The controlling of such periodicity imperfections is the most important issue for the fabrication of ceramic devices and materials. For example, electron or hole doping is necessary for semiconductor materials in order to achieve adequate conductivity and sufficient barrier height at the junction. However, doping causes imperfections in the crystal periodicity, which result in a low carrier mobility, low chemical stability, etc. Moreover, the defects and impurities cause structural distortion around the defects. For example, a defect with a lone pair of electrons causes a Jahn-Tellar distortion, which is an enhancement of the crystal periodicity imperfection by doping. In this subsection, structural relaxation around the defects in diamond will be introduced as an example of the application of the DV-Xα calculation.

Diamond is a representative of wide-gap semiconductors and thus, it is appropriate for use as a transparent semiconductor and switching devices in high-temperature environments. A lot of applications and devices using diamond have been proposed, but the difficulty of synthesizing n-type diamond is a great barrier in the fabrication of diamond-based semiconductor devices. Since diamond has the same crystal structure as silicon, it is expected that nitrogen and phosphorus in diamond form donor states in the same way as electron doping in a silicon crystal. In fact, boron in diamond forms a shallow acceptor state, and high hole conductivity is possible [4]. However, nitrogen in diamond is not an effective donor. It is believed that localization of doped electrons takes place due to lattice relaxation around nitrogen and/or segregation of impurities. For example, electron paramagnetic resonance (EPR) spectra measurements suggest that doped nitrogen has not T_d but C_{3v} symmetry and the nitrogen atoms tend to form a cluster of N_x in diamond [5]. Indeed, the relaxation of the local structure is essentially important in doping phenomena.

Several defects in diamond have been studied by molecular orbital (MO) calculation methods [6–11], because the MO calculation using cluster models is more convenient for calculating electronic states of localized defects than the band calculation methods based on Bloch's theorem. The electronic states of nitrogen [6–10] and phosphorus [11] in diamond have been calculated by ab initio or semi-empirical MO calculation methods. Most of those calculations have been performed to elucidate the electronic structure or mechanism of structural relaxation around those dopants. In this section, use of the DV-Xα method for the structural relaxation will be introduced.

The structural stability of solids is usually discussed on the basis of total energy calculation. However, the evaluation of total energy to discus structural stability requires a lot of time and computer resources. From the viewpoint of material design and property predictions, real-space images are sometimes more helpful than the k-space images given by band structure calculations. Taking real-space images into account, the use of the bond-overlap population (BOP) evaluated by the DV-Xα calculation is an efficient solution for material design. BOP is a measure of the strength of a chemical bond, since it corresponds to covalency. Hereafter, the use of the DV-Xα method is introduced as a tool for the prediction of structural relaxation.

4.2.3 Theory and Calculation

The nature of the chemical bonds in the cluster was quantitatively characterized by the BOP. The electron population of the lth MO, abbreviated to P_l, is given by

$$P_l = f_l \int |\Psi_l(r)|^2 \, \mathrm{d}\tau \,. \tag{4.3}$$

Here, $\Psi_l(r)$ and f_l represent the wave function and electron number occupied at the lth MO, respectively. The wave function $\Psi_l(r)$ is defined by a linear combination of atomic orbitals (LCAO) as

$$\Psi_l(r) = \sum_i \sum_\mu C_i^\mu(l) \chi_i^\mu \,. \tag{4.4}$$

Here, χ_i^μ denotes the μth atomic orbital in the ith atom. The orbital population P_l is thus re-written as

$$P_l = f_l \int \left| \sum_i \sum_\mu C_i^\mu(l) \chi_i^\mu \right|^2 \mathrm{d}\tau \,. \tag{4.5}$$

The overlap population between the jth and kth atoms in the lth MO, $O_{jk}(l)$, can be given by

$$O_{jk}(l) = f_l \sum_\nu \sum_\mu (C_j^\mu(l) \chi_j^\mu) \cdot (C_k^\nu(l) \chi_k^\nu) \,. \tag{4.6}$$

The BOP between the jth and kth atoms, B_{jk}, is defined as a summation of the overlap populations over all occupied MOs:

$$B_{jk} = \sum_l f_l \sum_{\nu,\mu} (C_j^\mu(l)\chi_j^\mu) \cdot (C_k^\nu(l)\chi_k^\nu) . \quad (4.7)$$

As found in (4.6), the sign of the overlap population O_{jk} corresponds to the character of the chemical bonding, i.e., antibonding for negative O_{jk} and bonding for positive O_{jk}. In a similar way, a chemical bond with a negative BOP $B_{jk} < 0$ is unstable, since such a chemical bond has, on the whole, an antibonding nature.

A cluster $[C_{29}H_{36}]$ of T_d symmetry, shown in Fig. 4.8, was used for the calculation of pure diamond. As for the impurity-doped diamond, a central carbon atom in the $[C_{29}H_{36}]$ cluster was substituted by nitrogen, boron, or phosphorus, i.e., $[AC_{28}H_{36}]$ (A = N, B, or P). Dangling bonds on the cluster surface were terminated by hydrogen atoms as shown in Fig. 4.8.

Lattice relaxation effects were taken into account by reducing the symmetry of the cluster from T_d to C_{3v} in the manner shown in Fig. 4.9: an A atom was displaced from its initial position to the position where the A–C(Ib) bond length elongates or shortens. The displacement of C(Ia) atoms was disregarded in this study.

The effective charge of each atom in a cluster was evaluated by the Mulliken population analysis. The stability of doped atoms was estimated from the BOP between a substituted A atom and an adjacent carbon atoms, C(Ia) and C(Ib).

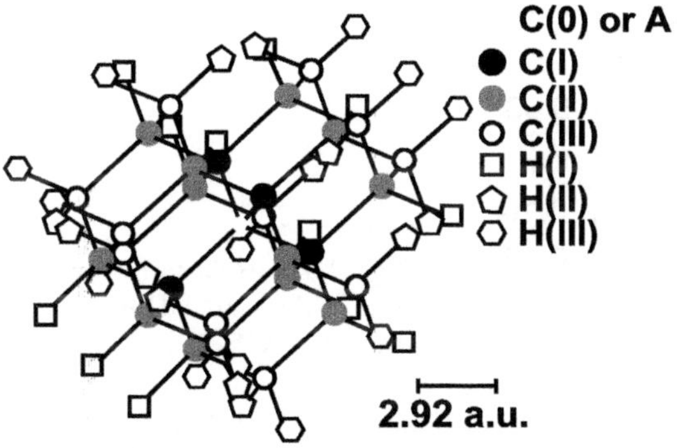

Fig. 4.8. A cluster model $[C_{29}H_{36}]$. A carbon atom, indicated by C(0) was substituted by a nitrogen, boron, or phosphorus atom

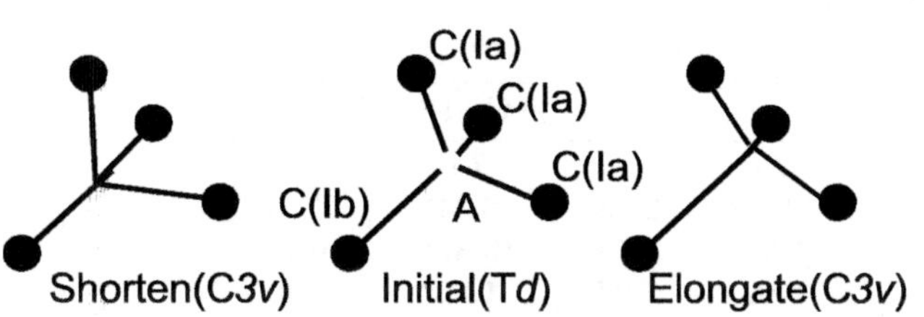

Fig. 4.9. Assumed relaxation with symmetry-lowering from T_d to C_{3v}

4.3 Results

The spin-orbital population for a [$NC_{28}H_{36}$] cluster evaluated by assuming C_{3v} symmetry is different from the result assuming T_d symmetry, though all of the atoms are fixed at the ideal position. For the cluster model of [$C_{29}H_{36}$], which corresponds to pure diamond, the evaluated population assuming C_{3v} symmetry agrees with the population assuming T_d symmetry. The BOP values for the A atom in [$AC_{28}H_{36}$] (see Figs. 4.8 and 4.9) are summarized in Table 4.1. These results are given for the ideal diamond structure without any atomic displacements. The BOP value is always anisotropic in the cluster [$C_{29}H_{36}$], even when C_{3v} symmetry is assumed. On the other hand, the BOPs for N–C bonds are anisotropic when assuming C_{3v} symmetry. This indicates that the nitrogen atom forms an anisotropic bond even though it is in an isotropic position. The formation of an anisotropic bond means that a nitrogen atom in a diamond lattice prefers to occupy a lower-symmetry site. Thus, examining the DV-Xα calculation is a conventional way to realize structural relaxation.

Figure 4.10 indicates a relation between lattice relaxation and the BOP of N–C bonds in the [$NC_{28}H_{36}$] cluster. As indicated in Fig. 4.10, the BOP for N–C(Ia) bonds is much larger than the BOP for N–C(Ib) bonds. This result suggests that a nitrogen atom makes chemical bonds with only three C(Ia) carbon atoms like an NH_3 molecule and has little interaction with other C(Ib) carbon atoms. The value of the BOP is saturated with an increase of

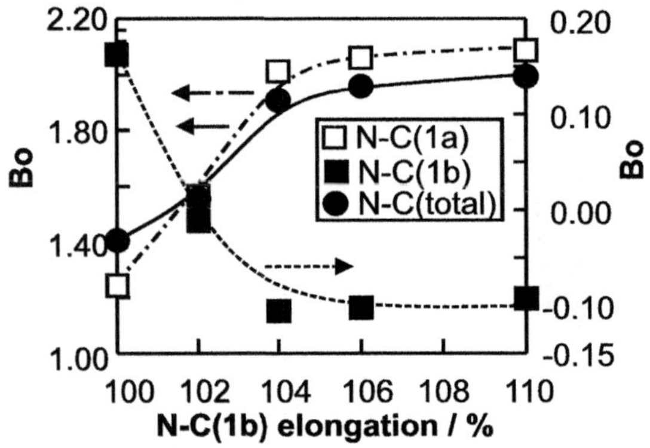

Fig. 4.10. Relation between bond order and lattice relaxation for [$NC_{28}H_{36}$]

Table 4.1. BOP for the A–C(Ia) and A–C(Ib) bonds in the [$AC_{28}H_{36}$] (A = C or N) clusters. "Sym" indicates the assumed symmetry of the MOs. All of the calculations were done assuming ideal atomic positions

Cluster	[$C_{29}H_{36}$]	[$C_{29}H_{36}$]	[$NC_{28}H_{36}$]	[$NC_{28}H_{36}$]
Sym.	T_d	C_{3v}	T_d	C_{3v}
A–C(Ia)	(2.27)	2.23	(0.94)	1.24
A–C(Ib)	(0.75)	0.74	(0.31)	0.16
Total	3.02	2.97	1.25	1.30

the N−C(Ib) distance and is constant over 106% elongation. The saturation is due to a simultaneous increase of bonding and antibonding components of N−C(Ia) bonds caused by the shortened N−C(Ia) distance. The elongation of the N−C(Ib) distance in diamond is estimated from EPR observations to be 120 – 128% [5]. However, a converged result is not achieved for a cluster of over 110% elongation. It is indicated that the displacement of the C(Ia) and/or C(Ib) atom must be taken into account to estimate the best arrangement of the nitrogen center.

As for the cluster [$BC_{28}H_{36}$], the MO calculation converges for T_d symmetry but not for C_{3v} symmetry, in which atom positions are the same as in T_d symmetry. This is due to the splitting of an impurity level by symmetry-lowering. The hole introduced by boron doping occupies the molecular orbital t_2 in T_d symmetry, while the t_2 orbital splits into a_1- and e-orbitals in C_{3v} symmetry. Since the energy splitting of the a_1- and e-orbitals is small, the hole transfer can take place between them, resulting in an oscillation in the calculated result. Therefore, the lattice relaxation is taken into account in the same way as the nitrogen-doped cluster.

The possibility of structural relaxation around a boron atom at a substitutional site was examined. The value of the BOP for B−C bonds shown in Fig. 4.10 is almost the same as the BOP value calculated for C−C bonds in the [$C_{29}H_{36}$] cluster and larger than the BOP value of N−C bonds.

As for Bo, the most stable position of boron in diamond could not be determined, since Bo for B−C bonds is almost independent of relaxation, as shown in Fig. 4.10. However, a hole is located in the a_1-orbital when the B−C(Ib) distance is shortened, while the hole is located in the e-orbital when the B−C(Ib) distance is elongated. As for the cluster with an elongated B−C(Ib) distance, the occurrence of Jahn–Tellar (J–T) distortion is expected, since a hole occupies the doubly degenerated e-orbital. It is expected that the symmetry of the boron center is reduced from C_{3v} to lower symmetry by elongation of the B−C(Ib) distance. The EPR spectra for a boron center are very broad in comparison with a nitrogen center in diamond. It was conjectured that the broadening of the EPR spectra is due to a dynamical J–T effect or to lattice relaxation. Taking the results of the DV-Xα calculation into consideration, it is expected that boron in diamond causes structural relaxation making the B−C(Ib) distance longer, since this type of displacement causes J–T distortion, as described above.

Remarks

Examples of the application of the BOP calculation for the evaluation of structural stability and prediction of lattice relaxation were introduced. The descriptions in this subsection were focused on the atomic arrangements around point defects in solid. However, the BOP is also a measure of the structural stability of phases. For example, the prediction of the stability of alloy is also possible using this technique. This BOP technique is appropriate for metals, organic compounds, as well as ceramics as a measure of structural stability and is one of the most useful functions of the DV-Xα method.

4.3.1 Assignments for the ESR Spectra Using Electron Density Calculations by DV-Xα

Introduction

Electron spin resonance (ESR) is one of the useful techniques for the detection and analysis of defects in materials. Several examples for ESR analysis were introduced in the previous subsection. In this subsection, simulation of metal impurities in diamond is introduced as an example of the application of the DV-Xα calculations for magnetic interactions [12].

Synthetic diamond is grown by the temperature-gradient method [13,14] under high-pressure conditions, e.g., over 5 GPa, or by the chemical vapor deposition method [15]. In the former method, diamond crystals are grown using metal solvents such as Fe, Ni, Co, Mn, or their alloys. Among these transition metal elements, only nickel ions in diamond give electron paramagnetic resonance (EPR) spectra [16]. According to Isoya et al. [17], two kinds of Ni centers have been identified by measuring EPR and electron nuclear double-resonance (ENDOR) spectra. They used a ^{13}C source for diamond synthesis, since observation of the super-hyperfine interaction between ^{13}C nuclei and electrons give rise to a precise understanding of the electronic state of defects. One of the Ni centers is the substitutional Ni^- ion, which has T_d symmetry and an effective spin of $S = 3/2$: this center is hereafter designated as Ni_s^-, and the other center is assigned to interstitial Ni^+ ions with an electronic configuration of $3d^9$ [18]. For the substitutional Ni^- ion, it was conjectured that the four electrons, $4s4p^3$ formed sp^3 hybrid orbitals to form chemical bonds with the four nearest-neighbor carbons [17].

Theory and Calculation

The DV-Xα method is an all-electron method; it is able to find the wave functions of all of the electrons in the assumed cluster model. This feature is important for the simulation of super-hyperfine (SHF) tensors.

The cluster used for the calculation of a Ni ion at a substitutional site is $[NiC_{28}H_{36}]^-$ in which an A-site of the cluster model shown in Fig. 4.8 is substituted by a Ni ion. The bond distance between a Ni ion and nearest-neighbor carbon atoms is fixed to the same value as the C$-$C bond length, d_{C-C}, of diamond except for the calculation to elucidate the effect of structural relaxation. The structural relaxation examined in the present study is that due to the displacement of the C(I) atom to elongate the Ni$-$C(I) bond length: positions of the other atoms are kept fixed without displacement, and the cluster is assumed to have T_d symmetry.

In the EPR study, the spin delocalization onto nearest-neighbor carbons was estimated from the ^{13}C SHF interaction ($A_{//} = 1.339\,\mathrm{mT}$ and $A_{\perp} = 0.340\,\mathrm{mT}$) by using the LCAO approximation [17,19]. According to the LCAO approximation, which neglects both the delocalization onto carbons, beyond the nearest neighbors and the overlap between metal d-orbitals and carbon orbitals. The wave functions of the unpaired electrons in the Ni_s^- center are given by

$$\begin{aligned}\Psi_{\mathrm{xy}} &= \alpha d_{\mathrm{xy}} + \frac{1}{2}\beta(s_1 - s_2 - s_3 + s_4) + \frac{1}{2}\gamma(\sigma_1 - \sigma_2 - \sigma_3 + \sigma_4) \\ &+ \frac{1}{4}\delta[(-\pi_{1\mathrm{x}} + \pi_{2\mathrm{x}} + \pi_{3\mathrm{x}} - \pi_{4\mathrm{x}}) + \sqrt{3}(\pi_{1\mathrm{y}} - \pi_{2\mathrm{y}} - \pi_{3\mathrm{y}} + \pi_{4\mathrm{y}})]\,, \quad (4.8\mathrm{a})\end{aligned}$$

$$\begin{aligned}\Psi_{\mathrm{xy}} &= \alpha d_{\mathrm{xy}} + \frac{1}{2}\beta(s_1 - s_2 + s_3 - s_4) + \frac{1}{2}\gamma(\sigma_1 - \sigma_2 + \sigma_{3} - \sigma_4) \\ &+ \frac{1}{4}\delta[(-\pi_{1\mathrm{x}} + \pi_{2\mathrm{x}} - \pi_{3\mathrm{x}} + \pi_{4\mathrm{x}}) + \sqrt{3}(\pi_{1\mathrm{y}} - \pi_{2\mathrm{y}} + \pi_{3\mathrm{y}} - \pi_{4\mathrm{y}})]\,, \quad (4.8\mathrm{b})\end{aligned}$$

$$\begin{aligned}\Psi_{\mathrm{xy}} &= \alpha d_{\mathrm{xy}} + \frac{1}{2}\beta(s_1 + s_2 - s_3 - s_4) + \frac{1}{2}\gamma(\sigma_1 + \sigma_2 - \sigma_3 - \sigma_4) \\ &+ \frac{1}{2}\delta[(\pi_{1\mathrm{x}} + \pi_{2\mathrm{x}} - \pi_{3\mathrm{x}} - \pi_{4\mathrm{x}})]\,. \quad (4.8\mathrm{c})\end{aligned}$$

Here, 1, 2, 3, and 4 refer to the nearest-neighbor carbons located in the directions of [111], $[\bar{1}1\bar{1}]$, $[1\bar{1}\bar{1}]$, and $[\bar{1}\bar{1}1]$, respectively, and the wave functions involve spin functions. The orbitals designated by σ are C-2p orbitals, which have lobes pointing toward the central Ni ion, while the orbitals designated by π are C-2p orbitals with lobes perpendicular to the carbon-nickel axis [20]. The value of $\beta^2 + \gamma^2 + \delta^2$ gives the sum of the spin densities on the four nearest-neighbor carbons. The ^{13}C SHF couplings of the nearest carbons, $A_{//}$ and $A_{\perp}$, are given by

$$A_{//} = A_{\mathrm{s}} + 2(A_{\mathrm{d}} + A_{\sigma} - A_{\pi})\,, \quad (4.9\mathrm{a})$$

$$A_{\perp} = A_{\mathrm{s}} - 2(A_{\mathrm{d}} + A_{\sigma} - A_{\pi})\,, \quad (4.9\mathrm{b})$$

where

$$A_{\mathrm{s}} = \frac{1}{2S}\frac{3}{4}\beta^2 g\mu_{\mathrm{b}} a_0\,, \quad (4.9\mathrm{c})$$

$$A_{\mathrm{d}} = \frac{1}{2S}3\alpha^2 g\mu_{\mathrm{b}} b_{\mathrm{dd}}\,, \quad (4.9\mathrm{d})$$

$$A_{\sigma} = \frac{1}{2S}\frac{3}{4}\gamma^2 g\mu_{\mathrm{b}} b_0\,, \quad (4.9\mathrm{e})$$

$$A_{\pi} = \frac{1}{2S}\frac{3}{8}\delta^2 g\mu_{b} b_0\,, \quad (4.9\mathrm{f})$$

$$a_0 = \frac{2}{3}\mu_0 g_{\mathrm{n}}\mu_{\mathrm{n}} \left|\Psi(0)\right|^2_{2\mathrm{s}}\,, \quad (4.9\mathrm{g})$$

$$b_0 = \frac{2}{5}\frac{\mu_0}{4\pi} g_{\mathrm{n}}\mu_{\mathrm{n}} \left\langle r^{-3}\right\rangle_{2p}\,, \quad (4.9\mathrm{h})$$

$$b_{dd} = \frac{\mu_0}{4\pi} g_{\mathrm{n}}\mu_{\mathrm{n}} R^{-3}\,. \quad (4.9\mathrm{i})$$

Table 4.2. LCAO parameters of the Ni_s^- center estimated from the ^{13}C superhyperfine interaction of the EPR spectra from [17]. The spin delocalization onto carbons beyond the nearest neighbors was neglected ($\alpha^2 + g\beta^2 + g\gamma^2 = 1$), and it was taken to be $\delta^2 = 0$

Atom	Orbital		Population	
			$R = 1.54$ Å	$R = 1.60$ Å
Ni	$3d$	(α^2)	0.746	0.721
C(I)	$2s$	(β^2)	0.024	0.024
	$2p$	(γ^2)	0.230	0.255

Here, S, $\Psi(0)$, r, R, g, and g_n are the total spin $S = 3/2$, the wave function at the nucleus, the electron-nucleus distance, the Ni–C bond length, the g-value for the electron, and the g-value for the nucleus, respectively.

The dipolar interaction between the magnetic moment of the electrons on the central Ni ion and the ^{13}C nuclear magnetic moment, A_d, depends on both the spin density of the Ni ion, α^2, and the Ni−C bond length, which might be increased from the normal C−C distance by structural relaxation. Since A_σ and A_π, which cannot be separated experimentally, counteract each other, only the minimum value of the spin density of C-$2p$ orbitals is estimated. The spin delocalization estimated in [17] is listed in Table 4.2.

The ^{13}C SHF couplings of the twelve second-nearest-neighbor carbons ($A_1 = 0.382\,\mathrm{mT}$, $A_2 = 0.382\,\mathrm{mT}$, $A_3 = 0.382\,\mathrm{mT}$ with the principal value A_3 associated with the principal axis along [110] and $A_\perp = 0.340\,\mathrm{mT}$) have been reported [6]. From the observed value of the isotropic part (0.308 mT), the total spin density on C-$2s$ orbitals of the second-nearest-neighbors was estimated to be 0.033.

The spin delocalization onto the nearest carbons cannot be determined uniquely from the ^{13}C SHF couplings observed, since the LCAO contains parameters that cannot be separated experimentally. On the other hand, the spin delocalization onto the nearest carbons is obtained from the Mulliken population analysis in the MO calculation. The ^{13}C SHF couplings are calculated based on the Mulliken population analysis and are compared with those obtained experimentally. In the calculation, the spin density, e.g., $n_{[C2s]}$, is used instead of β^2 in Eq. 4.9c, since the overlap population between Ni and nearest-neighbor carbons is not negligibly small. The parameters, a_0, b_0, and b_{dd}, are taken from [17] and [21].

4.3.2 Results

Figure 4.12 is the energy-level diagram for the $[NiC_{28}H_{36}]^-$ cluster. As marked by "a" and "b," two types of levels appear in the mid-gap. The energy level "a" belongs to a t_2-class of T_d symmetry, which is triply degenerate and is split into spin-up and spin-down levels. The lower energy level, with spin-up of $20t_2$, is fully occupied by electrons, and another higher level, with

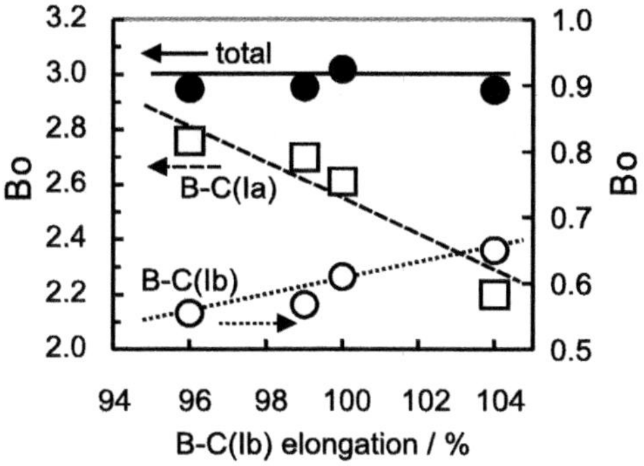

Fig. 4.11. Relation between Bo and lattice relaxation. The BOP for 100% elongation and other elongation values were calculated for T_d and C_{3v} symmetries, respectively

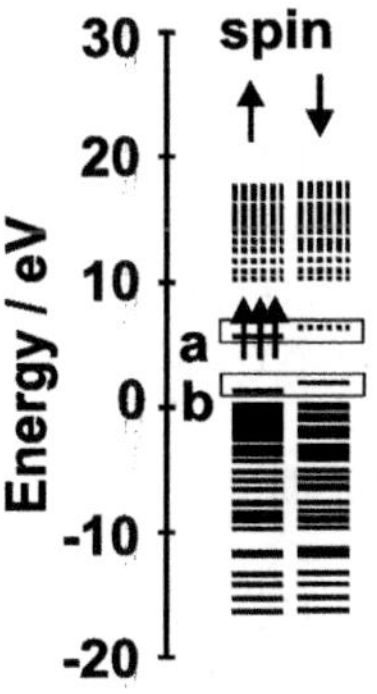

Fig. 4.12. Energy-level diagram for $[NiC_{28}H_{36}]^-$ cluster

spin-down of $20t_2$, is empty. The spin-up level of $20t_2$ is consequently the unpaired electrons of the Ni_s^- center with $S = 3/2$, which could be observed by EPR [17]. The level "b" belongs to an e-class of T_d symmetry and is also split into spin-up and spin-down levels ($9e$ levels), which are fully occupied by electrons. Therefore, the electron configuration $e^4t_2^3$ is obtained from the DV-Xα calculation for the cluster $[NiC_{28}H_{36}]^-$ and is consistent with the identification of the W8 EPR spectrum arising from the Ni_s^- center [17].

The ^{13}C SHF couplings of nearest-neighbor carbons, i.e., $A_{//}$ and $A_\perp$, were calculated: in the calculation, the value of $a_0 = \frac{2}{3}\mu_0 g_n \mu_n \left|\Psi(0)\right|^2_{2s}$ for the C-2s orbital of C(I) and $b_0 = \frac{2}{5}\frac{\mu_0}{4\pi} g_n \mu_n \langle r^{-3}\rangle_{2p}$ for the C-2p orbital of C(I) are cited from [17] and [21]. $A_{//}$ and $A_\perp$ were found to be 1.72 and 1.06 mT, respectively, while they were determined by the EPR measurements to be 1.339 and 0.347 mT, respectively. One of the possible reasons for the discrepancy between the observed and calculated SHF couplings is the structural relaxation associated with Ni ion substitution.

The structural relaxation is incorporated into the calculation by simply varying the Ni$-$C(I) distance. The DV-Xα calculation is well-defined for the cases in which $d_{Ni-C(I)}$ is less than $1.06 \times d_{C-C}$; further displacement of the C(I) atoms prevents convergence for the cluster model being studied, due partly to the relaxation of the next-nearest-neighbors. Figure 4.13, shows the SHF

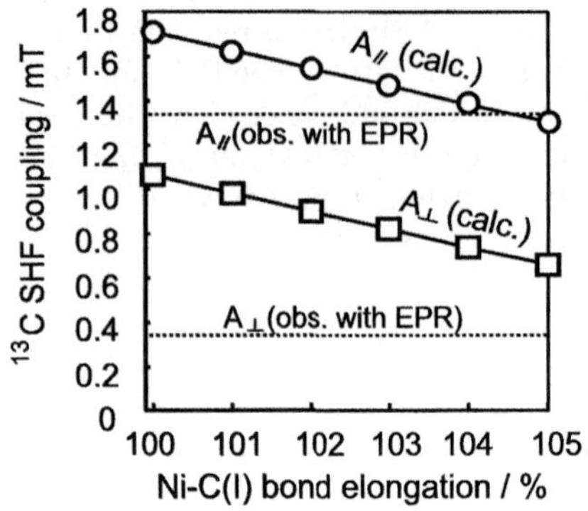

Fig. 4.13. Calculated $^{1}3$C super-hyperfine (SHF) coupling constant as a function of Ni-C(I) bond length. *Dotted line* shows the experimentally evaluated SHF constants

constants as a function of $d_{\mathrm{Ni-C}}$ distance. As shown in Fig. 4.13, the SHF constants approach the values (broken lines) determined via EPR measurement. This result suggests that the Ni−C(I) distance is elongated in comparison with the C−C distance in diamond. On the other hand, the bond order of the Ni−C(I) bond increases with an increase in the Ni−C(I) distance from 1.31 for $d_{\mathrm{Ni-C}} = d_{\mathrm{C-C}}$ to 1.50 for $d_{\mathrm{Ni-C}} = 1.05 \times d_{\mathrm{C-C}}$. This also suggests that the Ni_s^- center can be stabilized by elongating the Ni−C(I) distance.

A slight discrepancy between the observed and calculated SHF couplings remains even after the structural relaxation is taken into account. The most important reason is probably that the relaxation of the second nearest-neighbor is not taken into account in the calculation considered here: the relaxation of the C(I) site plausibly makes C(I)−C(II) bonds unstable unless the relaxation of the C(II) site takes place.

4.4 Property and Structure Predictions for Ceramics Using DV-Xα

Recent progress in computer power and computing techniques has enabled us to design new materials using computer simulations based on quantum mechanics. These simulations have been intensively applied to semiconductors, metals, and organic molecules. Nonetheless, the property calculation of inorganic crystals such as $Pb(Ti,Zr)O_3$, $BaTiO_3$, TiO_2, and SnO_2 have been limited though they have various interesting, useful functions such as piezoelectricity, opto-electronic effects, extremely high dielectric constants, opto-catalytic effects, and so on. This is partly due to the fact that these materials contain heavy metal ions and have a complex structure and that many important properties are closely related to the dynamic motion of ions, which requires high computation costs, including time and memory capacity. Therefore, these approaches have been carried out using molecular simulations such as lattice dynamics and molecular dynamics methods based on classical mechanics and use with empirical parameters or empirical interatomic potentials [22–24].

The status in quantum calculations for complex inorganic materials has been improved because of the rapid improvement in computation capability, which helps to solve the above difficulties. However, the materials that we are interested in have been getting much more complex. Development in ultra-thin film fabrication technologies has allowed the fabrication of new materials such as "artificial superlattices" and mesoscopic quantum structures. This means that we can create new materials at atomic scales to achieve improved and/or new functions. We may predict their properties using computer simulations; however, their specific dimensions are in mesoscopic scales which are quite large for typical quantum calculation methods even if the computation capability is drastically improved. As for empirical molecular simulations, they can be applied to such complex systems; however, they require empirical parameters, which may change according to material structure. This means that we cannot guarantee the accuracy of the calculation results if we apply the empirical techniques to new materials with structures that are significantly different from conventional materials.

Thus, we are still required to develop new methods to calculate the properties of these materials. In the following sections, we introduce some non-empirical approaches, including (i) a method to calculate structural and electronic properties, (ii) a fast electronic band structure calculation, and (iii) the indirect prediction of piezoelectric properties of inorganic solid solution materials. A fundamental, common idea is to remove empirical parameters and/or procedures from conventional calculation procedures. We mainly use the DV-Xα method because it is one of novel non-empirical techniques to calculate the electronic structures of clusters. It requires some trivial empirical parameters as input data, but these are not essential. One of its distinguishing features is that it requires rather inexpensive computer resources than other ab initio approaches because it adopts Slater's Xα approximation and a numerical approach using actual atomic wave functions in the clusters, which allows us to develop new calculation techniques without sacrificing the advantages of the above empirical approaches.

4.4.1 Calculation of Structural and Dielectric Properties of Inorganic Crystals Using DV-Xα Basis Functions

Introduction

In this section, we introduce a calculation method for structural and dielectric properties based on an ab initio approach. We have to know the total energy of crystals to calculate these properties (see [25] for a practical approach); however, the DV-Xα method usually does not provide the total energy value to sufficient degree of precision because it uses three-dimensional numerical integration with a finite number of integration points for the calculation of matrix elements required to solve a proper equation. Yamamoto et al. improved the accuracy of the numerical integrations [26]. In contrast, the ab initio techniques such as Gaussian03 [27], WIEN2k [28, 29] and Crystal03 [30, 31] can calculate total energy precisely. However, the calculated results depend on the selection of the basis functions, contracted Gaussian-type orbitals (CGTOs) [32]. Also, the optimal CGTOs may differ according to the nature of chemical bonding. In this section, we apply the DV-Xα method to determine appropriate CGTOs for subsequent ab initio calculations to determine structural and electronic properties.

Calculation

Figure 4.14 illustrates the structure of rock-salt-type $A^{n+}X^{n-}$ crystals. It can be seen that they are composed of $(AX_6)^{5n-}$ clusters with O_h symmetry. We calculated the atomic wave functions in the constituent cluster by the DV-Xα method. We simulated the Madelung potential, incorporating point-charges in the range of ± 3.5 lattices.

The atomic wave functions obtained, $\Psi(r)$, were expanded with Gaussian-type functions, $g_i(\alpha_i; r)$ [32]:

$$\psi(r) = \sum_i d_i g_i(\alpha_i; r) .$$

The parameters α_i and d_i are determined so as to minimize the following function by a least-squares method:

$$S = \sum_l \left(\psi(r_l) - \sum_i d_i g_i(\alpha_i; r_l) \right)^2 .$$

Crystal structure, and dielectric and elastic properties were calculated using the program Crystal88 [31, 33] with the CGTOs obtained above. Bulk modulus, C_V, is calculated from its definition using the total energy U:

$$\mathrm{d}U = \frac{1}{2} C_V \left(\frac{\mathrm{d}V}{V} \right)^2 ,$$

where $\mathrm{d}U$ is the total energy density change invoked by the volume change $\mathrm{d}V$.

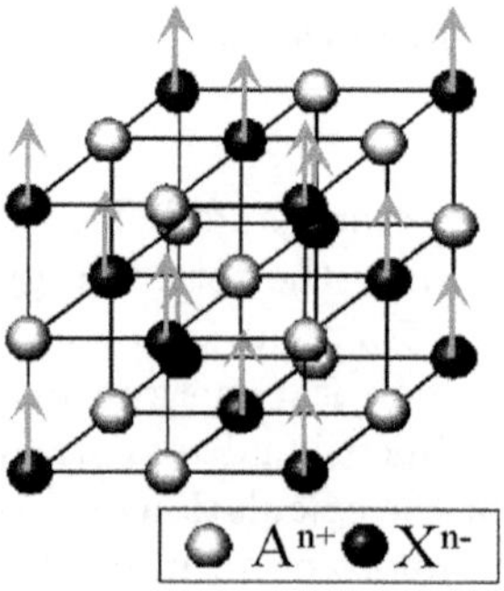

Fig. 4.14. Schematic illustration of the structure of a rock-salt-type $A^{n-}X^{n+}$ crystal. *Arrows* represent the ionic displacement pattern of the optical phonon

Similarly, dielectric permittivity ε is calculated from

$$dU = \frac{1}{2}\chi^{-1}P^2 ,$$

where χ is electric susceptibility and P is the electric polarization change invoked by the ionic displacement (a typical displacement pattern is illustrated in Fig. 4.14 with arrows).

Results

Figure 4.15 shows the O-2s and 2p orbitals calculated by the DV-Xα method and the fitted results with CGTOs. It demonstrates that O-2s and 2p orbitals calculated by the DV-Xα method can be reproduced by combination of CGTOs, when 7 GTOs are used for O-2s and 9 GTOs for O-2p. It was also confirmed that 5 to 9 GTOs could reproduce other basis functions as well.

The total energy of the crystals were calculated with Crystal88 using the CGTO obtained. Figure 4.16 shows the variation of the total energy density of MgO crystal with respect to the lattice volume. The total energy has a minimum value at the lattice parameter 4.197 Å, which agrees well with the experimental value of 4.21 Å measured at room temperature. The bulk modulus calculated at the room-temperature value for the lattice parameter is 2.16×10^{11} Pa, which is $\sim$ 30% larger than the experimental value of 1.53×10^{11} Pa.

Table 4.3 summarizes the calculation results obtained for MgO, NaCl, and NaF crystals. We can see that this technique of using the DV-Xα basis functions can provide good agreement with experimental values. The last line in Table 4.3 (indicated as MgO*) shows the result calculated using the re-optimized CGTOs after the determination using the DV-Xα basis functions. The variational principle tells us that the optimal basis function must give the minimum energy, which means that the MgO* basis should be better than the

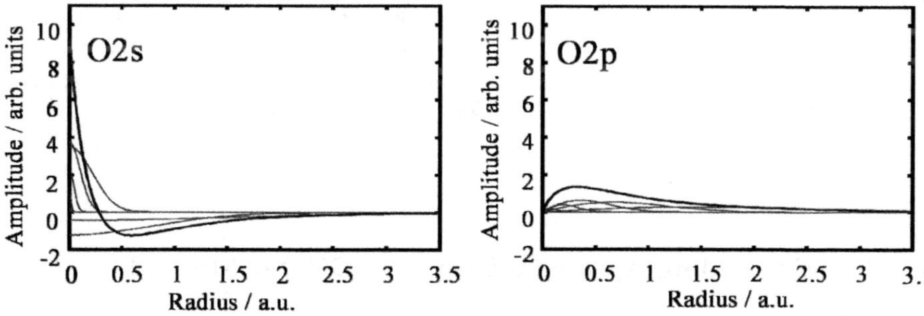

Fig. 4.15. O-2s and O-2p basis functions calculated by the DV-Xα method and the results fitted with CGTOs. The calculated and the fitted results are overlapped as thick lines, and thin lines show primitive GTOs used for the fitting

Table 4.3. Lattice parameter, bulk modulus, and dielectric constant measured and calculated using the DV-Xα bases. In the case of MgO*, the re-optimized CGTOs are used for the calculation. "Calc." and "exp." indicate the calculated and experimental values, respectively

	Lattice constant (Å)		Bulk modulus (Pa)		Dielectric permittivity	
	calc.	exp.	calc.	exp.	calc.	exp.
NaCl	5.622	5.64	26.7×10^{11}	24.6×10^{11}	4.93	5.62
NaF	4.580	4.62	57.4×10^{11}		4.18	5.05
MgO	4.197	4.21	2.16×10^{11}	1.53×10^{11}	7.98	9.77
MgO*	4.193	4.21	1.97×10^{11}	1.53×10^{11}	10.3	9.77

MgO basis determined by the DV-Xα method. In fact, the MgO* basis gives slightly better results. Nevertheless, the CGTOs found via the DV-Xα method are still good basis functions for these calculations.

We did not describe other techniques to calculate the above properties in which the DV-Xα method was not used because we think that we should concentrate on the actual application of the DV-Xα method in this chapter. We should note that similar ideas can be used for the calculation of piezoelectric constants and applied to the molecular dynamics [25,34,35]. Other approaches using a combination of the DV-Xα method and ab initio techniques to find the determinant factor of the battery voltage of $Li(Ti,Ni)O_2$ have been reported [36].

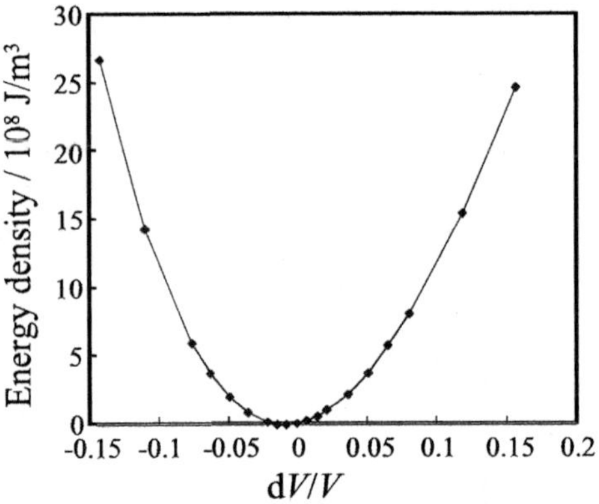

Fig. 4.16. Energy density calculated as a function of cell volume. The origin of the volume change dV is taken at the experimental lattice parameter at room temperature, 4.21 Å

4.4.2 Tight-binding Approach Using the DV-Xα Method

Introduction

In this section, we introduce a fast electronic band structure calculation technique [37]. The tight-binding method [38,39], the popular semi-empirical technique, is used in combination with the DV-Xα method to remove empirical parameters required in the procedure. Finally, we apply it to find the electronic structure of virtual artificial superlattices.

Calculation

In the DV-Xα method, the one-electron approximation and the linear combinations of atomic orbitals (LCAO) are used for convenience in practical calculations. In the LCAO approach, the molecular orbitals $|\Psi\rangle$ are expressed as follows, using the ith atomic wave function $|\Psi_i\rangle$:

$$|\Psi\rangle = \sum_i C_i |\Psi_i\rangle . \tag{4.10}$$

The parameters C_i are determined by the variational principle as

$$HC = \varepsilon SC , \tag{4.11}$$

where $H_{ij} = \langle \Psi_j | H | \Psi_i \rangle$ and $S_{ij} = \langle \Psi_j \mid \Psi_i \rangle$.

For periodic systems like perfect crystals, the Bloch's theorem reduces the number of electrons required for the calculation to those in the unit lattice. In the tight-binding approach, the following function is used to represent a basis function $|\Psi_i\rangle$ as a function of wave vector $\mathbf{k}$:

$$|\Psi_i(\mathbf{k})\rangle = \frac{1}{\sqrt{N_i}} \sum_{\substack{\text{all lattice} \\ \text{vectors } \mathbf{R}}} e^{i\mathbf{k}\cdot\mathbf{r_j}} |\Psi_i(\mathbf{r} - \mathbf{r}_j + \mathbf{R})\rangle , \tag{4.12}$$

where $\mathbf{R}$ is a lattice translation vector, $|\Psi_i(\mathbf{r} - \mathbf{r}_j)\rangle$ is the ith atomic wave function centered at the jth atomic site $\mathbf{r}_j$, and N_i is the number of wave functions used in (4.12). Finally, the following proper equation gives energy eigenvalues ε and crystal orbitals C as functions of $\mathbf{k}$.

$$H(k)C = \varepsilon S(k)C \tag{4.13}$$

$$H_{ij}(\mathbf{k}) = \frac{1}{\sqrt{N_i N_j}} \sum_{i',j'} e^{i\mathbf{k}\cdot(\mathbf{r}_{i'} - \mathbf{r}_{j'})} H_{i'j'} ,$$

$$S_{ij}(\mathbf{k}) = \frac{1}{\sqrt{N_i N_j}} \sum_{i',j'} e^{i\mathbf{k}\cdot(\mathbf{r}_{i'} - \mathbf{r}_{j'})} S_{i'j'} , \tag{4.14}$$

in which i' and j' are summed up for all equivalent atoms to the ith and jth atoms.

The tight-binding method simplifies the calculation of the integrations S_{ij} and H_{ij} using empirical parameters. We can obtain all S_{ij} and H_{ij} without empirical parameters by calculating the electronic structure of the constituent clusters using the DV-Xα method. In order to fix the origin of the energy for various clusters, which can't be determined absolutely because the clusters have finite net charge, we assume that the O-1s energy in each cluster has the same value. Subsequently, S_{ij} and H_{ij} are normalized using the energy shifts ε_0 as

$$(H + \varepsilon_0 S)\, C = (\varepsilon + \varepsilon_0)\, S \,. \tag{4.15}$$

That is, the S_{ij} and H_{ij} in (4.13) and (4.14) should be replaced with $(H + \varepsilon_0 S)$ and S, respectively.

4.4.3 Results

We start from the calculations for rutile-type crystals, TiO_2 and SnO_2, because these have a simple structure composed of a cluster $(MO_6)^{8-}$ (M = Ti, Sn). The crystal structure is illustrated in Fig. 4.17. Atomic basis functions used in the DV-Xα calculation are: orbitals up to 2p for O, orbitals up to 4p for Ti, and orbitals up to 5p (except for 4f) for Sn.

Figure 4.18 shows the density of states (DOS) near the highest occupied molecular orbital (HOMO) with the tight-binding model in comparison with the X-ray photoemission spectra (XPS) measured. We can see that the calculated DOS agrees well with the XPS spectra, which guarantees that the technique presented here can be used to calculate the electronic structure of the valence band to a good accuracy. Compared with the DOS calculated for the $[MO_6]^{8-}$ clusters (M = Ti^{4+}, Sn^{4+}), the cluster calculation (see Fig. 4.19) gives the result that the DOS peaks in the $[SnO_6]^{8-}$ cluster at the highest energy and at the middle energy have almost the same height (Figure 4.19), while the highest-energy peak is the strongest in the tight-binding result, demonstrating that the tight-binding calculation gives better agreement with the XPS spectra. This result also implies that the small $[SnO_6]^{8-}$ cluster is not enough to describe the electronic structure in the crystal.

Now, we proceed to the calculation of more complex oxides. We study perovskite-type crystals, $SrTiO_3$, $BaTiO_3$, and $SrZrO_3$ because these crystals have various interesting electronic and optical properties. Figure 4.20 shows the crystal structure of $SrTiO_3$, which is a cubic perovskite composed

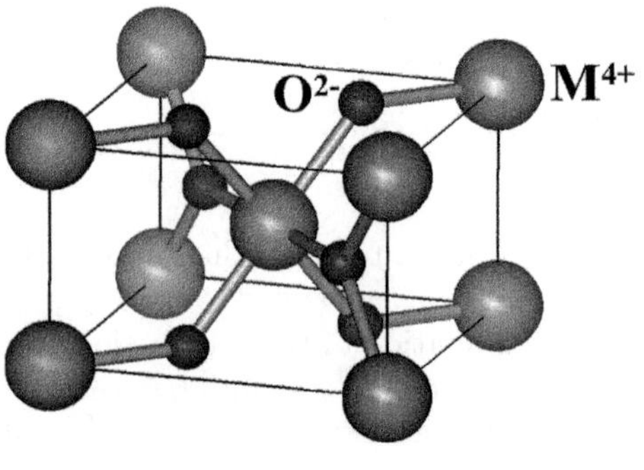

Fig. 4.17. Schematic illustration of the structure of a rutile-type crystal, MO_2

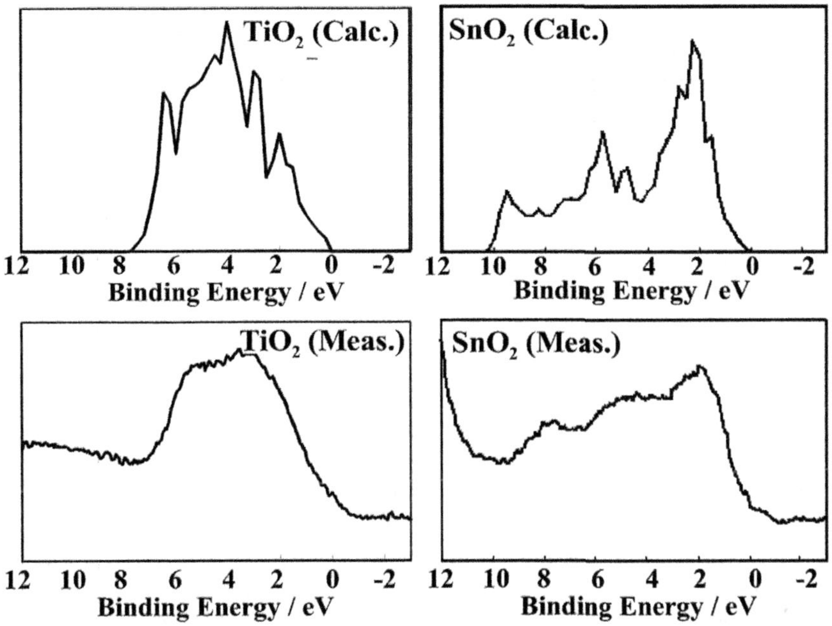

Fig. 4.18. DOS calculated (top) by the tight-binding method and XPS spectra measured (bottom) in the vicinity of valence band for TiO_2 (**a**) and SnO_2 (**b**). The HOMO energy is normalized to 0 eV

of $[TiO_6]^{8-}$ and $[SrO_{12}]^{22-}$ clusters. In the case of $BaTiO_3$, which is a ferroelectric crystal at room temperature, we need to substitute Sr for Ba and deform the lattice to be tetragonal.

Thus, we calculate the electronic structures in the $[TiO_6]^{8-}$, $[ZrO_6]^{8-}$, $[SrO_{12}]^{22-}$, and $[BaO_{12}]^{22-}$ clusters using the DV-Xα method. The Madelung potentials are calculated from point-charges in ± 3.6 lattices.

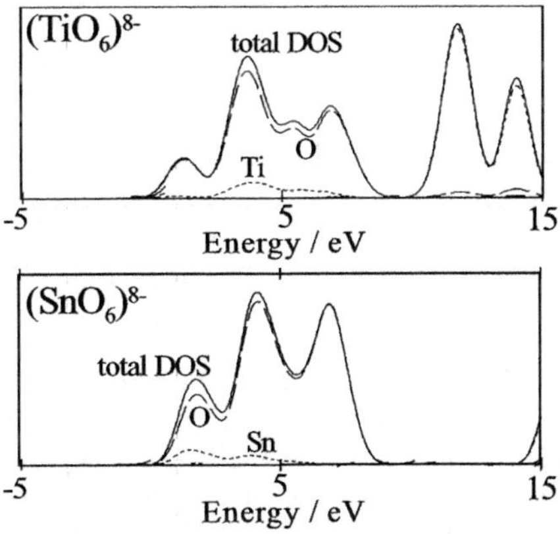

Fig. 4.19. DOS of $[TiO_6]^{8-}$ and $[SnO_6]^{8-}$ clusters calculated using the DV-Xα method. HOMO is assigned to be 0 eV

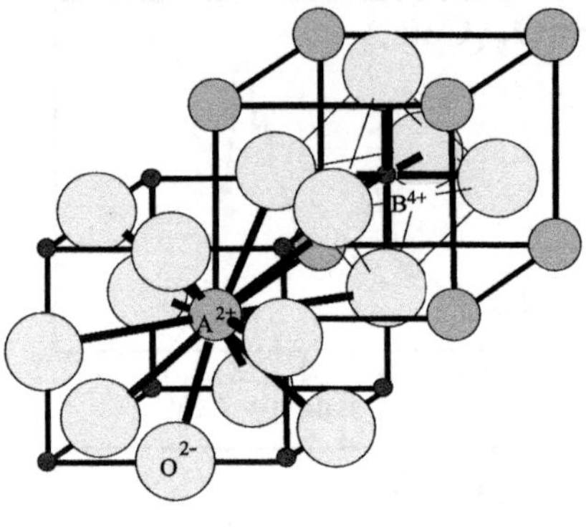

Fig. 4.20. Schematic illustrations of the structure of perovskite-type crystal ABO_3. Constituent clusters, $[BO_6]^{8-}$ and $[AO_{12}]^{22-}$, are explicitly shown

First we calculated the band structure of $SrTiO_3$ either by neglecting or by incorporating Sr (Fig. 4.21a,b, respectively) to extract the effect of the Sr on the electronic structure. We can see that the band structures near the band gap (i.e., around 0 eV) do not change remarkably, and the result agrees well with the results calculated by the self-consistent OLCAO method [40] even if the Sr is not considered in the calculation, suggesting that the Sr does not heavily affect the electronic structure near the band gap. This result is consistent with that reported by Tsukada et al. using the DV-Xα cluster calculation [41].

In contrast, we found that the incorporation of the Ba is necessary for $BaTiO_3$, indicating that the Ba-5p is strongly hybridized with O-2p, which is thought to be related to the appearance of ferroelectricity in $BaTiO_3$. Miura et al. also reported the similar conclusion for $PbTiO_3$ [42, 43]. Comparing

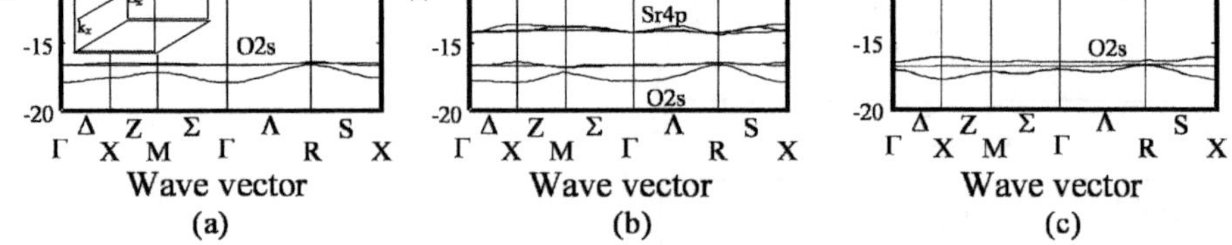

Fig. 4.21. Band structures calculated for cubic $SrTiO_3$ (**a** and **b**) and tetragonal $BaTiO_3$ (**c**). For results shown in (**a**) Sr is neglected in the tight-binding calculation. The *inset* in (**a**) shows the notation of symmetry points in the Brillouin zone. The band structure is not significantly changed by incorporation of Sr ions for SrTiO3, while incorporation of Ba changes the valence band structure largely in BaTiO3.

the band structure of cubic and tetragonal $BaTiO_3$, the energy levels in the tetragonal $BaTiO_3$ are split because it has lower crystal symmetry. Consequently, it has a bandgap smaller by $\sim 0.1\,eV$ than that of cubic $BaTiO_3$. As for the XPS spectra (inset in Fig. 4.22), the shape of the O-2p bands is different between $SrTiO_3$ and $BaTiO_3$(Fig. 4.22). The $SrTiO_3$ has two strong peaks while the $BaTiO_3$ has three peaks that are strongly overlapped. The third peak appearing in $BaTiO_3$ was not observed when the band structure was calculated without Ba. Also, the effective mass at the Γ point at the HOMO along the Δ direction becomes larger when the Ba is incorporated. These results indicate that the difference in the XPS spectra originates from the difference in the chemical bonding nature of Sr and Ba.

Now let us come back to the band structure of $SrTiO_3$. It shows an indirect transition between the Γ point and M or R points. Figure 4.23 shows the band structures of $SrTiO_3/SrZrO_3$ superlattices along the [001], [011], and [111] directions, respectively. Note that the [001], [011], and [111] superlattices have different symmetries, i.e., tetragonal, orthorhombic, and hexagonal, respectively. The different symmetries split the electronic levels that are degenerate in cubic $SrTiO_3$. It is also found that the energy dispersion in the conduction band becomes less than that of $SrTiO_3$, especially when the [111] superlattice is formed. In the superlattices, Zr replaces half of Ti, which breaks the Ti−O−Ti bonding network along the superlattice direction. It makes the interaction between Ti orbitals weak, resulting in less dispersion of the conduction bands (i.e., a larger effective mass for the electron).

The [001] superlattice still shows an indirect transition similar to the case of the original lattice, while the band structures of the [011] and [111] superlattices show a direct transition at the Γ point. These results arise from the fact that the Brillouin zone (BZ) of the double-period superlattices is half the size of the BZ of the original lattice, and the band structure folds up on the new boundary of the superlattice's BZ.

We should note that the transition probability at the direct band gap in the [011] and [111] superlattices may not be large, although the band structure changes to a direct transition, because the HOMO is almost the O-2p orbital,

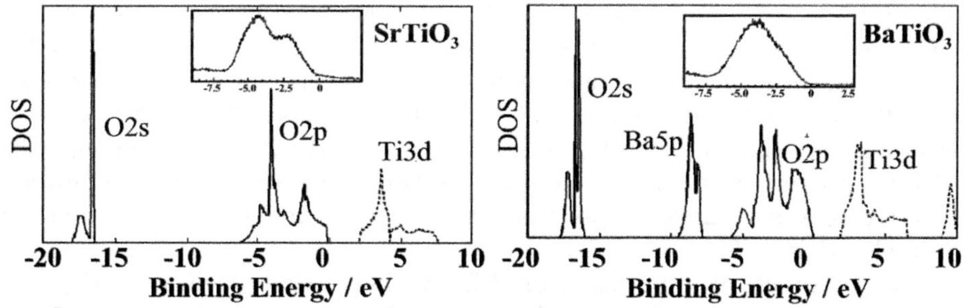

Fig. 4.22. DOS calculated by the tight-binding method for $SrTiO_3$ and $BaTiO_3$. *Insets* show the corresponding XPS spectra measured

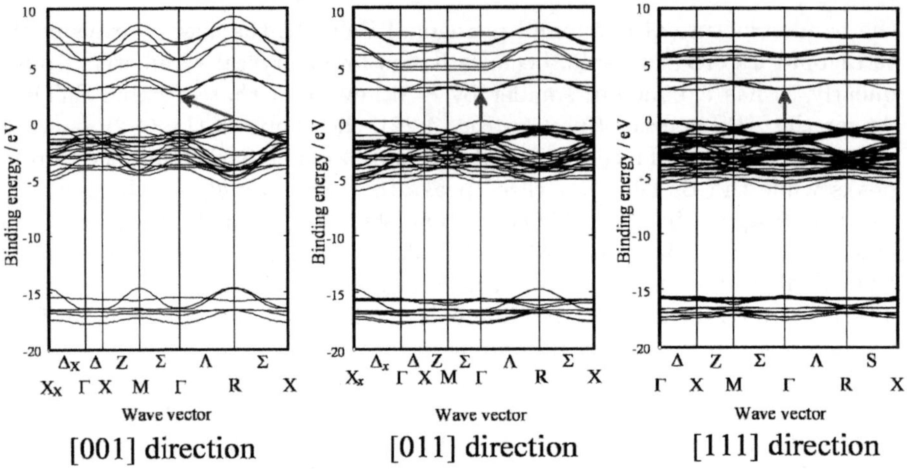

Fig. 4.23. Band structures of $SrTiO_3/SrZrO_3$ superlattices along [001], [011], and [111] directions

which has the same period as the original $SrTiO_3$ lattice. Thus, for practical design of new photo-electronic functional materials, we need to choose a more suitable superlattice structure so as to achieve a smaller effective mass at the conduction band and to modify the valence band for improved carrier mobility and transition probability. The present tight-binding approach combined with the DV-Xα method would help the procedure.

4.4.4 Indirect Prediction of the Piezoelectric Property Change of $Pb(Zr,Ti)O_3$ Induced by the Addition of Impurities

Introduction

As previously described, it is usually hard to apply quantum calculation methods to dielectric and piezoelectric properties in which the dynamic motion of ions play an essential role. We have reported the calculation of these properties via ab initio band calculations for AlN and ZnO [44]. However, it is very difficult to apply the same approach to solid-solutions such as $Pb(Zr_xTi_{1-x})O_3$ (PZT), which is one of the most popular piezoelectric materials, because Zr and Ti are distributed randomly in the B-sites of the perovskite lattice, and it is also known that the ordering or disordering of these B-site ions changes the electric properties completely [45]. In addition, cooperative lattice vibration referred to as soft-mode phonons play an important role in the appearance of ferroelectricity [46].

In the practical application of PZT, various oxides such as MnO_2 and Nb_2O_5 are used as additives in order to control its electronic properties [47,48]. We found that the electromechanical coupling factor of $Pb(Zr_{0.5}Ti_{0.5})O_3$ was increased by the addition of MnO_2 [49]; this was caused by the decrease in the axial c/a ratio, which represents how much the crystal lattice deforms from tetragonal. It is also known that the c/a ratio affects the Curie temperatures of perovskite ferroelectrics [50]. The Curie temperature is a measure of the ferroelectric and piezoelectric properties of perovskite-type oxides, and are closely related to the crystal structure, especially the axis c/a ratio. This leads us to expect that we can know the effect of additives on piezoelectric properties if we can estimate the c/a ratio change by computer simulation. Another indirect approach to evaluate the dielectric property change has been carried out by Fujita et al. using the bond overlap population as an indicator [51].

In this section, we apply the DV-Xα method to estimate the effect of additives in PZT on electronic properties through the estimation of the crystal structure change due to the addition of impurities. The relationship between the results of the DV-Xα calculations and the axis c/a ratio was investigated.

Calculation

DV-Xα calculations are performed with cubic and tetragonal clusters $[MO_6]^{n-}$ (M = Ti^{4+}, Zr^{4+}, Mn^{4+}, Nb^{5+}, Mg^{2+}, Al^{3+}, Cr^{3+}, Fe^{3+}, Co^{2+}, Ni^{2+}), as illustrated in Fig. 4.24. The configuration of the tetragonal cluster is extracted from the actual crystal structures of impurity-doped PZT determined by X-ray diffraction. The configuration of the cubic clusters is supposed to have a lattice parameter of 0.4064 nm. These clusters were placed in the Coulomb potential formed by surrounding ions in the range of ± 2.6 lattices to approximately simulate the Madelung potential. The formal charges are assumed to be $+2e$, $+4e$, $-2e$ (e is the elemenary electric charge, 1.602×10^{-19} C) for Pb, B-site cations, and O anions, respectively.

To discuss the stability of the tetragonal lattice, it is better to compare the total energy difference between tetragonal and cubic $[MO_6]^{n-}$ clusters. However, it is difficult to determine an absolute value of the energy level

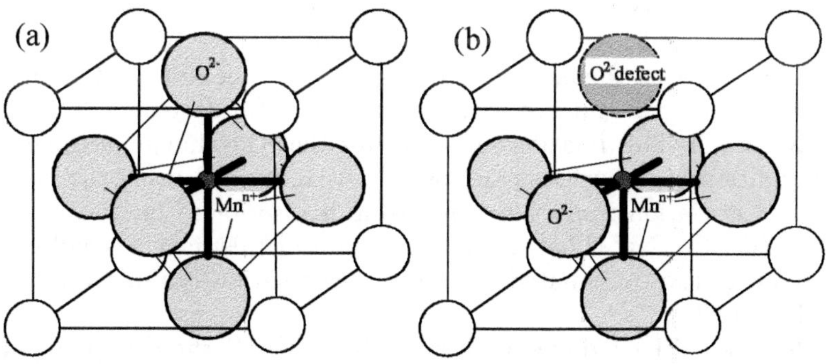

Fig. 4.24. Cluster models used for the calculation of electronic states of impurities in PZT. **a** MO_6 cluster. **b** MO_5 cluster with an oxygen vacancy

using the DV-Xα method because a cluster has a finite net charge. We used O-1s energy to normalize energy levels in different clusters. In addition, we introduce the following index ΔE to estimate the stability of a tetragonal lattice because it is hard to obtain total energy at enough precision using the DV-Xα method:

$$\Delta E = \sum_{E_{\mathrm{tet}} > E_{\mathrm{cub,HOMO}}} n \cdot \left(E_{\mathrm{tet}} - E_{\mathrm{cub,HOMO}}\right) ,$$

where n is the number of electrons occupied, $E_{\mathrm{cub,HOMO}}$ represents the energy level of the HOMO in the cubic cluster, and E_{tet} represents the energy level of the orbital occupied in the tetragonal cluster. Note that a lower ΔE value indicates that the tetragonal lattice is more stable.

Results

Figure 4.25 shows the DOS for cubic and tetragonal $[MO_6]^{n-}$ clusters normalized using O-1s energy levels. In this figure, we assumed a Gaussian-type distribution of 0.05 eV FWHM for each energy level. The broad bands near 0 eV are attributed to the hybridized orbitals made of O-2p and d-orbitals in transition metals. The O-2p bands in the tetragonal clusters are broader than those in the cubic clusters because the energy levels degenerate in the cubic clusters are split in the tetragonal clusters. The d electrons are in high spin states for all transition metal clusters, i.e., those of M = Fe^{3+}, Cr^{3+}, Co^{2+}, Mn^{4+}, and Ni^{2+}.

Figure 4.25(b) shows that the HOMO of the cubic clusters is lower than that of the tetragonal clusters, though the tetragonal lattice is actually more stable than the cubic lattice for PZT. This discrepancy could be caused by the fact that the long-range Coulomb potential is not considered. However, a considerable difference can be found in the DOS near the HOMO between the tetragonal and the cubic clusters, leading us to introduce the ΔE value to estimate the stability of the tetragonal lattice.

Figure 4.26 shows the relationship between the ΔE value and the change of the c/a ratio resulting from the addition of impurities. We can find a certain relationship between them, indicated with a solid line, for clusters of closed-shell cations, i.e., Cr and Mn cations; the c/a ratio decreases with increase in the ΔE value. This relationship indicates that the change in the c/a ratio due to addition of impurities can possibly be estimated using the ΔE value. However, a deviation from the relationship is also found in transition metal additives (Fe, Co, Ni). That is, the transition metals have a small c/a ratio compared with those estimated from the above relationship shown with a solid line in Fig. 4.26.

In the cases of PZT doped with Fe^{3+}, Ni^{2+}, Co^{2+}, there are two potential causes which may affect the c/a ratio; one is the coexistence of cations with different valence states, and the other is the formation of oxygen vacancies

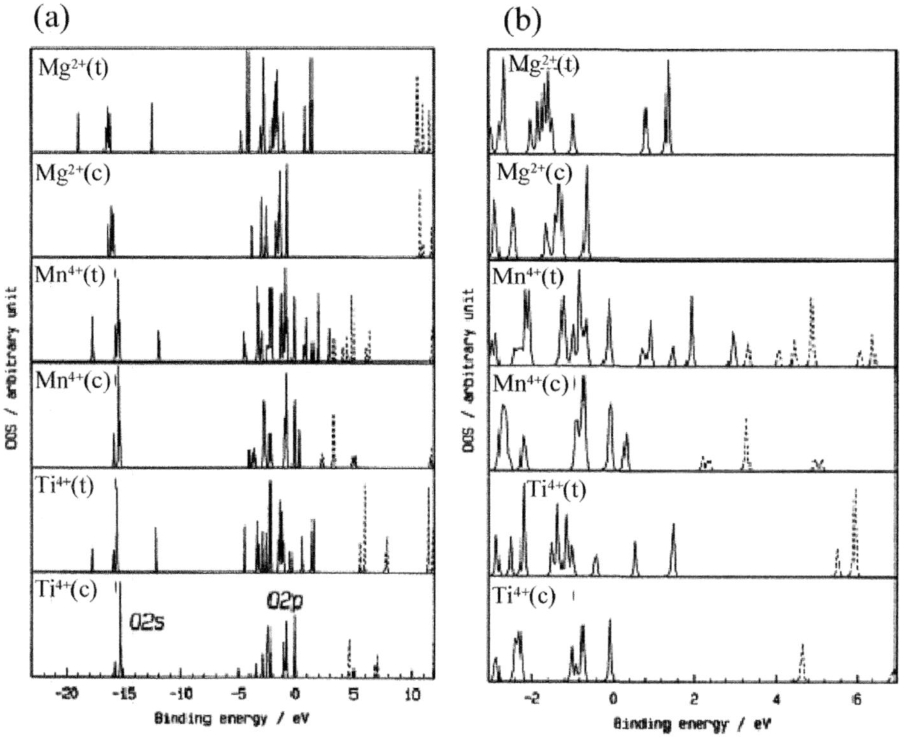

Fig. 4.25. Calculated DOS of $[MO_6]^{n-}$ clusters. *Solid lines* denote occupied states; *dashed lines* denote unoccupied states. *Right side* (**b**) is an expanded view of *left side* (**a**) in the vicinity of the HOMO levels. (t) and (c) represent tetragonal and cubic clusters, respectively. Symmetries of clusters used in the calculation are tetragonal and cubic and are indicated in the figures by (t) and (c), respectively.

introduced when an acceptor-type ion is doped. Fig. 4.27 superimposes the results calculated for clusters with different valency cations on Fig. 4.26. We can see that the decrease in the valency tends to the decrease in the ΔE value, which corresponds to an increase in the c/a ratio (note that the ΔE value of Fe^{2+} is negative). For Co, the change of the ΔE value is too small to explain the deviation. Thus, a different valency model cannot explain the deviation found in Fig. 4.26 for transition-metal clusters.

Next, we calculate the ΔE values for clusters with an oxygen defect using the cluster shown in Fig. 4.24(b). In $[FeO_5]^{7-}$, the Mulliken population analysis shows that about two electrons occupy the Fe-5p orbital but no electron occupies the Fe-5s orbital. The deficit of an oxygen forms a strong electric field at the Fe site and attracts electrons into the defect site. Therefore, electrons in the defect cluster occupy the polarized Fe-5p levels, which are not occupied in the cluster without the defect. As a result, it is found that the ΔE values for $[FeO_5]^{7-}$, $[NiO_5]^{8-}$, and $[CoO_5]^{8-}$ clusters are much larger than the solid

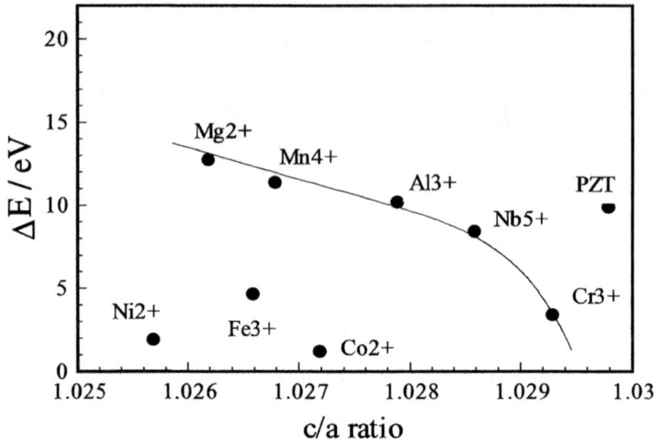

Fig. 4.26. Relationship between the axis c/a ratio and the ΔE value for $Pb(Zr_{0.5}Ti_{0.5})O_3 + 1.2\%MO_n$. Names in the figure indicate additive cations to $Pb(Zr_{0.5}Ti_{0.5})O_3$

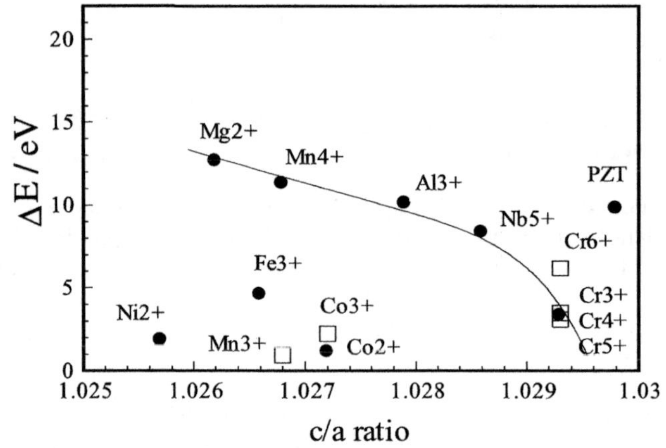

Fig. 4.27. Relationship between the axis c/a ratio and the ΔE value for $Pb(Zr_{0.5}Ti_{0.5})O_3 + 1.2\%MO_n$ in which additive cations have different valencies

line in Fig. 4.28. This suggests that the additions of Ni, Fe, or Co increase the c/a ratio if no oxygen vacancy is formed; however, the formation of an oxygen defect destabilizes the tetragonal lattice, resulting in a small axis c/a ratio in actual crystals.

In order to confirm the effect of oxygen vacancies on the c/a ratio, we experimentally investigate the effect of counter-doping with a donor to suppress the formation of oxygen vacancies introduced by acceptor doping. Figure 4.29 shows the c/a ratio of $PbTiO_3$ (PT) doped with Fe and La, in which Fe and La act as acceptor and donor, respectively. The c/a ratio decreases when either Fe or La is doped. The c/a ratio increases at an Fe content of $< 1.2\%$ and

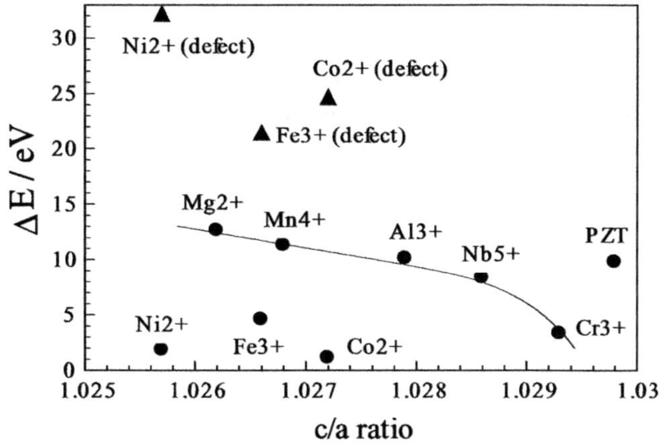

Fig. 4.28. Relationship between the axis c/a ratio and the ΔE value for $Pb(Zr_{0.5}Ti_{0.5})O_3$ + 1.2%MO_n compared with the ΔE value of the clusters with an oxygen defect

decreases at an Fe content of > 2.0% when Fe is doped to $(Pb_{0.98}La_{0.02})TiO_3$ (PLT). Oxygen vacancies are formed when Fe is added to pure PT, while they are suppressed when Fe is added to PLT due to charge compensation. Taking the condition of charge neutrality into consideration, oxygen vacancies are introduced when Fe content is larger than La content. This is why the c/a ratio begins to decrease when Fe content is larger than 2.0%. In addition, the previous discussion of the relationship between the valency of additive ions and the ΔE value suggests that Fe addition to PLT would also form a large

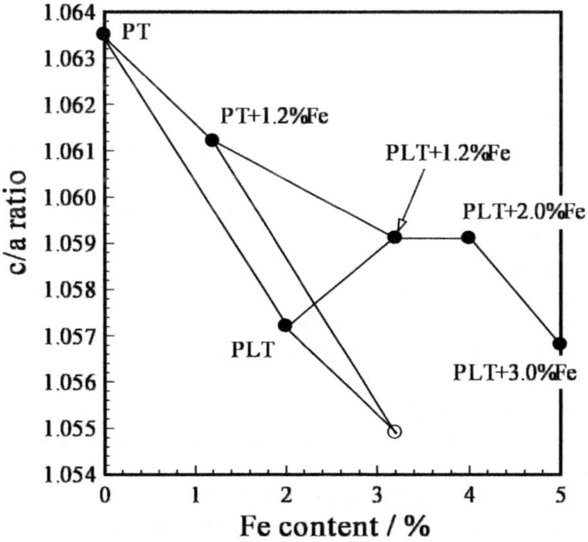

Fig. 4.29. Effect of counter-doping of Fe and La on the axial c/a ratio

amount of Fe^{2+}, which may increase the c/a ratio. It was also confirmed that the c/a ratio decreases when Fe, Co, or Ni doped in PZT is reduced in vacuum, demonstrating that the introduction of oxygen vacancies makes the c/a ratio small. Similar results are obtained for Nb- and La-doped PZT.

The above results suggest that we may control the axis c/a ratio by choosing an additive with an appropriate valency and defect structure. In particular, the combination of donor and acceptor doping would provide a flexible way to modify structure, which would subsequently lead to modification piezoelectric properties in a wide range.

4.5 Remarks

Examples of DV-Xα calculations in the field of ceramics research were introduced. It is needless to say that we cannot survey all of the DV-Xα papers on ceramics. Some of the examples were done at the beginning of 1990s. Quantum chemical calculations were state-of-the-art in 1980s because of a lack of computer resources. Nowadays, the power of calculators, especially personal computers, has increased, and most of these examples can be done with the PCs in any laboratory or office. The computer simulation of ceramics is like a "simplified cluster model with a simplified theory." However, the developments of both hardware and software resources enable us to compute a "complicated cluster model with a simplified theory" or a "simplified cluster model with a complicated theory." We described the prediction of material structure and properties via computer simulation. As a conclusion to this chapter, we predict that real material design with PC simulations will be realized in the near future.

References

1. R. Weissmann, K. Muller, Surf. Sci. Rep. **105**, 251 (1981)
2. S. Tanaka, C. Akita, N. Ohashi, J. Kawai, H. Haneda, J. Tanaka, J. Solid State Chem. **105**, 36 (1993)
3. N. Ohashi, S. Tanaka, C. Akita, J. Kawai, H. Adachi, O. Fukunaga, J. Tanaka, in *Computer Aided Innovation of New Materials II*, ed. by M. Doyama, J. Kihara, M. Tanaka, R. Yamamoto (Elsevier, Amsterdam, 1993), p. 1557
4. E. Yasu et al., Dia. Rel. Mat. **4**, 59 1994)
5. O.D. Tucker et al., Phys. Rev. B **50**, 15586 (1994)
6. R.P. Messmer, G.D. Watkins, Phys. Rev. B **7**, 2568 (1973)
7. A. Mainwood, J. Phys. C **11**, 2703, 2543 (1979)
8. M. Lanoo, Phys. Rev. B **25**, 2987 (1982)
9. H.W.L. Alves, A.N. Borges, J.L.A. Alves, J.R. Leite, A. Dal Pinto, Solid State Commun. **51**, 939 (1983)
10. S.J. Sferco, M.C.G. Paseggi, Solid State Commun. **62**, 517 (1987)
11. K. Jackson, M.R. Pederson, J.G. Harrison, Phys Rev. B **41**, 12641 (1990)

12. N. Ohashi, O. Fukunaga, J. Isoya, J. Tanaka, Japan. J. Appl. Phys. **36**, 1126 (1997)
13. R.H. Wentorf Jr., J. Phys. Chem. **75**, 1833 (1971)
14. H.M. Strong, R.M. Chrenko, J. Phys. Chem. **75**, 1838 (1971)
15. J.C. Angus, N.C. Gardner, D.J. Poferl, S. Chauhan, Pei Sung: Russ. Sint. Almazy **6**, 38 (1971)
16. H.N. Loubser, W.P. van Ryneveld, Nature **211**, 517 (1966)
17. J. Isoya, H. Kanda, J.R. Norris, J. Tang, M.K. Bowman, Phys. Rev. B **41**, 3905 (1990)
18. J. Isoya, H. Kanda, Y. Uchida, Phys. Rev. B **42**, 9843 (1990)
19. D.A. van Wezep, R. van Kemp, E.G. Sieverts, C.A.J. Ammerlaan, Phys. Rev. B **32**, 7129 (1985)
20. J. Owen, J.M.H. Thornley, Rep. Prog. Phys. **29**, 675 (1966)
21. B.A. Goodman, J.B. Raynor, in *Advance in Inorganic Chemistry, Radio Chemistry*, Vol. 13, ed. by H.J. Emeleus, A.G. Sharpe (Academic, New York, 1970), pp. 135–362
22. A. Khalal, D. Khatib, B. Jannot, Physica B **271**, 343 (1999)
23. T. Kamiya, Japan. J. Appl. Phys. **35**, 3688 (1996)
24. P. Ghosez, E. Cockayne, U.V. Waghmare, K.M. Rabe, Phys. Rev. B **60**, 836 (1999)
25. T. Kamiya, Japan. J. Appl. Phys. **35**, 4421 (1996)
26. K. Yamamoto, H. Ishikawa, K. Fujima, M. Iwasawa, J. Chem. Phys. **106**, 8769 (1997)
27. http://www.Gaussian.com/ http://www.wien2k.at
28. http://www.tuwien.ac.at/theochem/wien97 http://www.cse.clr.ac.uk/cmg/CRYSTAL/
29. P. Blaha, K. Schwarz, J. Luitz, *WIEN97, A full potential linearized augmented plane wave package for calculating crystal properties* (Tech. Univ. Wien, Vienna, 1999)
30. http://www.dl.ac.uk/TCSC/Software/CRYSTAL
31. C. Pisani, R. Dovesi, C. Roetti, *Hartree-Fock ab-initio treatment of crystalline systems, Lecture Notes in Chemistry*, Vol. 48 (Springer, Berlin Heidelberg New York, 1988)
32. S. Huzinaga, *Gaussian basis sets for molecular calculations* (Elsevier, Amsterdam, 1984)
33. R. Dovesi, C. Pisani, C. Roetti, M. Caus, in *Proceedings of the Quantum Chemistry Program Exchange*, Program Number 577 R.Dovesi, C.Pisani, C.Roetti, M.Caus, Crystal88 provided by the Quantum Chemicstry Program Exchange (QCPE). Program Number 577. http://qcpe.chem.indiana.edu/
34. T. Kamiya, Japan. J. Appl. Phys. **35**, 3688 (1996)
35. A.D. Corso, M. Posternak, R. Resta, A. Baldereschi, Phys. Rev. B **50**, 10715 (1994)
36. Y. Koyama, I. Tanaka, Y.-S. Kim, S.R. Nishitani, H. Adachi, Japan. J. Appl. Phys. **38**, 4804 (1999)
37. T. Kamiya, T. Tanaka, T. Tsurumi, M. Daimon, Japan. J. Appl. Phys. **33**, 3965 (1994)
38. J.C. Slater, G.F. Koster, Phys. Rev. **94**, 1498 (1954)
39. W.A. Harrison, *Electronic Structure, the Properties of Solids. The Physics of the Chemical Bond* (Freeman, San Francisco, 1980)

40. Y.-N. Xu, W.Y. Ching, R.H. French, Ferroelectrics **111**, 23 (1990)
41. M. Tsukada, C. Satoko, H. Adachi, J. Phys. Soc. Japan **48**, 200 (1980)
42. K. Miura, M. Tanaka, Japan. J. Appl. Phys. **37**, 6451 (1998)
43. K. Miura, M. Tanaka, Japan. J. Appl. Phys. **37**, 6460 (1998)
44. T. Kamiya, Japan. J. Appl. Phys. **35**, 4421 (1996)
45. T. Takahashi, Japan. J. Appl. Phys. **39**, 5637 (2000)
46. W. Cohran, Phys. Rev. Lett. **3**, 412 (1959)
47. T. Sakai, H. Kawamoto, J. Ceram. Soc. Japan **106**, 792 (1998)
48. K. Takahashi, M. Nishida, H. Hase, Japan. J. Appl. Phys. **37**, 5285 (1998)
49. T. Kamiya, T. Suzuki, T. Tsurumi, M. Daimon, Japan. J. Appl. Phys. **31**, 3058 (1992)
50. K. Singh, D.K. Bopardikar, Ferroelectrics **61**, 281 (1984)
51. M. Fujita, R. Sekine, S. Sugihara, Japan. J. Appl. Phys. **38**, 5664 (1999)

5

Magnetic Properties

Kimichika Fukushima

The DV-Xα method has been successfully applied to study of a variety of materials. This chapter focuses on magnetic properties – in particular, antiferromagnetic properties. It has thus far not been fully reported whether or not antiferromagnetic properties may be properly represented by the local density approximation. It is shown that, even within the scheme of the conventional Xα method, the molecular orbital method properly describes the antiferromagnetic state with the use of localized atomic orbitals. We present calculated antiferromagnetic properties for artificial hydrogen clusters, which have a rather long interatomic distance, as well as for Cu oxides.

5.1 Introduction

The DV-Xα method has been successfully used for the understanding of the properties and designs of a variety of materials, such as metals, semiconductors, ceramics, and so on, including a recently reported thermoelectric material with an anomalous high thermoelectric power [1]. Electronic structures of metallic, semiconducting, as well as ferromagnetic states are properly described by means of the local density approximation (LDA), which is developed based on the Slater's Xα method. Concerning the antiferromagnetic state, a ground-breaking neutron diffraction experiment was performed for MnO around 1951 [2], and the spin structure of the antiferromagnetic state was observed for the first time. After discovery, the description of the antiferromagnetic state became an important task in solid state physics.

In 1963, Hubbard succeeded in presenting a semi-empirical theory to describe the antiferromagnetic state based on the band structure model [3]. This theory is now called the Hubbard model and has a Hamiltonian expressed as

$$\mathcal{H} = -t \left(\sum_{\langle i,j \rangle} \sum_{\sigma} c_{i\sigma}^{\dagger} c_{j\sigma} \right) + U \sum_{i} n_{i\uparrow} n_{i\downarrow} \ , \qquad (5.1)$$

where $c_{i\sigma}^{\dagger}$ and $c_{i\sigma}$ are creation and annihilation operators for an electron with spin σ ($\sigma = \uparrow, \downarrow$) on the atomic site i, respectively, and $n_{i\sigma}$ is the number of electrons. The quantities t and U are the transfer integral and on-site Coulomb repulsion, respectively. The summation with $\langle i, j \rangle$ is carried out over a set of nearest-neighbor sites. In the case where the number of localized conductive electrons per atom is unity, and the on-site Coulomb repulsion U is weak with respect to the transfer integral t, the Hubbard model leads to a gapless band, with a bandwidth proportional to t. However, where the on-site Coulomb repulsion is strong with respect to the transfer integral, two electrons cannot exist on the same atomic site, and an antiferromagnetic state with an energy gap becomes stable in this model. The energy gap in this case is caused by the Coulomb repulsion between electrons on the same atomic site.

It has thus far not been fully reported whether or not the antiferromagnetic state may be described properly via first-principles by the local density approximation. For band structure calculations, some improvements have been proposed [4–6]: one is a modification using the self-interaction correction, and another is the LDA+U model, with U being a strong on-site Coulomb repulsion. These methods describe the antiferromagnetic state to some extent, but there remain problems. Concerning a molecular orbital method, it has not yet been shown whether or not the antiferromagnetic state may be represented properly by the LDA.

This chapter aims to show that with the use of localized atomic orbitals, the molecular orbital method can properly describe the antiferromagnetic state even by means of the conventional Xα method. The spin-polarized molecular orbital method is used to derive the antiferromagnetic state as follows: we examine a system where the number of localized conductive electrons is the same as that of atoms, and atomic orbitals with up and down spins have a different energy from each other, as shown in Fig. 5.1.

Then, the antibonding molecular orbital orbitals can be localized. Roughly speaking, there is no spin-up electron on an atomic site next to a site without a spin-down electron in the antiferromagnetic state.

From the point of view mentioned above, we performed molecular orbital calculations for artificial hydrogen clusters H_2, layered H_8 and H_{18}, which have a rather long interatomic distance, and the Cu_2O_{11} cluster composed of two octahedra. In hydrogen clusters, localized conductive electrons occupy

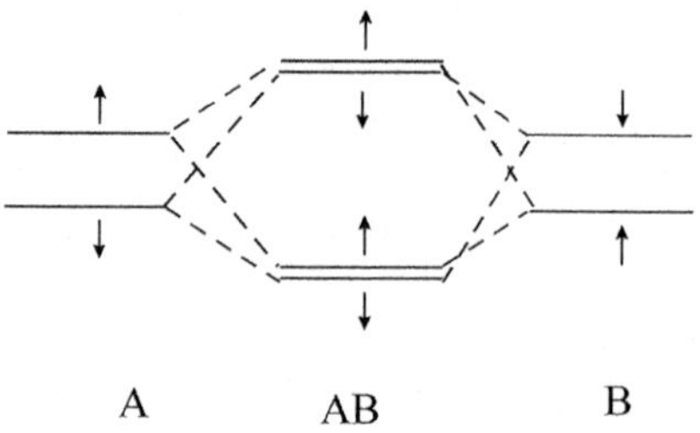

Fig. 5.1. Energy level structure for the antiferromagnetic state

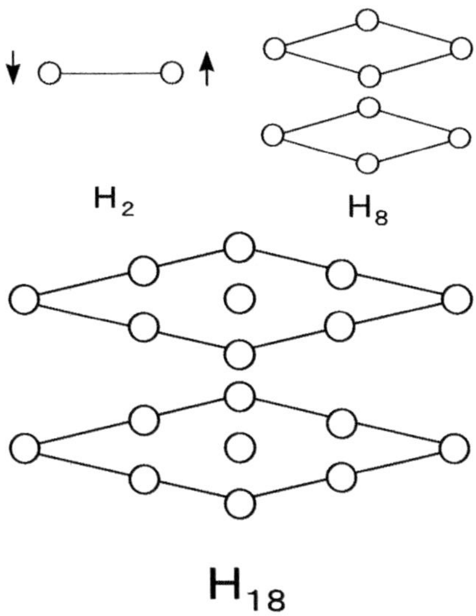

Fig. 5.2. Hydrogen clusters with a rather long interatomic distance

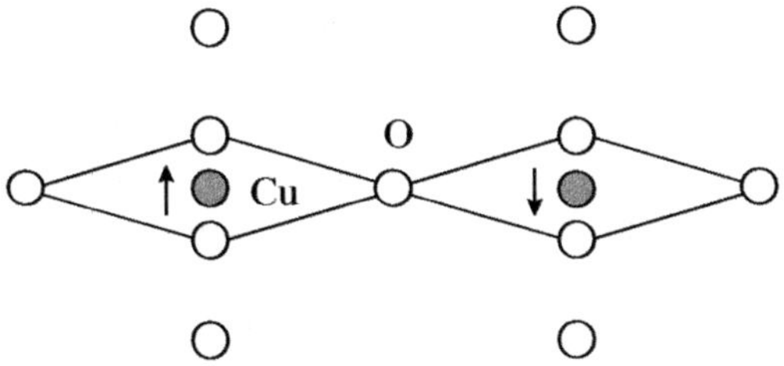

Fig. 5.3. Model cluster for Cu oxide

the states comprising the 1*s* atomic orbital, where the number of electrons per atom is unity. In the case of transition metal oxides, the interatomic distance between metal atoms is long. It is then reasonable to expect that the hydrogen clusters with a rather long interatomic separation will become antiferromagnetic.

This chapter is organized as follows: Sect. 5.2 presents the computational method and the model clusters; Sect. 5.3 describes the calculated results for the hydrogen clusters and Cu-oxide clusters; and Sect. 5.4 offers some conclusions.

5.2 Computational Method and Models

We carried out electronic structure calculations by using the spin-polarized molecular orbital theory where the Xα method was used. The value of exchange-correlation parameter used was 0.7 except in specific cases. The basis atomic orbitals were numerically calculated, and they were composed of 1*s* for the H atom; 1*s*, 2*s*, 2*p*, 3*s*, 3*p*, and 3*d* for the Cu atom; and 1*s*, 2*s*, and 2*p* for the O atom. For the H atom, a well potential was added, where the

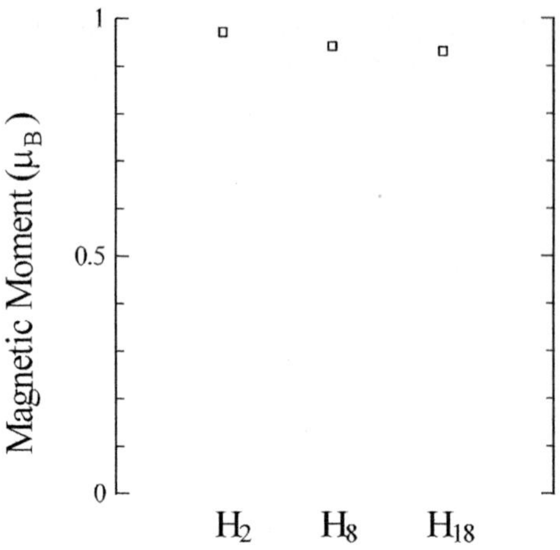

Fig. 5.4. Magnetic moment per atom for hydrogen clusters

radius of the well was 0.8 times the interatomic separation with a depth of −2.0 a.u. For the O^{2-} ion, which is not stable in a vacuum but is stabilized in a crystal, it is not easy to present atomic basis functions. We used two types trial functions as a basis set for O^{2-}. One, which we call the type-I basis function, was derived by using a well potential with a radius of 0.8 times the distance between the Cu atoms and a depth of −0.5 a.u. The other, which we call the type-II basis function, was calculated without a well potential, but the basis function has the same number of nodes and the same properties near the nucleus and at the region away from nucleus as the atomic orbital. In this case, discontinuity in the derivative was seen at a connecting point between the inner solution and the outer solutions. However, the solution used had a smooth derivative because it was interpolated by a polynomial. As for the Cu^{2+} ion, basis functions were easily calculated without well potential. The initial charges in self-consistent iterations were $+2e$ and $-2e$ for the Cu and O atoms, respectively.

Atomic configurations for calculated clusters were as follows, as shown in Fig. 5.2: H atoms for H_2 were at $(\pm a_H, 0, 0)$; those for layered H_8 were at $(\pm a_H, \pm a_H, \pm 1.25 a_H)$; and those for layered H_{18} were at $(b_H,\ b'_H,\ c_H)$, where $b_H,\ b'_H = \pm 2a_H$, 0 and $c_H = \pm 1.25 a_H$, and a_H was set to 0.126 nm.

As for the cluster Cu_2O_{11} shown in Fig. 5.3, positions for the Cu atoms were $(\pm a_C, 0, 0)$, and those for the O atoms were $(\pm a_C, \pm a_C, 0)$, $(\pm 2a_C, 0, 0)$, $(0, 0, 0)$, and $(\pm a_C, 0, \pm c_C)$, where a_C and c_C were set to 0.189 nm and 0.243 nm, respectively.

Although symmetry orbitals were not produced for the H clusters, the configuration of electrons for the atoms on the equivalent site was taken to be identical. For Cu_2O_{11}, we considered only the symmetry with respect to the xy-plane (two independent Cu atoms were on the x-axis) and adopted C_s as the point group. The O atoms on the symmetry orbital in the xy-plane and at the apex were regarded as two equivalent atoms.

In the integration for the matrix elements of the Hamiltonian and overlap integral, 500 sampling points per atom for the H clusters were used. We carried out calculations for Cu_2O_{11} by using the type-I and type-II basis functions at 9000 and 18 000 sampling points, respectively. The errors we estimated in the orbital energy were 0.01 eV for the former case and 0.04 eV for the latter case. In the calculation for Cu_2O_{11}, the ionic interaction effect from the surrounding ions were considered by adding the Madelung potential produced by the atoms in the K_2NiF_4 structure within a volume of $4.02a_K \times 3.02a_K \times 2.74c_K$ in the x-, y-, and z-directions, with a_K and c_K being the unit cell length along the a- and c-axes, and c_K being 1.33 nm.

5.3 Results and Discussion

We first investigated a dependence of cluster size on the antiferromagnetic state. Figure 5.4 shows the calculated results for the H clusters and indicates that the magnetic moments averaged on the H atom were 0.97, 0.94, and 0.93 μ_B for H_2, H_8, and H_{18}, respectively.

The antiferromagnetic configuration was seen between the nearest H atoms. The antiferromagnetic state is stable within the volume of the clusters above. The spin-unpolarized calculations were also performed for the same clusters with the same electronic configurations, and it was found that the energy gap between the top of the bonding states occupied by electrons (that is, the highest occupied molecular orbital, or HOMO) and the bottom of the unoccupied antibonding states (that is, the lowest unoccupied molecular orbital, or LUMO) were 1.2, 0.4, and 0.3 eV for H_2, H_8, and H_{18}, respectively, as shown in Fig. 5.5.

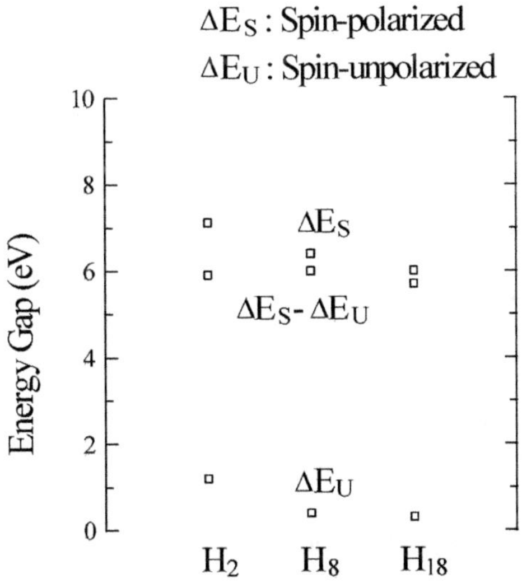

Fig. 5.5. Energy gap for hydrogen clusters

This suggests that the spin-unpolarized calculation may lead to quite a small energy gap for a large system, whereas the spin-polarized calculation shows that the energy gap between the HOMO and LUMO were 7.1, 6.4, and 6.0 eV for H_2, H_8, and H_{18}, respectively. The differences in the energy gap seen between the spin-polarized and spin-unpolarized calculations were 5.9, 6.0, and 5.7 eV for H_2, H_8, and H_{18}, respectively. Half of this energy difference is the localization energy, which will probably remain finite in an infinitely large system. We have determined that there is a discernible correlation between the electronic structure and the size and atomic configuration of H clusters. When we derive electronic structures for an infinitely large system by the molecular orbital method, the result coincides with that of a band structure calculation at the Γ point taking into account a corresponding supercell. We note that the set of increased α and the advanced renormalization of the mass enhancement – for example, by an electron-phonon interaction – result in the stabilization of the antiferromagnetic state. Additionally, in this chapter, we attempt to set an unrealistic value beyond unity for α for the specific case mentioned below.

For the Cu_2O_{11} cluster, the magnetic moment averaged on the Cu atom was 0.59 μ_B in the case where 0.7 was set to α and a type-I basis function was used. We can also see the antiferromagnetic state in this case. The energy gap between HOMO and LUMO was 0.83 eV for $\alpha = 0.7$, and the gap for $\alpha = 0.9$ was 1.2 times larger compared to that for $\alpha = 0.7$. As the value of α increased, the magnetic moment of the Cu atom changed little. With a type-II basis function, the magnetic moment averaged on the Cu atom was 0.98 μ_B for $\alpha = 0.9$, and the antiferromagnetic state was more stabilized compared to the case where a type-I basis function was used. Figure 5.6 depicts the energy level structure near the HOMO.

The origin of the energy for molecular orbitals is the energy of the HOMO. In the figure, the states occupied by electrons comprise the Cu-3d and O-2p atomic orbitals. The unoccupied states are antibonding states between the

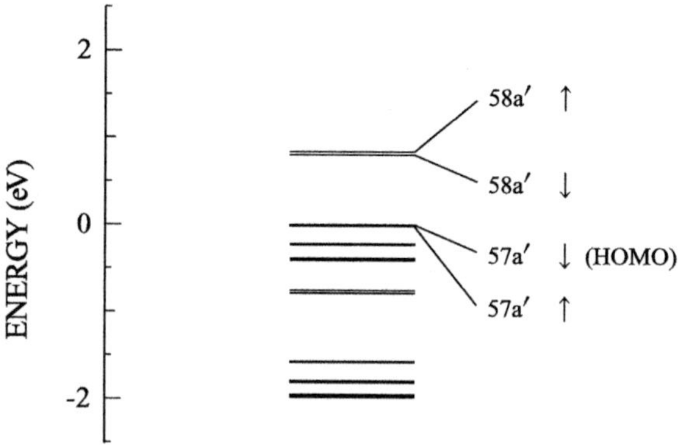

Fig. 5.6. Energy levels for Cu oxide

Cu-3*d* and O-2*p* atomic orbitals as well as between Cu-3*d* atomic orbitals. For an unrealistic value of $\alpha = 1.4$, the energy gap between HOMO and LUMO was 1.2 eV and the magnetic moment averaged on Cu atom was $0.77\,\mu_B$. (For H_2, the energy gap was 14.6 eV with $\alpha = 1.4$.)

5.4 Conclusions

This chapter has described an approach to modeling magnetic properties using the Xα method which has been successfully applied to a variety of materials; in particular, we focused on antiferromagnetic properties. It has thus far not been fully reported whether or not antiferromagnetic properties may be properly described by the local density approximation. It was shown that the molecular orbital method, with the use of the conventional Xα method, properly represents the antiferromagnetic state by means of localized atomic orbitals. Additionally, a good description of the antiferromagnetic state in Cu oxides needs proper calculations of the O^{2-} state. For instance, the basis function, which behaves like a wave function of a free electron at the region away from nucleus, may be suitable for transition-metal oxides.

References

1. K. Fukushima, N. Kondo, Japan J. Appl. Phys. **40**, 3226 (2001)
2. C.G. Shull, W.A. Strauser, E.O. Wollan, Phys. Rev. **83**, 333 (1951)
3. J. Hubbard, Proc. Royal Soc. **A276**, 238 (1963)
4. A. Svane, O. Gunnarsson, Phys. Rev. Lett. **65**, 1148 (1990)
5. V.I. Anisimov, J. Zaanen, O.K. Andersen, Phys. Rev. B **44**, 943 (1991)
6. M. Arai, T. Fujiwara, Phys. Rev. B **51**, 1477 (1995)

6

Optical Materials

Kazuyoshi Ogasawara and Hirohiko Adachi

6.1 Introduction

6.1.1 Optical Materials Based on Transition-Metal Ions

Transition-metal (TM) ions in transparent crystals are known to exhibit a variety of beautiful colors [1]. For example, a crystal of α-Al_2O_3 with Cr^{3+} ions has a beautiful red color and is widely known as a precious gemstone, ruby.

Similarly, a crystal of $Be_3Al_2(SiO_3)_6$ (beryl) with Cr^{3+} ions has a beautiful green color and is known as another gemstone, emerald. The colors of these gemstones originate from the absorption of light in the visible wavelength region due to the excitation of the impurity Cr^{3+} ions.

The absorption spectra of ruby and emerald are shown in Fig. 6.1 [2]. There is a characteristic spectral anisotropy and the spectra for different polarizations are shown together. All of them exhibit characteristic broad absorption peaks called the "U band" and the "Y band" in the visible region. In ruby, the U band mainly absorbs green light and the Y band mainly absorbs blue light. Accordingly, only red light is transmitted through the crystal, which is the phenomenological origin of the red color of ruby. In emerald, the shapes of the absorption spectra are quite similar. However, slightly lower peak energies give green color to the crystal instead of red because the U band mainly absorbs red light while the Y band still mainly absorbs blue light in this case. These peaks in the absorption spectra correspond to the energy levels of the impurity Cr^{3+} ions.

The eigenstates of a Cr^{3+} ion in crystals are basically dominated by the correlations among the three electrons occupying the Cr-3*d* orbitals and are generally called "multiplets," borrowing the terminology of atomic physics. However, the multiplet energy levels are strongly influenced by the properties of the host crystal through two factors: (1) crystal field splitting of the originally degenerate Cr-3*d* orbitals, and (2) partial delocalization of Cr-3*d* orbitals due to covalent interaction.

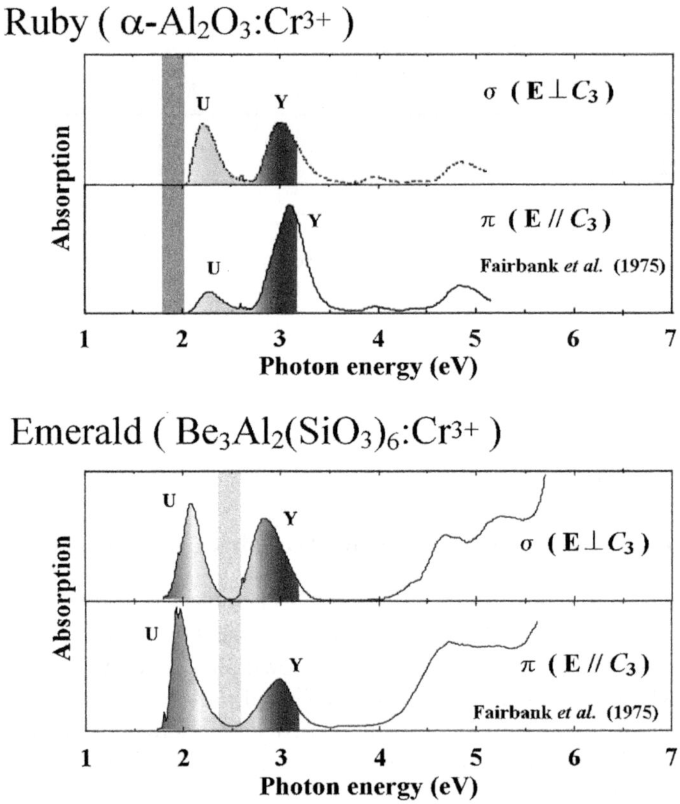

Fig. 6.1. Experimental absorption spectra of ruby and emerald for two different polarizations reported by Fairbank et al. [2]

Since the energy separations between multiplets of TM ions tend to be in the visible region, TM-doped crystals are frequently used as optical materials such as solid-state lasers, display materials, optical amplifiers, optical recording materials, etc., among which the first two applications are the most popular.

Ruby is actually famous as the first solid-state laser in history [3] and emerald is also known as an important tunable solid-state laser [4]. Other TM-doped tunable lasers such as Cr^{3+}:$BeAl_2O_4$ (alexandrite) [5], Cr^{4+}:Mg_2SiO_4 (Cr^{4+}:forsterite) [6], and Cr^{3+}:$LiSrAlF_6$ (Cr^{3+}:LiSAF) [7] are also quite useful and are commercially available.

On the other hand, semiconducting crystals with TM ions can be easily linked to already established electronics technology and are frequently used as display materials. There are many types of displays, which function according to a variety of excitation mechanisms. For example, cathodoluminescence (CL) is utilized in cathode-ray tube displays (CRTs), while electroluminescence (EL) is utilized in electroluminescent displays (ELDs). For ELD materials, full-color displays are currently realized with ZnS or SrS hosts doped with TM activators. The best phosphors currently used are a combination of (Cu, Ag)SrS and Mn:ZnS [8].

The multiplets are essentially many-electron correlation effects and cannot be reproduced by the ordinary electronic structure calculations based on the one-electron approximation.

6.1.2 Ligand-Field Theory

The multiplets of TM ions in crystals have traditionally been analyzed by a parameter-fitting method based on the ligand-field theory developed by Sugano et al. [1]. In the original formulation of ligand-field theory, only the octahedral (O_h) crystal field is considered through a parameter Δ describing the energy difference between the t_{2g} state and the e_g state. The effect of covalency is also taken into account through an electron–electron repulsion parameter, B. Accordingly, the multiplet energy levels can be expressed in terms of only two parameters, Δ and B. The energy levels of all d^N systems under an octahedral crystal field have been calculated as a function of Δ and summarized as the well-known Tanabe-Sugano diagrams [1], where both the energy and Δ are normalized with respect to B. These diagrams are quite popular for the analysis of multiplet structures of TM ions in crystals

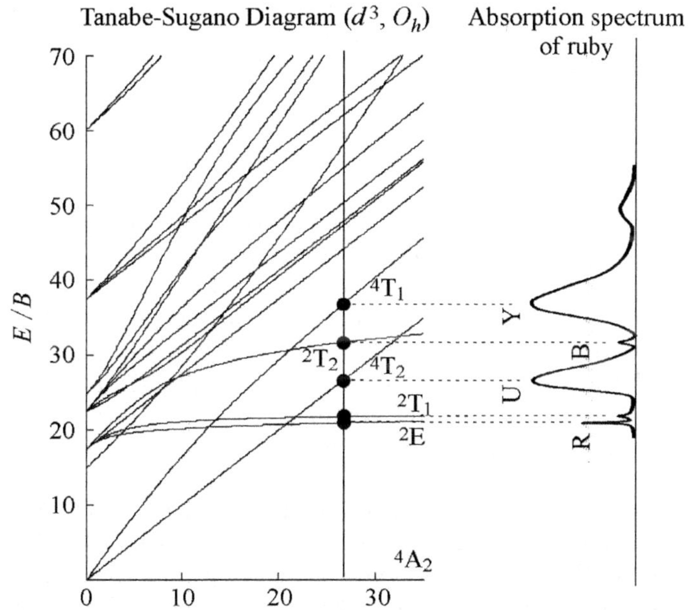

Fig. 6.2. Tanabe-Sugano diagram for the d^3 ion [1] and a schematic illustration to determine the values of B and Δ from the observed absorption spectrum of ruby

since the values of two parameters can be determined simply by comparing the observed spectrum with the corresponding diagram. The Tanabe-Sugano diagram for Cr^{3+} (d^3) is shown in Fig. 6.2, together with the corresponding analysis procedure for the absorption spectrum of ruby.

6.1.3 First-principle Calculation of Multiplets

Although ligand-field theory has been quite successful applied to the interpretation of the optical spectra of TM ions in crystals, it is an empirical method where the parameters are determined by the observed spectrum. Therefore, it cannot predict the optical properties of unknown or hypothetical materials. Since TM-doped crystals have quite important applications as optical materials such as solid-state lasers or display materials, non-empirical (first-principle or ab initio) prediction of multiplet structures of TM ions in crystals is quite important for theoretical design of new materials with improved properties.

Historically, the first non-empirical prediction of entire multiplet structures of TM ions in crystals was put forth by Watanabe and Kamimura based on a combination of the local density approximation (LDA) calculation and ligand-field theory [9–11]. They derived an analytic expression of the multiplet energies under T_{d} symmetry, where all multiplet energies are expressed in terms of ten independent electron–electron repulsion integrals. These integrals were calculated by cluster calculations based on LDA. They investigated the multiplet structures of TM ions in various semiconductors and reasonably reproduced the energies of the lower-lying multiplets. However, since the calculated multiplet energies tend to be overestimated, this type of calculation has not been conducted extensively.

A Hartree-Fock-type calculation combined with configuration interaction (CI) was recently performed by Seijo and Barandiarán et al. They calculated the multiplet structures of Ni^{2+} in MgO [12] and V^{3+} in some fluoride crystals [13, 14], and they achieved reasonable agreement with experiment. Although this approach is mathematically exact, an unfeasibly large number of Slater determinants are generally required for accurate calculation. Due to the computational requirements, only crystals with high (O_{h}) symmetry have been investigated so far. Density-functional-type calculations have been also utilized by taking the symmetry-reduction approach developed by Daul et al. [15], where an energy of a certain multiplet state is expressed in terms of energies of single-determinants. Although only the single-determinant energy can be obtained by a density-functional calculation, calculations for several different configurations can provide the multiplet energy levels. Moreno et al. calculated the multiplet structures of Cr^{4+} in some oxides by this approach and reproduced the energies of some lower-lying multiplets [16]. However, due to the absence of explicit many-electron wave functions, they could not calculate the transition probability between the multiplet states. Moreover, the symmetry-reduction procedure for the lower-symmetry systems is still somewhat ambiguous. Therefore, this approach has also only been applied to high-symmetry (O_h or T_d) systems.

6.1.4 DV-ME Method

Recently, we generalized and improved the approach by Watanabe and Kamimura and proposed a new hybrid method involving density-functional theory and a CI calculation [17–20]. Since the DV-Xα method is used for one-electron calculations, we called our method the discrete variational multielectron (DV-ME) method. In this approach, the many-electron Hamiltonian is slightly modified to include the electron correlation effects efficiently. However, this modification is made to satisfy the consistency between the one-electron density-functional calculation and the many-electron CI calculation without requiring any empirical parameters. Since the many-electron wave functions are obtained explicitly as a linear combination of Slater determinants, the transition probability between the multiplets can be also calculated numerically. In this chapter, we will show the first non-empirical calculation of absorption spectra of ruby [20] and Co^{2+}:ZnS [19] based on the DV-ME method, as examples of solid-state laser materials and display materials, respectively.

6.2 DV-ME Method

6.2.1 Configuration Interaction

In the discrete variational multielectron (DV-ME) method, molecular orbitals (MOs) are calculated based on ordinary DV-Xα cluster calculations [21]. Within a subspace spanned by the selected electrons and the selected orbitals, all of the possible Slater determinants are prepared and used as the basis-functions to describe the many-electron states. By diagonalizing the many-electron Hamiltonian, the many-electron wave functions are expressed as a linear combination of Slater determinants. The interactions between Slater determinants are generally called "configuration interactions" (CIs) in the terminology of quantum chemistry.

The many-electron Hamiltonian is constructed based on the ground state charge density obtained by an ordinary DV-Xα cluster calculation as follows:

$$H = \sum_{i=1}^{M} h(\boldsymbol{r}_i) + \sum_{i} \sum_{j>i} g(\boldsymbol{r}_i, \boldsymbol{r}_j) \ , \tag{6.1}$$

where M is the number of the selected electrons, and h and g are the one-electron and two-electron operators expressed as

$$\begin{cases} h(\boldsymbol{r}_i) &= -\dfrac{1}{2}\nabla_i^2 + V_{nuc}(\boldsymbol{r}_i) + V_0(\boldsymbol{r}_i) \\ g(\boldsymbol{r}_i, \boldsymbol{r}_j) &= \dfrac{1}{r_{ij}} \ . \end{cases} \tag{6.2}$$

Here, $\boldsymbol{r}_i$ is the position of the ith electron, r_{ij} is the distance between the ith electron and the jth electron, V_{nuc} is the potential due to the nuclei, and V_0 is the potential due to the other electrons (frozen electrons). The explicit form of V_0 was derived by Watanabe and Kamimura as

$$V_0(\boldsymbol{r}) = \int \frac{\rho_0(\boldsymbol{r}')}{|\boldsymbol{r}-\boldsymbol{r}'|}\mathrm{d}\boldsymbol{r}' + \frac{3}{4}\left[\frac{\rho(\boldsymbol{r})V_{\text{xc}}\{\rho(\boldsymbol{r})\} - \rho_0(\boldsymbol{r})V_{\text{xc}}\{\rho_0(\boldsymbol{r})\}}{\rho_{\text{sel}}(\boldsymbol{r})} - V_{xc}\{\rho_{\text{sel}}(\boldsymbol{r})\}\right], \tag{6.3}$$

where ρ, ρ_{sel}, and ρ_0 represent the charge density of all electrons, the charge density of the selected electrons, and the charge density of the other electrons, respectively, and V_{xc} is the Slater's Xα potential [22].

The matrix elements of H can be generally expanded as

$$\begin{aligned} H_{pq} &= \langle \Phi_p \mid H \mid \Phi_q \rangle \qquad (6.4)\\ &= \sum_{i=1}^{L}\sum_{j=1}^{L} A_{ij}^{pq}\langle i \mid h \mid j\rangle + \sum_{i=1}^{L}\sum_{j=1}^{L}\sum_{k=1}^{L}\sum_{l=1}^{L} B_{ijkl}^{pq}\langle ij \mid g \mid kl\rangle, \end{aligned}$$

where L is the number of selected orbitals, and A_{ij}^{pq} and B_{ijkl}^{pq} are coefficients obtained by the expansion of the Slater determinants. Here, $\langle i \mid h \mid j\rangle$ and $\langle ij \mid g \mid kl\rangle$ are defined by

$$\langle i \mid h \mid j\rangle = \int \phi_i^*(\boldsymbol{r})h(\boldsymbol{r})\phi_j(\boldsymbol{r})\,\mathrm{d}\boldsymbol{r}, \tag{6.5}$$

and

$$\langle ij \mid g \mid kl\rangle = \iint \phi_i^*(\boldsymbol{r}_1)\phi_j^*(\boldsymbol{r}_2)\frac{1}{r_{12}}\phi_k(\boldsymbol{r}_1)\phi_l(\boldsymbol{r}_2)\,\mathrm{d}\boldsymbol{r}_1\,\mathrm{d}\boldsymbol{r}_2, \tag{6.6}$$

respectively, where ϕ denotes the selected orbitals obtained by the cluster calculation.

These matrix elements are calculated numerically using the DV integration scheme [23] as follows:

$$\langle i \mid h \mid j\rangle = \sum_{\alpha} \omega(\boldsymbol{r}_\alpha)\phi_i^*(\boldsymbol{r}_\alpha)h(\boldsymbol{r}_\alpha)\phi_j(\boldsymbol{r}_\alpha), \tag{6.7}$$

and

$$\langle ij \mid g \mid kl\rangle = \sum_{\alpha} \omega(\boldsymbol{r}_\alpha)\phi_i^*(\boldsymbol{r}_\alpha)\left[\sum_{\beta(\neq\alpha)} \omega(\boldsymbol{r}_\beta)\phi_j^*(\boldsymbol{r}_\beta)\frac{1}{r_{\alpha\beta}}\phi_l(\boldsymbol{r}_\beta)\right]\phi_k(\boldsymbol{r}_\alpha), \tag{6.8}$$

where $\boldsymbol{r}_\alpha$ is a sampling point, and ω is a weight function proportional to the inverse of the sampling point density. The multiplet energy levels can be easily obtained by diagonalization of the Hamiltonian matrix.

6.2.2 CDC Approach

In the DV-ME method, all Slater determinants are constructed using the ground-state MOs, and the many-electron Hamiltonian is also constructed based on the ground-state charge density. However, the relaxation of the electronic structure during transition is sometimes not negligible. In fact, direct diagonalization of (6.1) tends to result in an overestimation of crystal-field splitting typically by about 20% [20]. On the other hand, the energy difference between different configurations can be well estimated by one-electron calculations using Slater's transition-state method [22]. Therefore, we introduce a configuration-dependent correction (CDC) approach, where a configuration-dependent correction term is added to the diagonal matrix elements of the many-electron Hamiltonian. This correction term is determined from the consistency between the one-electron MO calculation and the CI calculation without introducing any empirical parameters [20].

For example, in the case of d^N ions under O_h symmetry, the Slater determinants can be classified into several configurations specified by $(t_{2g})^m(e_g)^n$ $(m+n=N)$. In this case, the matrix elements can be expressed as

$$\begin{aligned} H_{pq} &= \langle \Phi_p \mid H \mid \Phi_q \rangle \\ &= \left[\sum_{i=1}^{L} \sum_{j=1}^{L} A_{ij}^{pq} \langle i \mid h \mid j \rangle + \delta_{pq} D_{\mathrm{CDC}}(m,n) \right] \\ &\quad + \sum_{i=1}^{L} \sum_{j=1}^{L} \sum_{k=1}^{L} \sum_{l=1}^{L} B_{ijkl}^{pq} \langle ij \mid g \mid kl \rangle , \end{aligned} \tag{6.9}$$

where $D_{\mathrm{CDC}}(m,n)$ is the CDC to the electronic configurations specified by $(t_{2g})^m(e_g)^n$. In order to determine the value of $D_{\mathrm{CDC}}(m,n)$, the contribution of the average value of the electron–electron interaction term should be subtracted. This procedure can be accomplished as follows.

According to the Slater's transition-state method, energy separation between adjacent configurations, $(t_{2g})^m(e_g)^n$ and $(t_{2g})^{m-1}(e_g)^{n+1}$, can be well-approximated by the one-electron energy difference, $\Delta(m,n;m-1,n+1) = \varepsilon_{eg} - \varepsilon_{t2g}$, in the Slater's transition state, $(t_{2g})^{m-0.5}(e_g)^{n+0.5}$. Then the value of $D_{\mathrm{CDC}}(m,n)$ can be determined by the relation

$$\begin{aligned} &\frac{1}{N(m,n)} \sum_{(m^p,n^p)=(m,n)} H_{pp} \\ &- \frac{1}{N(m-1,n+1)} \sum_{(m^p,n^p)=(m-1,n+1)} H_{pp} = \Delta(m,n;m-1,n+1) , \end{aligned} \tag{6.10}$$

where $N(m,n)$ is the number of Slater determinants belonging to the $(t_{2g})^m(e_g)^n$ configuration, and m^p and n^p are the numbers of electrons occupying the t_{2g} and e_g states in the pth Slater determinants, respectively.

In the actual calculation, the value of $\Delta(m,n,m-1,n+1)$ does not strongly depend on the configuration $(m,\ n)$. Therefore, in the work presented here, $\Delta(m,n,m-1,n+1)$ for all configurations is approximated by the value obtained in the ground state, i.e.,

$$\Delta(m,n;m-1,n+1) = e_{\mathrm{g}}^{\mathrm{GS}} - t_{2\mathrm{g}}^{\mathrm{GS}} = \Delta_{\mathrm{eff}} , \tag{6.11}$$

as has been done in [20] Then (6.11) can be simplified to,

$$\frac{1}{N(m,n)} \sum_{(m^p,n^p)=(m,n)} H_{pp} = n\Delta_{\mathrm{eff}} . \tag{6.12}$$

6.2.3 Correlation Correction

In the DV-ME method, we generally consider only electrons occupying the impurity state orbitals. However, in this subspace, the number of Slater determinants is insufficient to describe the effect of electron correlations accurately, and the electron–electron repulsion energies are overestimated significantly. Mathematically, the remaining electron correlation effects can be taken into account by gradually increasing the number of Slater determinants. However, this approach is computationally quite inefficient. Therefore, we introduce a more efficient approach to take into account the remaining electron correlations. In this approach, all matrix elements of the electron–electron repulsion operator are multiplied by a certain reduction factor c (correlation correction factor). Then the matrix elements of H can be finally expressed as

$$\begin{aligned} H_{pq} &= \langle \Phi_p \mid H \mid \Phi_q \rangle \\ &= \left[\sum_{i=1}^{L} \sum_{j=1}^{L} A_{ij}^{pq} \langle i \mid h \mid j \rangle + \delta_{pq} \Delta_{\mathrm{CDC}}(m,n) \right] \\ &\quad + \sum_{i=1}^{L} \sum_{j=1}^{L} \sum_{k=1}^{L} \sum_{l=1}^{L} B_{ijkl}^{pq} c \langle ij \mid g \mid kl \rangle . \end{aligned} \tag{6.13}$$

The value of c can be also determined from the consistency between the one-electron MO calculation and the many-electron CI calculation. However, in this case, a spin-unrestricted (spin-polarized) calculation is utilized.

For example, in the case of a d^3 system under O_h symmetry, we consider a spin-flip transition within the $(t_{2\mathrm{g}})^3$ configuration, Within the one-electron approximation, this transition energy can be calculated by $\varepsilon_{t2\mathrm{g}\downarrow} - \varepsilon_{t2\mathrm{g}\uparrow}$ in the Slater's transition state, $(t_{2\mathrm{g}\uparrow})^{2.5}$ $(t_{2\mathrm{g}\downarrow})^{0.5}$. On the other hand, in the multiplet energy levels, the same transition corresponds to transitions from ${}^4A_2$ $(S_z = 3/2)$ to four different states: ${}^4A_2$, ${}^2E$, ${}^2T_2$, ${}^2T_1$ $(S_z = 1/2)$. By

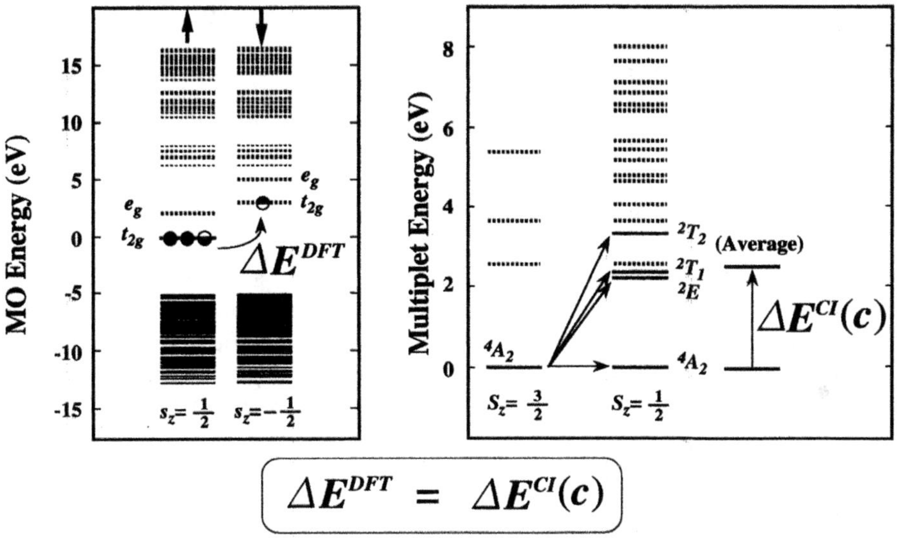

Fig. 6.3. Schematic illustration of a spin-flip transition energy within a $(t_{2g})^3$ configuration in the one-electron energy levels (*left*) and in the multiplet energy levels (*right*)

setting the average transition energy of these transitions equal to the value obtained by the transition-state calculations, we obtain

$$\frac{\sum_k g_k E_k^E(c)}{\sum_k g_k} - E^G(c) = \varepsilon_{t\downarrow}^{TS} - \varepsilon_{t\uparrow}^{TS}, \tag{6.14}$$

where g_k is the degeneracy of the state k. In this case, the sum of k is taken over $k = {}^4A_2$, ${}^2E$, ${}^2T_2$, ${}^2T_1$ ($S_z = 1/2$) with $g_k = 1, 2, 3, 3$, respectively. This condition is schematically shown in Fig. 6.3.

6.2.4 Transition Probability

Using the eigenvector obtained by the diagonalization of the many-electron Hamiltonian, the many-electron wave functions corresponding to each multiplet state can be obtained explicitly as a linear combination of the Slater determinants. Therefore, the oscillator strength for the electric-dipole transition between multiplets can be calculated directly using

$$I_{if} = 2(E_f - E_i) \mid \langle \psi_f \mid \sum \boldsymbol{r}_k \cdot \boldsymbol{e} \mid \psi_i \rangle \mid^2, \tag{6.15}$$

where ψ_i and ψ_f are the many-electron wave functions of the initial and final states, while E_i and E_f are the energy eigenvalues of these states. $\boldsymbol{r}_k$ denotes the position of the kth electron, and $\boldsymbol{e}$ denotes the unit vector parallel to the direction of the electric field.

6.3 Calculation of the Absorption Spectrum of Ruby

6.3.1 Model Cluster

In ruby, a chromium ion substitutes for an aluminum ion, and has six nearest-neighbor oxygen ions. This oxygen octahedron is slightly distorted, and the actual site symmetry is C_3. A model cluster composed of 63 atoms with C_3 symmetry (Fig. 6.4) was adopted for the calculation of the absorption spectrum of ruby. The impurity chromium ion is located at the center of the cluster, and 14 aluminum ions are considered. All of the nearest-neighbor oxygen ions to these aluminum ions are considered. In order to reproduce the effective Madelung potential, several thousand point-charges are located at atomic positions outside the cluster. The MO calculation was performed using a nearly minimal basis set (1*s*–4*p* for Cr, 1*s*–3*d* for Al, and 1*s*–2*p* for O), with 40 000 sampling points.

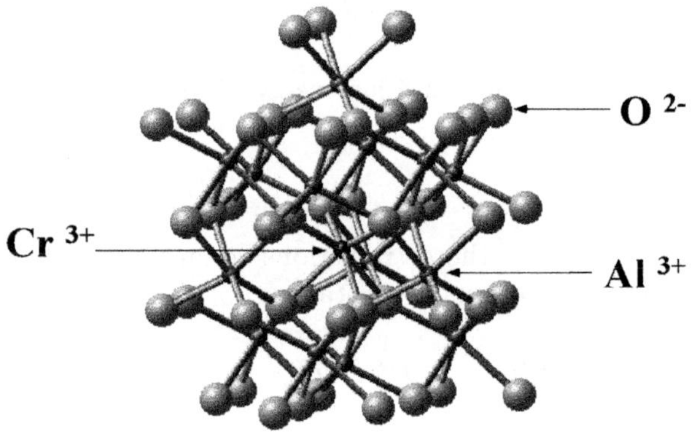

Fig. 6.4. Model cluster adopted for the calculation of the absorption spectrum of ruby

6.3.2 One-Electron Energy Level

The calculated one-electron energy level is shown in Fig. 6.5. The valence band mainly consists of O-2*p* orbitals and the conduction band mainly consists of Al-3*s*, 3*p*, and 3*d* orbitals. The impurity states mainly consisting of Cr-3*d* orbitals are formed within the band gap, which splits into the t_{2g} state and the e_g state due to the presence of the octahedral crystal field. Although a slight trigonal crystal field further splits the t_{2g} state into an a state and an e state, this splitting is quite small compared to the octahedral crystal field. The octahedral crystal field splitting, Δ, calculated by the difference between the energy of the e_g state and the average energy of the t_{2g} state was 2.11 eV.

6.3.3 Multiplet Energy Level

For the calculation of the multiplet structures of ruby, three electrons and ten impurity-state orbitals are considered. Therefore, the many-electron Hamil-

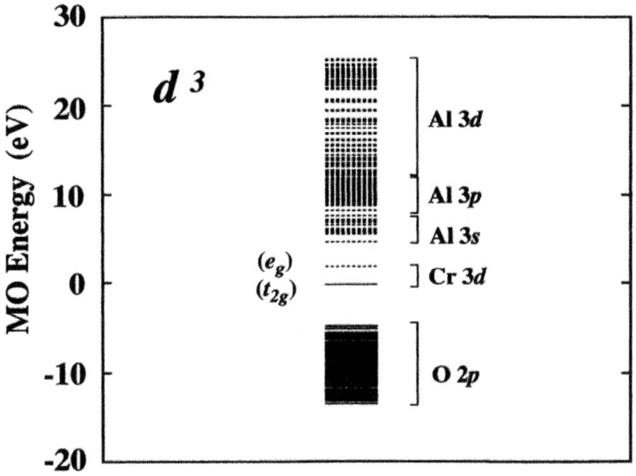

Fig. 6.5. Calculated one-electron molecular orbital (MO) energy levels of ruby. The energy of the highest occupied molecular orbital (HOMO) is set at the origin

tonian is expressed in terms of ${}_{10}C_3 = 120$ Slater determinants. Although there is a slight trigonal field, the multiplet energy levels are still dominated by the octahedral crystal field. Therefore, in the case being considered, the spin-flip transition in the $(t_{2g})^3$ configuration described in Fig. 6.3 is used for the determination of the correlation correction factor c. The obtained value of c is 0.844, indicating that the electron–electron repulsion energy is reduced by a factor of 0.844 due to the remaining effect of electron correlations.

The calculated multiplet energy levels with and without correlation correction (CC) are shown in Fig. 6.6 together with the peak positions in the observed absorption spectra [2]. Each multiplet energy level is labeled by the symmetry in the octahedral approximation. Due to the presence of the trigonal field, each T state splits into A and E states. When CC is taken into account, the theoretical multiplet energies decrease significantly, especially in doublets, and the experimental multiplet energy levels, including the trigonal splittings, are excellently reproduced without any empirical parameters.

6.3.4 Absorption Spectra

Under exact octahedral symmetry, the electric-dipole transitions between the multiplets (d–d transitions) are completely forbidden. However, in the case of ruby, the presence of a slight trigonal field allows electric-dipole transitions between multiplets through the trigonal distortion of the many-electron wave functions. In the DV-ME method, the trigonally distorted many-electron wave functions are treated explicitly. Thus, the transition probabilities for electric dipole transitions from the ground state to all excited states are calculated directly. For easy comparison between the theoretical transition probabilities

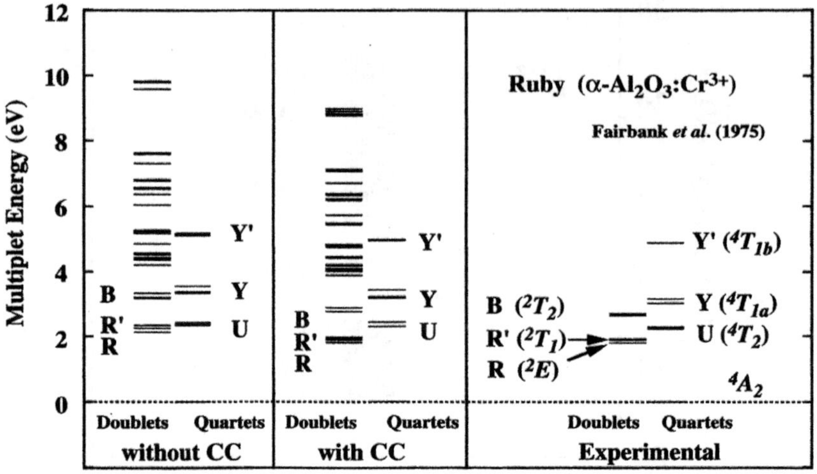

Fig. 6.6. Calculated multiplet energy levels of ruby without correlation correction (CC) (*left*) and with CC (*center*), together with the peak positions in the experimental absorption spectrum reported by Fairbank et al. [2] (*right*)

and the observed absorption spectra, each peak is broadened by a Gaussian function with a full-width at half-maximum (FWHM) of 0.3 eV, and the contributions from all excited states are summed up. The π spectrum is obtained by setting the electric field vector ($\boldsymbol{e}$) parallel to the C_3 axis, while the σ spectrum is obtained by setting $\boldsymbol{e}$ perpendicular to the C_3 axis. The final theoretical absorption spectra are shown in Fig. 6.7, together with the experimental spectra reported by Fairbank et al. [2]. In the experimental spectra, a characteristic anisotropy is observed, where the intensity of the U band is stronger in the σ spectrum while the intensity of the Y band is stronger in the

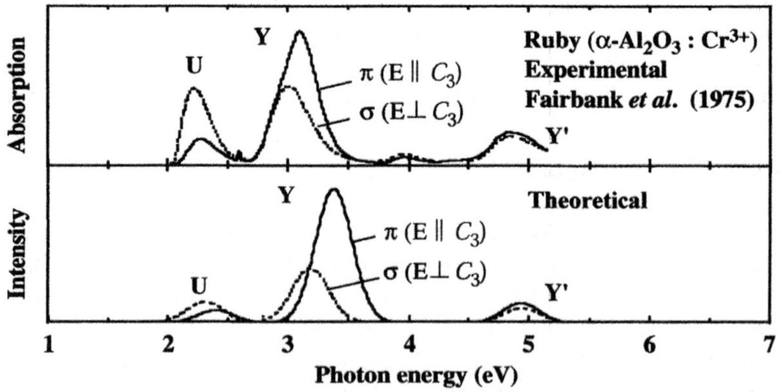

Fig. 6.7. Theoretical absorption spectra of ruby for two different polarizations, together with the experimental absorption spectra reported by Fairbank et al. [2]

π spectrum. The peak positions in the σ spectrum shift toward lower energies compared to the π spectrum. All of these features are clearly reproduced in the theoretical spectra shown in Fig. 6.7.

6.4 Calculation of the Absorption Spectrum of Co^{2+}:ZnS

6.4.1 Model Cluster

In Co^{2+}:ZnS, a cobalt ion substitutes for a zinc ion, and is tetrahedrally coordinated by four nearest-neighbor sulfur ions. A model cluster composed of 41 atoms with T_d symmetry (Fig. 6.8) was adopted for the calculation of the absorption spectrum of Co^{2+}:ZnS. In order to reproduce the effective Madelung potential, several thousand point charges are located at atomic positions outside the cluster. The MO calculation was performed using nearly minimal basis set ($1s$–$4p$ for Co and Zn, and $1s$–$3p$ for S), with 99 999 sampling points.

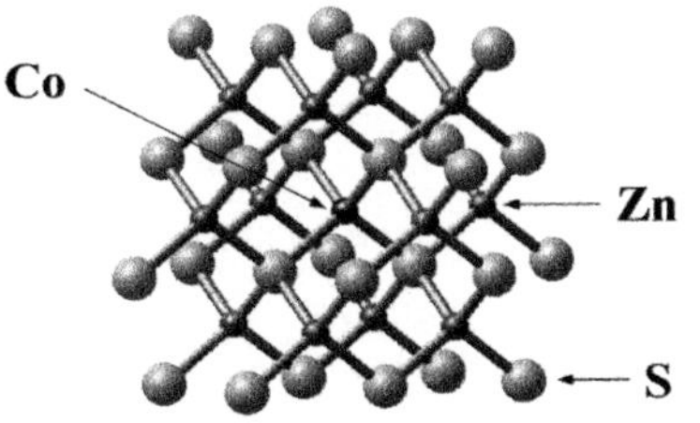

Fig. 6.8. Model cluster adopted for the calculation of the absorption spectrum of Co^{2+}:ZnS

6.4.2 One-Electron Energy Level

The calculated one-electron energy level is shown in Fig. 6.9. The valence band mainly consists of S-$3p$ and Zn-$3d$ orbitals, and the conduction band mainly consists of Zn-$4s$ and $4p$ orbitals. The impurity states mainly consisting of Co-$3d$ orbitals are formed within the band gap. Since cobalt has four nearest-neighbor sulfur ions, these impurity states split into the t_2 state and the e state due to the tetrahedral crystal field. The tetrahedral crystal field splitting, Δ, calculated by the difference between the energy of the e state and the energy of the t_2 state, is 0.62 eV.

6.4.3 Multiplet Energy Level

For the calculation of the multiplet structures of Co^{2+}:ZnS, seven electrons and ten impurity-state orbitals are considered. Therefore, the many-electron wave functions are expressed in terms of ${}_{10}C_7 = 120$ Slater determinants. In

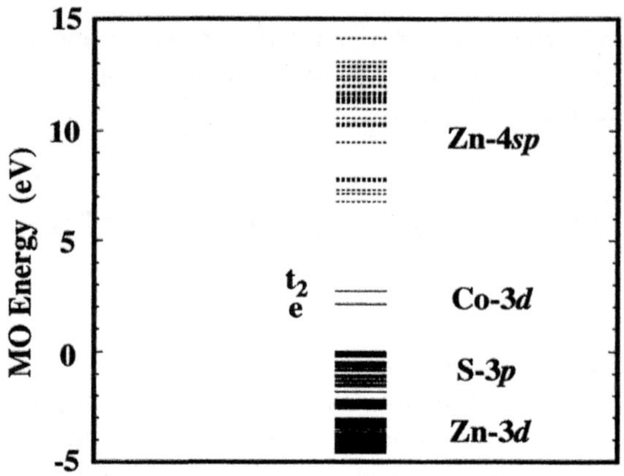

Fig. 6.9. Calculated one-electron molecular orbital (MO) energy levels of Co^{2+}:ZnS. The energy of the top of the valence band is set at the origin

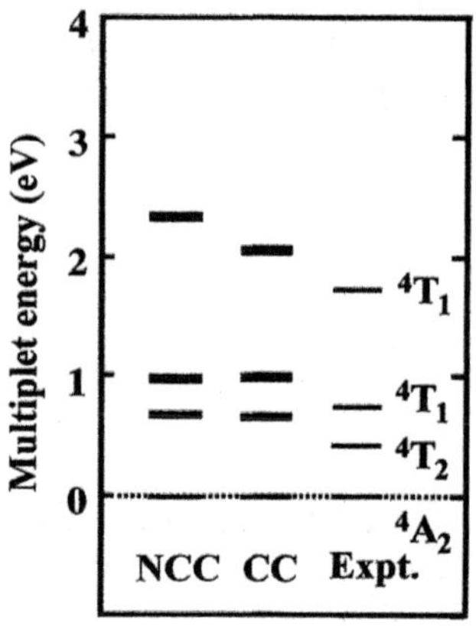

Fig. 6.10. Calculated multiplet energy levels of Co^{2+}:ZnS without correlation correction (CC) (*left*) and with CC (*center*), together with the peak positions in the experimental absorption spectrum of Co^{2+}:ZnS reported by Weakliem [24]

the case being considered, the spin-flip transition in the $(e)^4(t_2)^3$ configuration is used for the determination of the correlation correction factor c. The obtained value of c is 0.844, indicating that the electron–electron repulsion energy is reduced by a factor of 0.822 due to the remaining effect of electron correlations.

The calculated multiplet energy levels with and without CC are shown in Fig. 6.10 together with the peak positions in the observed absorption spectrum [24]. As shown in the figure, when CC is taken into account, the theoretical multiplet energy levels slightly decrease and approach the experimental values.

6.4.4 Absorption Spectrum

In the absence of covalency between the impurity cobalt ion and the host materials, the electric-dipole transitions between the multiplets (d–d transitions) are completely forbidden. However, in the case of Co^{2+}:ZnS, the covalent interaction with the tetrahedrally coordinated sulfur ions breaks the inversion symmetry and allows the electric-dipole transitions between the multiplets.

In the DV-ME method, the covalent effects are automatically taken into account by the use of molecular orbitals. Thus, the transition probabilities for electric-dipole transitions from the ground state to all excited states are calculated directly. For easy comparison between the theoretical transition probabilities and the observed absorption spectra, each peak is broadened by a Gaussian function with a FWHM of 0.1 eV, and the contributions from all excited states are summed up. The theoretical absorption spectrum is shown in Fig. 6.11, together with the experimental spectra reported by Weakliem [24]. In the experimental spectrum, there are two broad peaks. The analysis of the configuration interaction shows that the peak near 1 eV mainly consists of the $(e)^2(t_2)^5$ configuration (a two-electron excited state), while the peak near 2 eV mainly consists of $(e)^3(t_2)^4$ configuration (a one-electron excited state). Although the transition to the former is forbidden in the one-photon absorption process, a transition is possible in the presence of a CI due to the small composition of the one-electron excited state. In the theoretical spectrum, the relative intensities of these peaks are reproduced quite well, indicating that CI is included appropriately in our calculations.

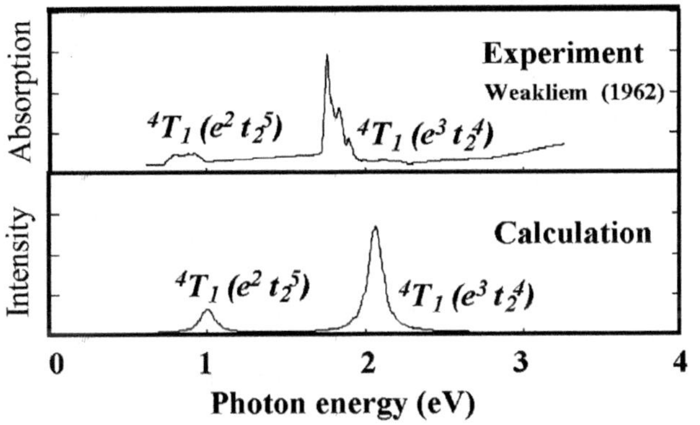

Fig. 6.11. Theoretical absorption spectrum of Co^{2+}:ZnS together with the experimental absorption spectrum reported by Weakliem [24]

6.5 Summary

In this chapter, we showed two examples of first-principle calculations of the absorption spectra of optical materials.

In the case of ruby, the transitions are caused by a combined effect of covalency and trigonal distortion of the ligand oxygen coordination, which results in the trigonally distorted molecular orbitals. The absorption spectrum of ruby has been calculated directly using trigonally distorted wave functions.

The theoretical spectra excellently reproduced the peak positions, their relative intensities, and the characteristic spectral anisotropy without any empirical parameters. This is actually the first non-empirical reproduction of the absorption spectrum of ruby.

In the case of Co^{2+}:ZnS, the transitions are purely caused by the covalent effects with tetrahedrally coordinated sulfur, which breaks the inversion symmetry of the system. The theoretical spectrum reproduced the relative intensities of two peaks excellently, indicating that the CI is taken into account appropriately in our calculations.

As optical materials, rare-earth ions are also widely utilized, as shown by the great success of the Nd:YAG laser or the erbium-doped fiber amplifier (EDFA). In these heavy elements, relativistic effects should be also taken into account. To this end, we have recently developed a relativistic version of the DV-ME method, and an analysis of rare-earth doped optical materials is also in progress.

Acknowledgement. This work was supported by a Grant-in-Aid for Scientific Research from the Ministry of Education, Science, Sports and Culture.

References

1. S. Sugano, Y. Tanabe, H. Kamimura, *Multiplets of Transition-Metal Ions in Crystals* (Academic Press, New York, 1970)
2. W.M. Fairbank, Jr, G.K. Klauminzer, A.L. Schawlow, Phys. Rev. **B11**, 60 (1975)
3. T.H. Maiman, Nature **187** (1960) 493
4. J. Buchert, A. Katz, R.R. Alfano, IEEE J. Quantum Electron. **19**, 1477 (1983)
5. J.C. Walling, O.G. Peterson, H.P. Jenssen, R.C. Morris, E.W. O'Dell, IEEE J. Quantum Electron. **16**, 1302 (1980)
6. V. Petricevic, S.K. Gayan, R.R. Alfano, K. Yamagishi, H. Anzai, Y. Yamagishi, Appl. Phys. Lett. **52**, 1040 (1988)
7. S.A. Payne, L.L. Chase, L.K. Smith, W.L. Kway, H.W. Newkirk, J. Appl. Phys. **66**, 1051 (1989)
8. S. Sun, Displays **19**, 145 (1999)
9. S. Watanabe, H. Kamimura, J. Phys. Soc. Japan **56**, 1078 (1987)
10. S. Watanabe, H. Kamimura, J. Phys. C: Solid State Phys. **20**, 4145 (1987)
11. S. Watanabe, H. Kamimura, Mater. Sci. Eng. **B3**, 313 (1989)
12. R. Llusar, M. Casarrubios, Z. Barandiarán, L. Seijo, J. Chem. Phys. **105**, 5321 (1996)
13. S.L. Moraza, Z. Barandiarán, J. Chem. Phys. **105**, 50 (1996)
14. S.L. Moraza, L. Seijo, Z. Barandiarán, Phys. Rev. **57**, 11974 (1998)
15. C. Daul, Int. J. Quantum Chem. **52**, 867 (1994)
16. K. Wissing, J.A. Aramburu, M.T. Barriuso, M. Moreno, Solid State Commun. **108**, 1001 (1998)
17. K. Ogasawara, T. Ishii, Y. Ito, H. Ida, I. Tanaka, H. Adachi, Japan. J. Appl. Phys. **37**, 4590 (1998)

18. K. Ogasawara, T. Ishii, I. Tanaka, H. Adachi, Mater. Trans. Japan. Inst. Met. **40**, 396 (1999)
19. K. Ogasawara, T. Ishii, F. Oba, I. Tanaka, H. Adachi, Mat. Res. Soc. Symp. Proc. **560**, 341 (1999)
20. K. Ogasawara, T. Ishii, I. Tanaka, H. Adachi, Phys. Rev. **62**, 143 (2000)
21. H. Adachi, M. Tsukada, C. Satoko, J. Phys. Soc. Japan **45**, 875 (1978)
22. J.C. Slater, *Quantum Theory of Molecules and Solids*, Vol. 4 (McGraw-Hill, New York, 1974)
23. D.E. Ellis, G.S. Painter, Phys. Rev. **B2**, 2887 (1970)
24. H.A. Weakliem, J. Chem. Phys. **36**, 2117 (1962)

7

Heavy Elements

Tomohiko Ishii, Masahiro Yamashita, Rika Sekine, and Toshiaki Enoki

7.1 Introduction

For heavy elements with atomic numbers greater than 50, such as gold, relativistic effects become significant in the electronic structure and in chemical bonding [1,2]. Namely, the direct relativistic effects due to the finite speed of light and spin-orbit interaction lead both to radial contraction and energetic stabilization of s and p shells and to the spin-orbit splitting of the molecular orbital (MO) levels. The contraction of the s and p orbitals increases the screening of nuclear charge for outer d and f electrons, resulting in the expansion of the radial distribution of d and f orbitals. The magnitude of the expansion depends upon the magnitude of contraction. These two kinds of relativistic effects affect the valence levels relevant to chemical bonding.

In this chapter, we will show an example of a molecular orbital calculation including relativistic effects, using $AuCl_3$ as the target complex, since it contains gold, a heavy element. Graphite intercalation compounds (GICs) have a variety of novel physical properties associated with graphite-intercalate-graphite sandwiched layered structures [3]. Transition-metal chloride GICs are known as typical acceptor GICs, with triple atomic layers of chlorine-metal-chlorine composite in graphitic galleries [4]. In the acceptor-type GICs, it has been reported that the magnitude of the charge transfer rate from the graphitic π-band is considerably small [5], reflecting the weak interaction between the graphitic layers and intercalates. Such a small charge transfer changes only slightly the Fermi level of the graphitic π-band with no mixing of the electronic states between graphitic layers and the intercalate, which validates the use of the rigid-band concept. Therefore, the electronic structure of the acceptor-type GICs in the vicinity of the Fermi energy is described in terms of the graphitic two-dimensional π-bands. In fact, the optical reflectance for the $SbCl_5$ and AsF_5 GICs are explained by the analysis of the electronic structure of the graphitic π-band model [6].

On the other hand, experimental investigation [7] reveals that $AuCl_3$ GICs have exceptional features in the electronic structure among acceptor-GICs.

A large plasma frequency ω_p in optical reflectance measurements suggests the existence of an overlap of the intercalate band and the graphitic π-band near the Fermi level. The possibility of the existence of the intercalate band is also suggested from the observation of the large Pauli paramagnetic susceptibility and electronic specific heat coefficient, which indicate an enhanced density of states at the Fermi level. In addition, the c-axis conductivity value for the first-stage compound reaches 23 $(\Omega\mathrm{cm})^{-1}$ at room temperature, which is the highest value ever observed in acceptor GICs [8], and which implies that the orbital overlapping between the graphitic π-band and the intercalate band enhances the conductivity along the c-axis. These experimental findings suggest a strong coupling between the graphitic π-electron system and the intercalate, which is related to the essence of the intercalated molecule Au_2Cl_6.

Meanwhile, in donor-type GICs, in which the charge transfer rate is considerable, the intercalate band is reported to overlap with the graphitic π^*-band in the vicinity of the Fermi level. The coexistence of the intercalate band and the graphitic π-band around at the Fermi energy is confirmed by strong interactions between them. The large charge transfer rate generated by the strong interaction between the graphitic layers and the intercalate in $AuCl_3$ GICs is reminiscent of the situation in donor-type GICs. In this chapter, we discuss the electronic properties of an intercalate Au_2Cl_6 molecule calculated by means of the DV-Xα molecular orbital method in order to clarify the electronic structure and the chemical character of the Au_2Cl_6 molecule, which will help us understand the peculiar features of $AuCl_3$ intercalate in graphitic galleries. The system being studied, Au_2Cl_6, includes gold, a heavy element. Therefore, in this calculation, we also use a relativistic version of the DV-Xα method, compare the results with those obtained by the non-relativistic calculation, and discuss the relativistic effects on the electronic structure of Au_2Cl_6.

7.2 Method of Calculation

In the analysis of the electronic structure of intercalate $AuCl_3$, we carried out an electronic structure calculation using the DV-Xα molecular orbital method. The computation details of both methods used in our work have been described elsewhere [9–12].

We apply the DV-Xα method to the calculation of the electronic structure of planar molecule Au_2Cl_6.

Figure 7.1 shows the structural model of a dimerized planar molecule of Au_2Cl_6 based on the pristine $AuCl_3$ crystal [13].

In our calculation, we assume that intercalate $AuCl_3$ has the same molecular structure as the Au_2Cl_6 molecule in the pristine crystal [14]. As shown in Fig. 7.1, there are two kinds of chlorine atoms – one is located between two gold atoms (bridge Cl) and the other is at the terminal of the molecule (terminal Cl). The four terminal chlorine atoms form an isosceles trapezoid,

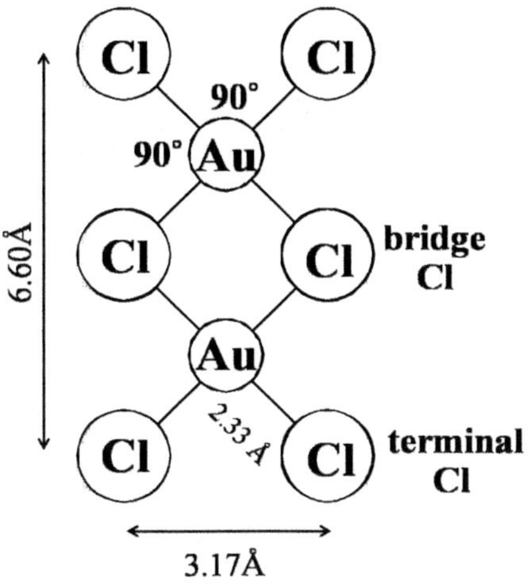

Fig. 7.1. The structural model of a dimerized planar molecule Au_2Cl_6 based on the $AuCl_3$ pristine crystal

which encloses two gold atoms. We look upon this trapezoidal structure as a rectangular one, because the difference in length between two sides is very small ($\sim 0.3\%$). On this assumption, the symmetry of this molecule can be regarded as a D_{2h} point group [15]. As the spin function is included, the D_{2h} group reduces to a D_{2h}^* double group in a calculation with relativistic effects. The symmetry orbitals corresponding to irreducible representations of the D_{2h}^* symmetry are constructed from the atomic orbitals using the standard projection operator method [16]. The relativistic and non-relativistic DV-Xα calculations are performed by means of a special program, which takes into account relativistic effects. The Slater exchange parameter, α, is set to 0.7 for all atoms, and 5000 DV sampling points are used, which provides valence electron energy eigenvalues to a precision of better than 0.1 eV. We employ basis functions up to the $3p$ orbital for the chlorine atom, up to the $6p$ orbital for the gold atom. In order to investigate the effect of the high-order shell orbits, calculations employing basis functions up to the $4p$ orbital for the chlorine atom are also done. The calculations are carried out self-consistently until the difference in orbital populations between the initial and final states of the iteration is less than 0.01.

In $AuCl_3$ GICs, we assume that the intercalate $AuCl_3$ has the same molecular structure as the Au_2Cl_6 molecule in a pristine crystal. This assumption is considered to be valid in view of the available electron microscope results on the in-plane superstructure of intercalate and the EXAFS results [14]. The charge transfer rate per carbon atom f_C is estimated to be ~ 0.017 from the Raman spectrum measurement performed on the first-stage $AuCl_3$-GIC [7]. About 0.5 electrons are considered to be transferred from 28 carbon atoms to a unit of Au_2Cl_6 in the first-stage compound of $C_{28}Au_2Cl_6$.

The electronic structure of $[Au_2Cl_6]^{-0.5}$ was calculated assuming the same structure as a pristine Au_2Cl_6 molecule, since the difference in structures before and after the electron transfer is small and negligible.

7.3 Results and Discussion

Table 7.1 summarizes the energy levels and the characteristics of the molecular orbitals of the pristine Au_2Cl_6 molecule, which are obtained by the non-relativistic DV-Xα calculation with the basis functions of the chlorine atom up to the $3p$ orbital.

In the HOMO level, which is assigned to the $17b_{1g}$ orbital in the D_{2h} symmetrical Au_2Cl_6 molecule, the terminal Cl-$3p$ atomic orbital (see Fig. 7.1) plays a major role in comparison with those of the bridging chlorine and the gold atoms. Meanwhile, about 23% of the Au-$5d$ atomic orbital contributes to the LUMO level assigned to the $21b_{2u}$ orbital, while the proportions of the terminal and bridge Cl-$3p$ atomic orbitals compared to the LUMO are about 54% and 20%, respectively. The HOMO–LUMO band gap is estimated to be 1.5 eV (i.e., −5.26 eV − (−6.75 eV)), which is consistent with the fact that the color of the $AuCl_3$ crystal consisting of the Au_2Cl_6 molecule is brownish red.

In order to make the calculation more exact, we employ the basis functions of the chlorine atom up to the $4p$ orbital. Comparing between this calculation and the one done previously, in which only basis functions up to the $3p$ orbital were considered, we find that the essential features of the electronic

Table 7.1. The energy levels and the characteristics of the molecular orbitals of the Au_2Cl_6 molecule near the HOMO and LUMO levels obtained by means of the non-relativistic DV-Xα calculation. The basis functions of the chlorine atoms were used up to the $3p$ orbital. The character "e" in the table denotes the electron occupation number

				Orbital population(%)				
		Energy		Au			Cl (bridge)	Cl (terminal)
	Orbital	(eV)	e	$5d$	$6s$	$6p$	$3p$	$3p$
(HOMO)	$11b_{2g}$	−7.55	2.0	0.0420		0.0723		0.8857
	$29a_g$	−7.26	2.0	0.0829	0.0018	0.0007	0.0404	0.8745
	$13b_{1u}$	−7.20	2.0	0.0747		0.0103	0.1719	0.7431
	$26b_{3u}$	−7.19	2.0	0.0437	0.0010	0.0001	0.0465	0.9089
	$20b_{2u}$	−7.09	2.0	0.0171		0.0244	0.0235	0.9312
	$6a_u$	−7.05	2.0	0.0359				0.9641
	$8b_{3g}$	−7.04	2.0	0.0456			0.0146	0.9397
	$17b_{1g}$	−6.76	2.0	0.0000		0.0004	0.0381	0.9619
(LUMO)	$21b_{2u}$	−5.26	0.0	0.2312		0.0150	0.2029	0.5423
	$18b_{1g}$	−4.93	0.0	0.2267		−0.0004	0.2501	0.5221
	$14b_{1u}$	−0.54	0.0	0.0001		0.8843	0.0634	0.0521
	$12b_{2g}$	−0.32	0.0	0.0005		0.9284		0.0711
	$30a_g$	0.12	0.0	0.0250	0.6719	0.0538	0.1056	0.1050
	$27b_{3u}$	1.21	0.0	0.0193	0.7473	−0.0223	0.1311	0.1115
	$22b_{2u}$	2.95	0.0	0.0023		0.8949	0.0003	0.0634

structure are almost the same, while the absolute value of each energy level is slightly different between the two cases. The values in the higher-order case are shifted by $\sim 0.3\,\mathrm{eV}$ from the values in Table 7.1 with little change in the electronic structure. This implies that the use of basis functions up to the $4p$ orbital for the chlorine atom is not necessary, and that the calculation using basis functions up to the $3p$ orbital is valid enough to discuss the electronic properties of the Au_2Cl_6 planar molecule.

Next, we introduce the DV-Xα calculation that takes into consideration relativistic effects, in which we employ the basis functions of the Cl atom up to the $3p$ orbital, taking into account the fact that the the relativistic DV-Xα calculation using basis functions up to the Cl-$4p$ orbital does not converge. Table 7.2 summarizes the energy levels and the characteristics of the molecular orbitals of the pristine Au_2Cl_6 molecule, which are obtained by the relativistic DV-Xα calculation with the basis functions of the chlorine atom up to the $3p$ orbital.

In the region below the HOMO level assigned to the $65\gamma_{5g}$ orbital in the D_{2h}^* double group, the terminal Cl-$3p$ atomic orbital plays a major role, similar to the case calculated by the non-relativistic DV-Xα method. The most

Table 7.2. The energy levels and the characteristics of the molecular orbitals of the Au_2Cl_6 molecule near the HOMO and LUMO levels obtained by the DV-Xα calculation involving relativistic effects. The basis functions of the chlorine atom were used up to the $3p$ orbital. The character "e" in this table denotes the electron occupation number. The atomic orbitals $5d$, $6s$, $6p$, and $3p$ correspond to $5d_{3/2}$ or $5d_{5/2}$, $6s_{1/2}$, $6p_{1/2}$, or $6p_{3/2}$, and $3p_{1/2}$ or $3p_{3/2}$, respectively

				Orbital population(%)				
		Energy		Au			Cl (bridge)	Cl (terminal)
	Orbital	(eV)	e	$5d$	$6s$	$6p$	$3p$	$3p$
(HOMO)	$62\gamma_{5g}$	−7.55	2.0	0.078	0.002	0.069	0.005	0.846
	$62\gamma_{5u}$	−7.27	2.0	0.068	0.002	0.009	0.076	0.846
	$63\gamma_{5g}$	−7.19	2.0	0.119	0.000	0.003	0.040	0.839
	$63\gamma_{5u}$	−7.15	2.0	0.067	0.000	0.008	0.039	0.886
	$64\gamma_{5u}$	−7.10	2.0	0.070	0.001	0.014	0.078	0.837
	$65\gamma_{5u}$	−6.98	2.0	0.105	0.001	0.002	0.028	0.863
	$64\gamma_{5g}$	−6.96	2.0	0.113	0.000	0.001	0.035	0.851
	$65\gamma_{5g}$	−6.85	2.0	0.006	0.000	0.000	0.035	0.959
(LUMO)	$66\gamma_{5u}$	−4.79	0.0	0.330	0.000	0.020	0.188	0.448
	$66\gamma_{5g}$	−4.49	0.0	0.309	0.000	0.000	0.229	0.457
	$67\gamma_{5g}$	−1.01	0.0	0.044	0.542	0.047	0.148	0.185
	$67\gamma_{5u}$	−0.78	0.0	0.007	0.099	0.731	0.093	0.069
	$68\gamma_{5u}$	−0.60	0.0	0.031	0.490	0.148	0.161	0.163
	$68\gamma_{5g}$	−0.59	0.0	0.001	0.011	0.905	0.001	0.082
	$69\gamma_{5u}$	2.91	0.0	0.003	0.001	0.882	0.002	0.078

important difference between the relativistic and the non-relativistic results is the ratio of the Au-5d atomic orbital contribution to the HOMO.

In the LUMO level assigned to the $66\gamma_{5u}$ orbital, there is a contribution of about 33% of the Au-5d atomic orbital, while the proportions of the terminal and bridge Cl-3p atomic orbitals to the LUMO are about 45% and 19%, respectively. These values are considerably different from the values obtained from the non-relativistic calculation, suggesting that the relativistic effect is very important for the orbital of Au_2Cl_6 molecule, in which the heavy Au atom is included.

Figures 7.2, 7.3, and 7.4 show the basis functions $rR(r)$ of the Au-6s, 6p and 5d atomic orbitals, respectively, obtained by the non-relativistic and relativistic calculations. The relativistic results show the energy-splitting due to the spin-orbit interaction.

Comparing the non-relativistic basis functions with the relativistic ones, one can see that the $5d_{3/2}$ and $5d_{5/2}$ orbitals are more extended due to the indirect relativistic effect, and the $6s_{1/2}$ and $6p_{1/2}$ orbitals show remarkable contraction due to the direct relativistic effects. The $6p_{3/2}$ orbital is almost the same as the non-relativistic 6p orbital. The basis function $rR(r)$ of the

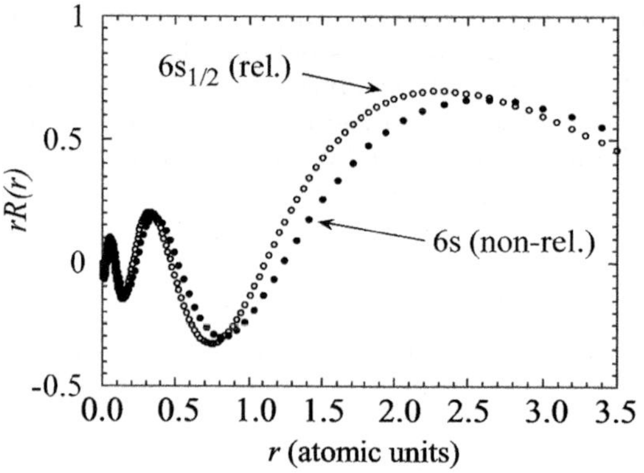

Fig. 7.2. Comparison of the basis functions $rR(r)$ of the Au-6s atomic orbital obtained by the non-relativistic method (*filled circles*) with those obtained by the relativistic method (*open circles*)

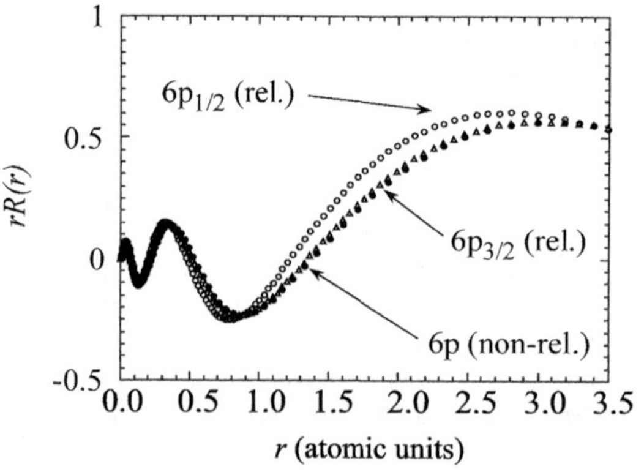

Fig. 7.3. Comparison of the basis functions $rR(r)$ of the Au-6p atomic orbital obtained by the non-relativistic method (*filled circles*) with those obtained by the relativistic method (*open circles* and *triangles*)

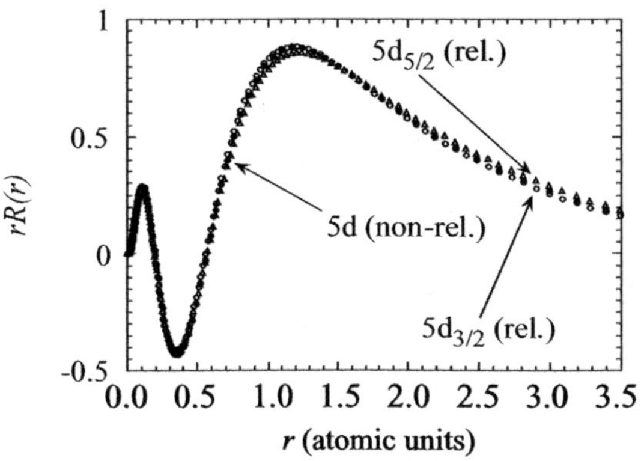

Fig. 7.4. Comparison of the basis functions $rR(r)$ of the Au-5d atomic orbital obtained by the non-relativistic method (*filled circles*) with those obtained by the relativistic method (*open circles and triangles*)

chlorine 3s atomic orbital obtained by the non-relativistic method is also compared with those found via the relativistic calculation. It is revealed that the relativistic effect is not apparent for a light atom like chlorine, since the energy difference of core levels between the relativistic and non-relativistic models is relatively small for chlorine. These results suggest that the contribution of the Au-5d atomic orbital at the LUMO level changes from 23% to 33% mainly due to the indirect relativistic expansion.

Figure 7.5 shows the molecular orbital energy levels of the Au_2Cl_6 molecule with contributions from the consistent atomic orbitals. Generally, the energy levels of the gold 5d atomic orbital are located around -12 eV, while those of the chlorine 3p atomic orbitals are located around -7 eV.

The LUMO ($66\gamma_{5u}$) level of the Au_2Cl_6 molecule is calculated to be -4.79 eV, as summarized in Table 7.2. This suggests that the Au_2Cl_6 molecule works as an acceptor to the pristine graphite, taking into account the fact that the electron affinity level of graphite is reported to be -4.43 eV [17], which is consistent with the results of the in-plane Raman E_{2g2} mode [8].

Figure 7.6 shows contour maps of the HOMO ($17b_{1g}$) and LUMO ($21b_{2u}$) levels of the Au_2Cl_6 molecule for the cross section cut by the plane involving

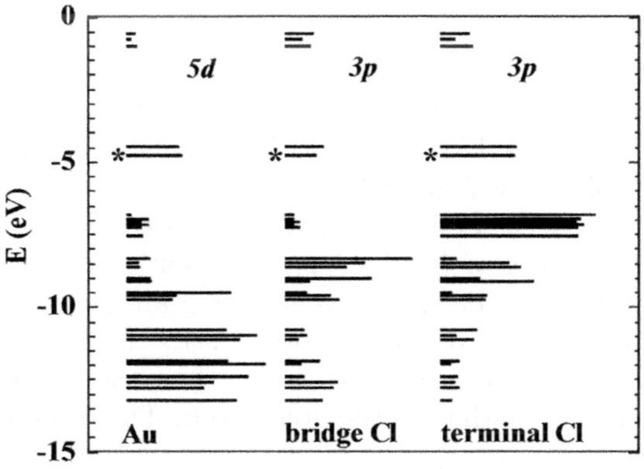

Fig. 7.5. The electronic level structure of Au_2Cl_6. The length of each bar denotes the atomic orbital composition of each molecular orbital. The levels indicated by *asterisks* (∗) are the LUMO levels

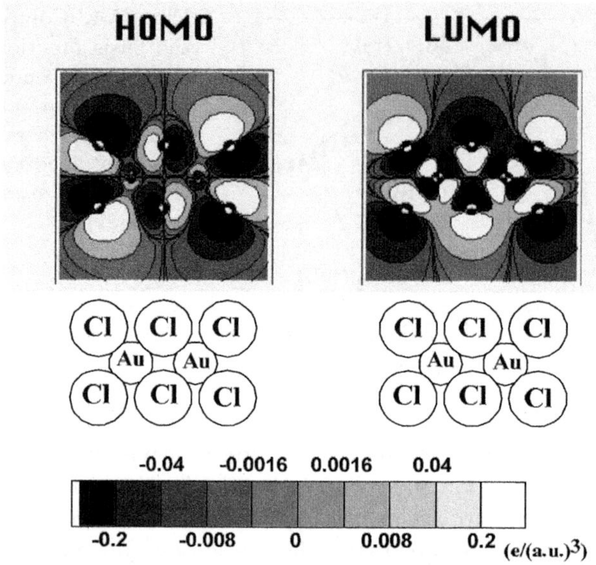

Fig. 7.6. Contour maps of the HOMO (*left*) and LUMO (*right*) of the Au_2Cl_6 molecule, for the cross-section cut by the molecular plane of an Au_2Cl_6. The *white* and the *black regions* denote the positive and the negative signs of the wave functions, respectively, ranging from -0.2 to 0.2 electrons per $(a.u.)^3$

the molecular plane of Au_2Cl_6. Here, we draw the MOs for the non-relativistic calculation since it is difficult to figure out the MOs for the relativistic calculation.

In the HOMO, the orbitals of the terminal chlorine atoms play a larger role in comparison with those of the bridge chlorine and the gold atoms, as mentioned before. Meanwhile, the atomic orbitals of gold atoms make an important contribution to the LUMO in addition to those of the terminal chlorine atoms. These contour maps (Fig. 7.6) suggest that all of the bonds between neighbor atoms have an anti-bonding character in both the HOMO and LUMO. In the figure, white and black denote positive and negative signs of the atomic orbital phase, respectively. This implies that the bond lengths will be elongated by a charge transfer from the graphitic π-band, which weakens the bonds when Au_2Cl_6 molecules are intercalated in graphitic galleries.

Figure 7.7 shows contour maps of the LUMO of the Au_2Cl_6 molecule, for the vertical sections cut by the plane perpendicular to the planar molecule Au_2Cl_6. Figure 7.7a shows the cross-section involving two terminal Cl atoms and one bridging Cl atom parallel to the molecular long axis, while Fig. 7.7b shows the cross-section perpendicular to the long axis involving an Au atom. The contribution of the terminal chlorine $3p$ atomic orbitals to the LUMO wave function is spread out along the vertical planes wider than the contributions of the bridge chlorine atom and that of the gold atom. This finding

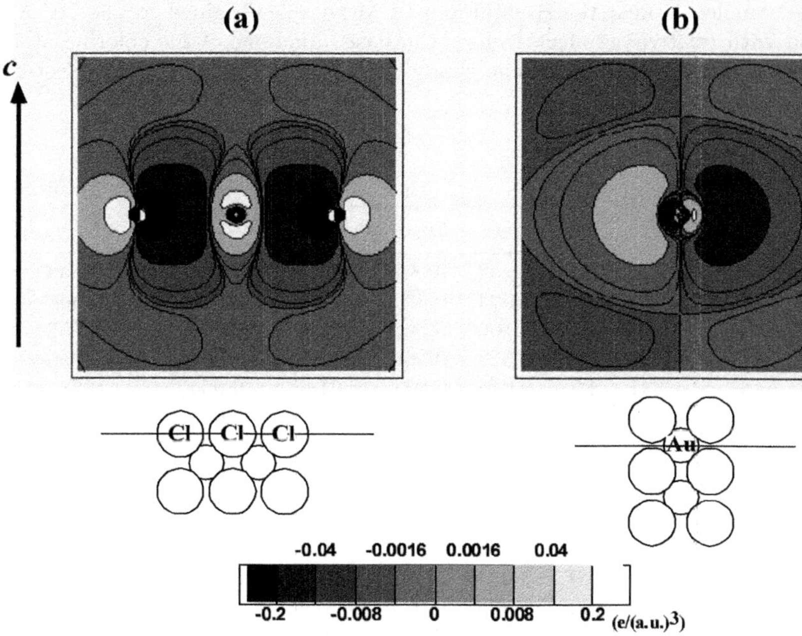

Fig. 7.7. Contour maps of the LUMO of the Au_2Cl_6 molecule for the vertical sections cut by the plane perpendicular to the molecular plane. **a** The cross-section involving two terminal Cl atoms and one bridging Cl atom parallel to the molecular long axis. **b** The cross-section perpendicular to the long axis involving an Au atom. The *white* and the *black regions* denote the positive and the negative signs of the wave functions, respectively, ranging from -0.2 to 0.2 electrons per $(a.u.)^3$. The "c" indicates the direction of the *c*-axis, which is perpendicular to the Au_2Cl_6 plane

suggests that the wave function of the terminal chlorine atoms will interact strongly with the graphitic π-band along the *c*-axis when the Au_2Cl_6 molecule is intercalated in a graphitic gallery.

Next, we discuss the effect of the charge transfer on the electronic structure of the Au_2Cl_6 intercalate in $AuCl_3$ GICs. As mentioned before, the amount of electrons transferred from the graphitic π-band is estimated at 0.5 per Au_2Cl_6 molecular unit in the case of a first-stage compound. Table 7.3 summarizes the energy levels and the characteristics of the molecular orbitals after the transfer of 0.5 electrons from the graphitic layer to the Au_2Cl_6 molecule (i.e., $[Au_2Cl_6]^{-0.5}$) near the HOMO and LUMO levels.

In this calculation, we employed the basis functions of the chlorine atom up to the $3p$ orbital with relativistic effects, since consideration of the Cl-$4p$ orbital yields a divergent result for the Au_2Cl_6 molecule, as mentioned before. The LUMO ($66\gamma_{5u}$) level of the Au_2Cl_6 molecule is partially occupied by 0.5 electrons transferred from the graphitic π-band, as shown in Table 7.3. By

Table 7.3. The energy levels and the characteristics of the molecular orbitals of the $[Au_2Cl_6]^{-0.5}$ molecule near the HOMO and LUMO levels obtained by the DV-Xα calculation with relativistic effects, where the basis functions of the chlorine atom up to the $3p$ orbital is employed. The character "e" in this table denotes electron occupation number

				Orbital population(%)				
		Energy		Au			Cl (bridge)	Cl (terminal)
	Orbital	(eV)	e	$5d_{3/2,5/2}$	$6s_{1/2}$	$6p_{1/2,3/2}$	$3p_{1/2,3/2}$	$3p_{1/2,3/2}$
(HOMO)	$62\gamma_{5g}$	−5.86	2.0	0.068	0.002	0.060	0.004	0.866
	$62\gamma_{5u}$	−5.59	2.0	0.064	0.002	0.007	0.079	0.848
	$63\gamma_{5g}$	−5.53	2.0	0.106	0.000	0.002	0.039	0.855
	$63\gamma_{5u}$	−5.48	2.0	0.072	0.000	0.005	0.058	0.865
	$64\gamma_{5u}$	−5.41	2.0	0.046	0.001	0.014	0.054	0.884
	$65\gamma_{5u}$	−5.32	2.0	0.094	0.001	0.002	0.029	0.874
	$64\gamma_{5g}$	−5.31	2.0	0.103	0.001	0.001	0.035	0.861
	$65\gamma_{5g}$	−5.15	2.0	0.002	0.000	0.000	0.033	0.964
	$66\gamma_{5u}$	−3.20	0.5	0.319	0.000	0.016	0.191	0.459
(LUMO)	$66\gamma_{5g}$	−2.90	0.0	0.299	0.000	0.000	0.237	0.460
	$67\gamma_{5g}$	0.51	0.0	0.044	0.572	0.040	0.137	0.177
	$67\gamma_{5u}$	0.82	0.0	0.005	0.076	0.786	0.076	0.056
	$68\gamma_{5u}$	1.00	0.0	0.030	0.537	0.103	0.163	0.160
	$68\gamma_{5g}$	1.03	0.0	0.001	0.007	0.922	0.000	0.069
	$69\gamma_{5u}$	4.16	0.0	0.003	0.000	0.910	0.001	0.061

comparing Tables 7.2 and 7.3, one can see that the characteristics of the molecular orbitals of $[Au_2Cl_6]^{-0.5}$ are not changed very much. The minor change in the features of the molecular orbitals due to the charge transfer proves the validity of the rigid-band concept in discussing the electronic structure of the Au_2Cl_6 molecule.

Now, we discuss the density of states (DOS) at the Fermi level of the $AuCl_3$ GIC involving $[Au_2Cl_6]^{-0.5}$.

Figure 7.8 shows the DOS of the $[Au_2Cl_6]^{-0.5}$ molecule near the HOMO level.

Since the integral of a Lorentzian function with respect to energy from $-\infty$ to ∞ is estimated to be π, a bandwidth of $1/\pi(= 0.31)$ eV was used in the calculation of the DOS as a normalized coefficient of the Lorentzian function in which both the peak height and the total number of electrons are normalized to each level.

In this figure, the position of the Fermi level of the π-band in the $AuCl_3$ GIC is adjusted to coincide with the vacuum level. The electron affinity level, which is -4.43 eV in pristine graphite, is shifted down by -0.97 to -5.4 eV after intercalating Au_2Cl_6, according to the results of Raman spectra experiments [8]. Meanwhile, the small charge transfer of $\sim$ 0.5 electrons elevates the Fermi energy of the Au_2Cl_6 molecule from −4.79 to −3.20 eV, as summarized in Tables 7.2 and 7.3. These values are elevated due to the lack of consideration given to the Madelung energy generated between $C_{28}^{+0.5}$ and $[Au_2Cl_6]^{-0.5}$. The energy level of the HOMO in $[Au_2Cl_6]^{-0.5}$ appears to be lowered when $[Au_2Cl_6]^{-0.5}$ is intercalated between the positively charged graphitic layers through the Madelung energy. Moreover, the lack of information on the energy level of graphite after the intercalation makes the calculation of the Madelung energy impossible. Therefore, we adjust the HOMO level of $[Au_2Cl_6]^{-0.5}$ to the Fermi level of the graphitic π-band of the $AuCl_3$ GIC ($E_F = -0.97$ eV) [8].

In Fig. 7.8, the position of the Fermi energy at ~ -5.8 eV (indicated by the dashed line) is shifted up to −5.40 eV, (indicated by the solid line), as a result of the transfer of 0.5 electrons from the graphitic π-band. The local density of states per carbon atom for the graphitic π-band is estimated at $D_\pi(E_F) = 0.036$ states/(eV atom of C) (= 0.17 states/(eVcm^3)) [8]. The inset of Fig. 7.8 shows the electronic structure of the first-stage $AuCl_3$ GIC with the overlap of the graphitic π-band (solid line) and the Au_2Cl_6 intercalate band, where the value of the local DOS of the graphitic π-band is normalized by multiplying by a factor of 28, taking into account the fact that the composition of $AuCl_3$ GIC is considered to be $C_{28}Au_2Cl_6$.

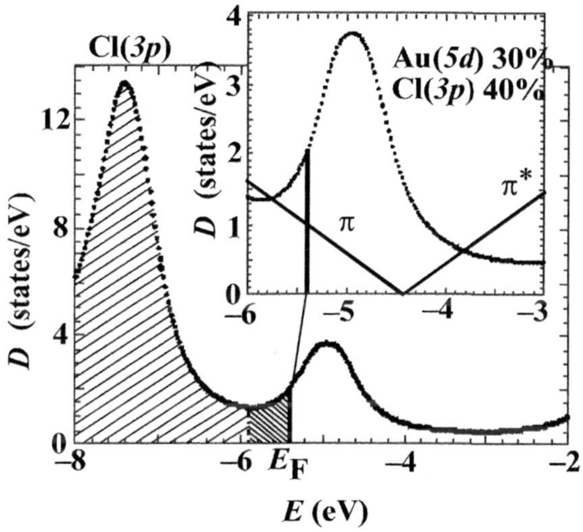

Fig. 7.8. The electronic density of states (DOS) of the $[Au_2Cl_6]^{-0.5}$ molecule near the HOMO level superimposed on the graphitic π-bands. In the calculation of the DOS, a Lorentzian function with a bandwidth of 0.32 eV is employed so that each level accommodates 2.0 electrons. *Dashed* and *solid lines* denote the Fermi levels before and after the transfer of 0.5 electrons from the graphitic π-band, respectively. The *inset* shows the band structure of the first-stage $AuCl_3$ GIC, showing the overlap of the graphitic π-band and the Au_2Cl_6 intercalate band, the latter of which has a mixture of Au-5d and Cl-3p states

The DOS plot in Fig. 7.8 exhibits a large contribution of the Au_2Cl_6 HOMO band, which is constructed mainly with 30% Au-5*d* and 40% terminal Cl-3*p* orbitals, as mentioned in Table 7.3, exists at the Fermi level superimposed on the graphitic linear π-band. These results also suggest that the total density of states at the Fermi level consisting of the graphitic π-band and the Au_2Cl_6 intercalate band is about three times larger than the value expected from the graphitic π-band. This is in good qualitative agreement with the optical reflectance, magnetic susceptibility, specific heat, and electrical conductivity [7,8], which prove the overlap of the graphitic π-band and the $AuCl_3$ intercalate band.

Next, we move on to the chemical features of the Au_2Cl_6 molecule in relation to the charge transfer. Tables 7.4 and 7.5 show the results of Mulliken population analysis [18], which show the orbital populations of the gold and chlorine valence atomic orbitals.

It is worth noting that the Au-6*p* atomic orbital has electrons in both Au_2Cl_6 and $[Au_2Cl_6]^{-0.5}$, although a free gold atom has no electrons in the 6*p* orbital. This result implies that the energy levels of the Au-6*p* atomic orbital is lowered in these molecules in comparison with the free gold atom [2] as a consequence of bond-formation with chlorine and/or a ligand effect. From the Mulliken orbital population, the valence of gold is calculated to be $Au^{+0.236}$, suggesting that the bonding between gold and chlorine has a rather covalent nature. In addition, the valence of the terminal chlorine is estimated to be $Cl^{-0.125}$, which is considerably smaller than the valance of an ordinary metal chloride where chlorine atom has an almost complete negative charge Cl^{-1}.

This result suggests the possibility that the chlorine atom plays the role of a strong acceptor to graphite, since the LUMO at the terminal Cl-3*p* site

Table 7.4. The Mulliken orbital population of the gold atom in Au_2Cl_6 and $[Au_2Cl_6]^{-0.5}$

	Au			
Atom or molecule	5*d*	6*s*	6*p*	Valence electron
Au	10	1	0	Au^0
Au_2Cl_6	9.274	0.846	0.643	$Au^{+0.236}$
$[Au_2Cl_6]^{-0.5}$	9.378	0.814	0.498	$Au^{+0.311}$

Table 7.5. The Mulliken orbital population of the chlorine atom in Au_2Cl_6 and $[Au_2Cl_6]^{-0.5}$

	Cl (bridge)			Cl (terminal)		
Atom or molecule	3*s*	3*p*	Valence electron	3*s*	3*p*	Valence electron
Cl	2	5	Cl^0	2	5	Cl^0
Au_2Cl_6	1.934	5.051	$Cl^{+0.014}$	1.951	5.174	$Cl^{-0.125}$
$[Au_2Cl_6]^{-0.5}$	1.952	5.139	$Cl^{-0.091}$	1.962	5.272	$Cl^{-0.235}$

spreads widely along the c-axis when Au_2Cl_6 is intercalated in graphitic galleries, as mentioned in Fig. 7.7.

The valences of the gold and the chlorine atoms are calculated to be $Au^{+0.311}$ and $Cl^{-0.235}$, respectively, in the case of the $[Au_2Cl_6]^{-0.5}$ molecule. This means that the charge transfer from graphite makes the Au_2Cl_6 molecule more ionic than the pristine Au_2Cl_6 molecule. Taking into account the fact that all the bonds between neighbor atoms are anti-bonding in character, as mentioned in Fig. 7.6, this enhanced ionicity of the Au_2Cl_6 molecules results in the weakening of the bond strengths due to electron transfer. Experimental evidence supporting the proposition that the bond strength in Au_2Cl_6 becomes weak after intercalating into a graphitic gallery is that the a_g mode observed in Raman spectra (at 350 cm^{-1}), associated with the Au–Cl terminal and bridge stretching vibration [7], shows a red shift (to 342 cm^{-1}) in $AuCl_3$ GICs in comparison with the measurements for pristine Au_2Cl_6.

Finally, we discuss the results of the overlap population between gold atom and two kinds of chlorine neighbors. Table 7.6 summarizes the overlap population.

Comparison of the values before and after the transfer of 0.5 electrons per Au_2Cl_6 results in a 30% decrease in the overlap population of the $AuCl_3$ GIC, suggesting that the bond strength becomes weak by the charge transfer to the antibinding orbital. This result is also supported by the low energy shift of the intramolecular vibrational Raman mode measurement.

Table 7.6. The overlap population between gold and the two kinds of chlorine atoms in Au_2Cl_6 and $[Au_2Cl_6]^{-0.5}$

Molecule	Au–Cl(bridge)	Au–Cl (terminal)
Au_2Cl_6	0.274	0.404
$[Au_2Cl_6]^{-0.5}$	0.187	0.302

Acknowledgement. The authors wish to thank Profs. B. Fricke (Univ. of Kassel) and S. Sugiura (Shonan Inst. of Tech.), Drs. H. Nakamatsu (Kyoto Univ.), J. Onoe (Tokyo Inst. of Tech), M. Hirata (JAERI), M. Kurihara, and T. Bastug (Australia National University) for fruitful discussions, useful technical advice, and encouragement.

References

1. P. Pepper, B.E. Bursten, Chem. Rev. **91**, 719 (1991)
2. P. Pyykkö, Chem. Rev. **88**, 563 (1988)
3. M.S. Dresselhaus, G. Dresselhaus, Adv. Phys. **30**, 139 (1981)
4. K. Ohashi, I. Tsujikawa, Tanso **95**, 154 (1978)
5. H. Kamimura, Ann. Phys. **11**, 39 (1986)

6. J. Blinowski, Nguyen Hy Hau, C. Rigaux, J.P. Vieren, R. Le Toullec, G. Furdin, A. Hérold, J. Melin, J. Phys. (Paris) **41**, 47 (1980)
7. T. Ishii, Y. Komatsu, K. Suzuki, T. Enoki, A. Ugawa, K. Yakushi, S. Bandow, Mol. Cryst. Liq. Cryst. **245**, 1–6 (1994)
8. T. Ishii, M. Nakao, K. Suzuki, T. Enoki, R. Nishitani, Y. Nishina, Solid State Commun. **84**, 1055 (1992)
9. A. Rosén, D.E. Ellis, Chem. Phys. **62**, 3039 (1975)
10. H. Adachi, M. Tsukada, C. Satoko, J. Phys. Soc. Japan **45**, 875 (1978)
11. H. Nakamatsu, H. Adachi, T. Mukoyama, Bull. Inst. Chem. Res. Kyoto Univ. **70**, 16 (1992)
12. J. Onoe, K. Takeuchi, H. Nakamatsu, T. Mukoyama, R. Sekine, B.-I. Kim, H. Adachi, J. Chem. Phys. **99**, 6810 (1993)
13. S.C. Edward, H.T. David, Acta Cryst. **11**, 284 (1958)
14. J. Ehrich, P. Behrens, W. Metz, W. Niemann, Synth. Met. **34**, 217 (1989)
15. H. Nakamatsu, H. Adachi, T. Mukoyama, Bull. Inst. Chem. Res. Kyoto Univ. **69**, 342 (1991)
16. J. Meyer, W.D. Sepp, B. Fricke, A. Rosén, Comput. Phys. Commun. **54**, 55 (1989)
17. Landolt-Börnstein, Neue Serie II/11 (Springer, Berlin Heidelberg New York, 1981)
18. R.S. Mulliken, J. Chem. Phys. **23**, 1833 (1955)
19. R.S. Mulliken, J. Chem. Phys. **23**, 1841 (1955)
20. R.S. Mulliken, J. Chem. Phys. **23**, 2338 (1955)
21. R.S. Mulliken, J. Chem. Phys. **23**, 2343 (1955)

Part III

Spectroscopy

8

Radiative Transitions

Takeshi Mukoyama

The general theory of X-ray emission and absorption is described. The expressions for transition probabilities are presented in the dipole approximation, and the dipole matrix element is obtained for atoms. Theoretical models for molecular X-ray emission rates are explained. Various approximations used in the case of molecular X-ray emission are tested with the DV-Xα method. The chemical effect on the $K\beta/K\alpha$ X-ray intensity ratios for chemical compounds of 3d elements is investigated with the DV-Xα method. The calculated results are compared with the experimental data, and information about chemical states is obtained. The validity of the DV-Xα method to calculate X-ray spectra is demonstrated.

8.1 Introduction

Since early days of X-ray physics, X-ray spectroscopy has been widely used to investigate the electronic structures of materials. It is well-known that X-ray absorption and emission spectra are sensitive to the chemical environments of elements in molecules and solids [1–4]. Extensive experimental X-ray spectra for various molecules and solids have been reported. With recent advances in high-resolution X-ray spectrometers, high-power X-ray sources, and synchrotron radiation facilities, X-ray spectroscopy has become one of the most powerful experimental tools to study electronic structures. In particular, the advent of third-generation synchrotron radiation facilities has made it possible to produce strong energy-tuned photon beams and to obtain more elaborate experimental spectra. However, in order to analyze these measured data and to extract information about electronic structures, comparison with theoretical calculations based on molecular orbital (MO) methods or band theories is indispensable.

On the other hand, in the case of molecules and solids, most theoretical studies have been performed only for transition energies and line profiles, and exact calculations of X-ray absorption and emission spectra are rather scarce. This can be attributed to two reasons: first, molecular orbital methods and band theories are usually developed for the ground states and are difficult to apply to excited states or to states with inner-shell vacancies. Second, in absorption and emission spectra of molecules, the matrix elements of electron transitions should be calculated with multi-center integrations by the use of molecular wave functions. Such calculations are tedious and time-consuming even for high-speed computers with large memories.

The discrete-variational (DV) Xα method [5] has several advantages over other theoretical methods for calculations of electron transition probabilities in molecules. In this method, numerical wave functions, obtained by solving the self-consistent field (SCF) equation, are used as the basis functions. This means that realistic molecular wave functions and molecular potentials are obtained with a relatively small number of basis functions compared to other methods. Furthermore, the matrix elements in the secular equations are evaluated with the DV integration method [6], i.e., as the weighted sum of the integrand values at the sampling points distributed randomly. This DV integration method can also be applied to calculate the transition matrix elements. In this way, we can avoid the difficulty inherent to multi-center integration. The DV-Xα method has been successfully applied to X-ray and electron spectroscopy [7].

The interaction between atoms (molecules) and an electromagnetic field can be classified into two processes: X-ray emission and absorption. The X-ray emission process corresponds to an electron transition between two bound states. When there exists an inner-shell vacancy, one of the outer-shell electrons makes a transition to the vacancy with the emission of electromagnetic radiation or another atomic electron. The latter process, called the Auger effect, is a non-radiative transition and is not discussed here.

On the other hand, in X-ray absorption, the electron is excited to an unoccupied bound state or is ionized to the continuum. At a certain energy, a sudden jump occurs in the absorption spectrum, and this energy is called the absorption edge. When the incident X-ray energy is higher than the ionization threshold, the electron is transferred to the continuum, and the photoelectron is ejected.

In the case of molecules and solids, there also exist fine structures on the high-energy side of the absorption edge; the X-ray absorption near-edge structure (XANES) is located in the energy region above the absorption threshold to $\sim 40\,\mathrm{eV}$, and the extended X-ray absorption fine structure (EXAFS) is in the energy region up to $\sim 1000\,\mathrm{eV}$. The latter is considered as a single scattering of photoelectrons by the surrounding atoms and contains information on the local structure of molecules. On the other hand, XANES is due to the multiple scattering of ejected electrons by neighboring atoms and is strongly sensitive to the electron density around the atom excited by X-rays.

Here, we confine ourselves to the case of electron transitions between two discrete states and are mostly concerned with X-ray emission spectra, because we need special techniques to investigate continuum states in the DV-Xα method. The calculations of the XANES spectra with the DV-Xα method are described in [8,9]. The DV-Xα calculations for X-ray photoelectron spectroscopy (XPS) are reported in [10].

In the following section, the general theory of the interaction between electromagnetic radiation and electrons will be presented. The X-ray emission and absorption probabilities in the dipole approximation are discussed. In Sect. 8.3, the dipole matrix element is estimated for wave functions in a central potential. The X-ray emission rates for molecules are discussed in Sect. 8.4, and the advantage of the DV-Xα method is elucidated. In Sect. 8.5, the computational methods used to calculate X-ray emission rates with the DV-Xα method, such as the DV integration method, the electronic relaxation effect, and the contribution from interatomic transitions, are tested. The subsequent sections are devoted to simple examples for the $K\beta/K\alpha$ X-ray intensity ratios in chemical compounds of $3d$ transition metals. Finally, it is demonstrated that the DV-Xα method is useful to calculate X-ray emission rates and spectra in molecules.

8.2 Transition Probability

The Hamiltonian for an electron of charge e and mass m in a static potential $V(r)$ and interacting with a radiation field is given by

$$H = \frac{1}{2m}\left(\boldsymbol{p} - \frac{e}{c}\boldsymbol{A}(\boldsymbol{r},t)\right)^2 + V(r) , \tag{8.1}$$

where $\boldsymbol{p}$ is the momentum of electron, c is the velocity of light, and the electromagnetic field is described by the vector potential $\boldsymbol{A}(\boldsymbol{r},t)$. This Hamiltonian can be expressed as

$$H = H_0 + H' , \tag{8.2}$$

where H_0 is the Hamiltonian in the absence of the electromagnetic field:

$$H_0 = \frac{\boldsymbol{p}^2}{2m} + V(r) . \tag{8.3}$$

When we employ the Coulomb gauge $\nabla \cdot \boldsymbol{A} = 0$, the interaction Hamiltonian is given by

$$H' = -\frac{e}{mc}\boldsymbol{p} \cdot \boldsymbol{A} . \tag{8.4}$$

Here, we neglect the small term proportional to $\boldsymbol{A}^2$.

When radiation field is expressed as the superposition of oscillators confined within volume V, $\boldsymbol{A}$ can be written as

$$\boldsymbol{A}(\boldsymbol{r},t) = \left(\frac{4\pi c^2}{V}\right)^{1/2} \sum_l \varepsilon_l \left(\boldsymbol{a}_l \mathrm{e}^{i(\boldsymbol{k}_l\cdot\boldsymbol{r}-\omega t)} + \boldsymbol{a}_l^\dagger \mathrm{e}^{-i(\boldsymbol{k}_l\cdot\boldsymbol{r}-\omega t)}\right) , \tag{8.5}$$

where $\boldsymbol{k}_l$ is the propagation vector, ε is the unit vector in the direction of polarization perpendicular to $\boldsymbol{k}$, ω is the frequency of the oscillator, and $\boldsymbol{a}$

and $\boldsymbol{a}^\dagger$ are the annihilation and creation operators for a photon, respectively. The matrix elements of the operators not equal to 0 are given by

$$\langle n-1|\boldsymbol{a}|n\rangle = \left(\frac{n\hbar}{2\omega}\right)^{1/2} , \tag{8.6}$$

$$\langle n+1|\boldsymbol{a}^\dagger|n\rangle = \left[\frac{(n+1)\hbar}{2\omega}\right]^{1/2} , \tag{8.7}$$

where n is the number of photons, and $\langle n+1|\boldsymbol{a}^\dagger|n\rangle$ means that there are $(n+1)$ photons in the final state and n photons in the initial state. In the case of X-ray emission, there is no photon ($n=0$) in the initial state and only the value of $\langle 1|\boldsymbol{a}^\dagger|0\rangle$ is allowed. On the other hand, only the value of $\langle 0|\boldsymbol{a}|1\rangle$ is not equal to zero for X-ray absorption.

First, we shall consider of the X-ray emission process. The matrix element for an electron that transfers from the state i to the state f can be written as

$$\langle \Psi_f|H'|\Psi_i\rangle = -\frac{e}{mc}\left(\frac{2\pi c^2\hbar}{V\omega}\right)^{1/2}\int \Psi_f^* \mathrm{e}^{-i\boldsymbol{k}\cdot\boldsymbol{r}}\,(\varepsilon\cdot\boldsymbol{p})\Psi_i\,\mathrm{d}\tau , \tag{8.8}$$

where Ψ_i is the electron wave function in the initial state, Ψ_f is the electron wave function in the final state, and we omit the subscript l. The factor $\mathrm{e}^{-i\boldsymbol{k}\cdot\boldsymbol{r}}$ in (8.8) is expanded as

$$\mathrm{e}^{-i\boldsymbol{k}\cdot\boldsymbol{r}} = 1 - i\boldsymbol{k}\cdot\boldsymbol{r} + \frac{1}{2!}(\boldsymbol{k}\cdot\boldsymbol{r})^2 - \cdots ,$$

where we call the term corresponding to $(\boldsymbol{k}\cdot\boldsymbol{r})^L$ the electric 2^{L+1}-pole transition. In the case of X-ray emission, we can assume $\boldsymbol{k}\cdot\boldsymbol{r} \ll 1$. When we retain only the first term with $L=0$ and replace $\mathrm{e}^{-i\boldsymbol{k}\cdot\boldsymbol{r}}$ by 1, this approximation is called the dipole approximation for the reason described below.

Writing the matrix element of $\boldsymbol{p}$ as $\boldsymbol{p}_{fi}$, considering $\boldsymbol{p} - m\boldsymbol{v} = m d\boldsymbol{r}/\mathrm{d}t$, and applying the Heisenberg equation of motion to $\boldsymbol{r}$ [11], we have

$$\frac{1}{m}\boldsymbol{p}_{fi} = \frac{\mathrm{d}}{\mathrm{d}t}\boldsymbol{r}_{fi} = \frac{1}{i\hbar}[\boldsymbol{r}, H_0] = \frac{1}{i\hbar}(E_i - E_f)\boldsymbol{r}_{fi} = i\omega_{fi}\boldsymbol{r}_{fi} ,$$

where [ ] is the Poisson bracket, $\hbar\omega_{fi} = E_i - E_f$, and E_i and E_f are the energies of the electron in the initial and final states, respectively.

The operator $e\boldsymbol{r}$ has a form equivalent to a classical electric dipole, and its matrix element $\boldsymbol{d}_{fi} = e\boldsymbol{r}_{fi}$ is called the dipole matrix element. Using this matrix element, we can write the matrix element of the dipole transition as follows:

$$H'_{fi} = -i\left(\frac{2\pi\hbar}{V\omega}\right)^{1/2}\omega_{fi}\varepsilon\cdot\boldsymbol{d}_{fi} . \tag{8.9}$$

According to the Fermi's golden rule, the probability of X-ray emission with an electron transition from the state i to the state f is given by

$$P_{fi} = \frac{2\pi}{\hbar}\left|H'_{fi}\right|^2\rho(E) = \frac{4\pi^2}{V\omega}\omega_{fi}^2\left|\varepsilon\cdot\boldsymbol{d}_{fi}\right|^2\rho(E) , \tag{8.10}$$

where $\rho(E)$ is the final density of states of the photons. When a photon is emitted into a solid angle $\mathrm{d}\Omega$, the density of states allowed in the energy interval between E and $E + \mathrm{d}E$ is written as

$$\rho(E) = \frac{\mathrm{d}}{\mathrm{d}E}\left(\frac{p^2\,\mathrm{d}p\,\mathrm{d}\Omega}{h^3/V}\right) = \frac{\mathrm{d}}{\mathrm{d}\hbar\omega}\left(\frac{V\omega^2\,\mathrm{d}\omega}{8\pi^3c^3}\right)\mathrm{d}\Omega = \frac{V\omega^2}{8\pi^3\hbar c^3}\,\mathrm{d}\Omega\,,$$

because of $E = \hbar\omega = pc$. Setting $\omega_{fi} = \omega$, we have

$$P_{fi} = \frac{\omega^3}{2\pi\hbar c^3}\int \left|\boldsymbol{\varepsilon}\cdot(e\boldsymbol{r})_{fi}\right|^2 \mathrm{d}\Omega\,. \tag{8.11}$$

For convenience, we assume that $\boldsymbol{k}$, $\boldsymbol{\varepsilon}$, and $\boldsymbol{r}_{fi}$ are on a plane, and we let the angle between $\boldsymbol{\varepsilon}$ and $\boldsymbol{r}_{fi}$ be $\theta = (\boldsymbol{\varepsilon}, \boldsymbol{r}_{fi})$. Then the direction of photon emission is expressed by the polar angle $\Theta = (\boldsymbol{k}, \boldsymbol{r}_{fi})$ and the azimuthal angle Φ. Since the direction of polarization is perpendicular to the direction of propagation, $(\boldsymbol{\varepsilon}, \boldsymbol{k}) = \pi/2$, we can write $\boldsymbol{\varepsilon}\cdot\boldsymbol{r}_{fi} = |\boldsymbol{r}_{fi}|\cos\theta = |\boldsymbol{r}_{fi}|\sin\Theta$. The probability P_{fi} is obtained as

$$P_{fi} = \int_0^{2\pi}\int_0^{\pi}\frac{\omega^3}{2\pi\hbar c^3}\left|(e\boldsymbol{r})_{fi}\right|^2\sin^2\Theta\,\mathrm{d}\Theta\,\mathrm{d}\Phi = \frac{4}{3}\frac{e^2\omega^3}{\hbar c^3}\left|\boldsymbol{r}_{fi}\right|^2. \tag{8.12}$$

It is convenient to introduce the oscillator strength

$$f = \frac{2m}{3\hbar}\omega\left|\boldsymbol{r}_{fi}\right|^2. \tag{8.13}$$

Then, (8.12) becomes

$$P_{fi} = 8.0\times 10^9\left[\frac{E_i - E_f}{Re}\right]^2 f \quad (\mathrm{sec}^{-1})\,, \tag{8.14}$$

where Re is the Rydberg constant.

In the case of X-ray absorption, the electron is excited from a lower energy state to a higher one. However, when we put $n = 1$ in (8.6), the matrix element for absorption is same as that for emission. In the dipole approximation, X-ray absorption probability per unit time can be written in a form similar to (8.10):

$$P_{fi} = \frac{4\pi^2}{V}\omega\left|\boldsymbol{\varepsilon}\cdot\boldsymbol{d}_{fi}\right|^2\delta(E_i + \hbar\omega - E_f)\,, \tag{8.15}$$

where $\delta(x)$ is the delta function, which takes care of energy conservation. It should be noted that the initial state for absorption (i.e., the ground state) corresponds to the final state for emission.

The cross-section σ for absorption of a photon with frequency ω is defined as the ratio of transition probability per unit time to the incident photon flux. When there is a single photon in a volume V, the photon flux is the density of photons multiplied by their speed and is given by c/V. This yields

$$\begin{aligned}\sigma(E) &= \frac{4\pi^2}{c}\omega\left|\boldsymbol{\varepsilon}\cdot(e\boldsymbol{r})_{fi}\right|^2\delta(E_i + \hbar\omega - E_f)\\ &= 4\pi^2\alpha E\left|\boldsymbol{\varepsilon}\cdot\boldsymbol{r}_{fi}\right|^2\delta(E_i + E - E_f)\,,\end{aligned} \tag{8.16}$$

where α is the fine-structure constant, and $E = \hbar\omega$ is the energy of incident X-rays.

When the incident X-ray beam is not polarized, (8.16) should be averaged with respect to the direction of $\boldsymbol{\varepsilon}$. In the absence of an external field, we can assume that the direction of the vector $\boldsymbol{r}$ is distributed uniformly in the coordinate space. Then the average over $\boldsymbol{\varepsilon}$ is replaced by the average over the direction of $\boldsymbol{r}$:

$$\overline{|\boldsymbol{\varepsilon}\cdot\boldsymbol{r}_{fi}|^2} = \frac{1}{3}\left|\boldsymbol{r}_{fi}\right|^2 .$$

The X-ray absorption cross-section is given as a function of E:

$$\sigma(E) = \frac{4\pi^2}{3}\alpha E\left|\boldsymbol{r}_{fi}\right|^2 . \tag{8.17}$$

8.3 Dipole Matrix Element

Equations (8.14) and (8.17) indicate that we can calculate X-ray emission rates and absorption cross-sections if the dipole matrix element $\boldsymbol{r}_{fi}$ is known. In this section, we assume an isolated atom and calculate the dipole matrix element.

The one-electron wave function of the isolated atom with principal quantum number n, orbital angular momentum quantum number l, and magnetic quantum number m is expressed in polar coordinates as

$$u_{nlm}(\boldsymbol{r}) = R_{nl}(r)\,Y_{lm}(\theta,\phi) , \tag{8.18}$$

where $R_{nl}(r)$ is the radial part of the solution to the Schrödinger equation for the Hamiltonian H_0, and $Y_{lm}(\theta,\phi)$ denotes the spherical harmonics. Since $z = r\cos\theta$, the z component of the matrix element corresponding to the transition from the state with quantum numbers nlm to the state with $n'l'm'$ is given by

$$\begin{aligned} z_{nlm}^{n'l'm'} &= \int u_{n'l'm'}^{*}\, z\, u_{nlm}\,\mathrm{d}\tau \\ &= \int_0^\infty r^3\,\mathrm{d}r R_{n'l'}(r)R_{nl}(r)\int Y_{l'm'}^{*}(\theta,\phi)\cos\theta Y_{lm}(\theta,\phi)\,\mathrm{d}\Omega . \end{aligned} \tag{8.19}$$

From the definition of the spherical harmonics, integration of the angular part is reduced to

$$I = \int Y_{l'm'}^{*}\cos\theta Y_{lm}\,\mathrm{d}\Omega = \sqrt{\frac{4\pi}{3}}\int Y_{l'm'}^{*}Y_{l0}Y_{lm}\,\mathrm{d}\Omega , \tag{8.20}$$

where the variables θ and ϕ are omitted for the sake of simplicity.

Using the formula [12]

$$\int Y^*_{l_3 m_3} Y_{l_2 m_2} Y_{l_1 m_1}\,\mathrm{d}\Omega = \left[\frac{(2l_1+1)(2l_2+1)}{4\pi(2l_3+1)}\right]^{1/2} C(l_1 l_2 l_3; m_1 m_2) C(l_1 l_2 l_3; 00)\delta_{m_1+m_2,m_3}\,,$$

the angular integration can be performed as follows:

$$I = \left[\frac{2l+1}{2l'+1}\right]^{1/2} C(l1l'; m0) C(l1l'; 00)\delta_{mm'}\,, \tag{8.21}$$

where $C(l_1 l_2 l_3; m_1 m_2)$ is the Clebsch–Gordan coefficient.

According to the properties of the Clebsch–Gordan coefficient, (8.21) differs from zero only when $l + 1 + l'$ is an even number, and $|l-1| \le l' \le l+1$. Thus, if $l \neq 0$, $l' = l-1$ or $l' = l+1$ is allowed. In the case of $l = 0$, $l' = 1$ is the only possible value. This is called the dipole selection rule.

Using the values of the Clebsch–Gordan coefficients [12], we obtain

$$z_{nlm}^{n'l+1m} = \left[\frac{(l+1-m)(l+1+m)}{(2l+1)(2l+3)}\right]^{1/2} R_{nl}^{n'l+1}\,, \tag{8.22}$$

$$z_{nlm}^{n'l-1m} = \left[\frac{(l-m)(l+m)}{(2l-1)(2l+1)}\right]^{1/2} R_{nl}^{n'l-1}\,, \tag{8.23}$$

$$z_{nlm}^{n'l'm'} = 0 \quad l' \neq l \pm 1\,,$$

with

$$R_{nl}^{n'l'} = \int R_{nl}(r) R_{n'l'}(r) r^3\,\mathrm{d}r\,.$$

The x and y components of the matrix element can be evaluated in a similar manner. Practically, it is more convenient to calculate the matrix elements of the linear combinations

$$x + iy = r\sin\theta\,\mathrm{e}^{i\phi} = -\sqrt{\frac{8\pi}{3}} Y_{11}(\theta,\phi)\,,$$

$$x - iy = r\sin\theta\,\mathrm{e}^{-i\phi} = \sqrt{\frac{8\pi}{3}} Y_{1-1}(\theta,\phi)\,.$$

Then, we obtain

$$(x+iy)_{nlm}^{n'l+1m+1} = \sqrt{\frac{(l+m+2)(l+m+1)}{(2l+3)(2l+1)}} R_{nl}^{n'l+1}\,, \tag{8.24}$$

$$(x-iy)_{nlm}^{n'l+1m-1} = -\sqrt{\frac{(l-m+2)(l-m+1)}{(2l+3)(2l+1)}} R_{nl}^{n'l+1}\,, \tag{8.25}$$

$$(x+iy)_{nlm}^{n'l-1m+1} = -\sqrt{\frac{(l-m)(l-m-1)}{(2l+1)(2l-1)}} R_{nl}^{n'l-1} , \tag{8.26}$$

$$(x-iy)_{nlm}^{n'l-1m-1} = \sqrt{\frac{(l+m)(l+m-1)}{(2l+1)(2l-1)}} R_{nl}^{n'l-1} . \tag{8.27}$$

Finally, we have

$$\begin{aligned}\sum_{m'} \left| r_{nlm}^{n'l+1m'} \right|^2 &= \left| z_{nlm}^{n'l+1m} \right|^2 + \left| x_{nlm}^{n'l+1m+1} \right|^2 \\ &\quad + \left| x_{nlm}^{n'l+1m-1} \right|^2 + \left| y_{nlm}^{n'l+1m+1} \right|^2 + \left| y_{nlm}^{n'l+1m-1} \right|^2 \\ &= \frac{l+1}{2l+1} \left| R_{nl}^{n'l+1} \right|^2 . \end{aligned} \tag{8.28}$$

Similarly,

$$\sum_{m'} \left| r_{nlm}^{n'l-1m'} \right|^2 = \frac{l}{2l+1} \left| R_{nl}^{n'l-1} \right|^2 . \tag{8.29}$$

From (8.28) and (8.29), the general expression becomes

$$\sum_{m'} \left| r_{nlm}^{n'l'm'} \right|^2 = \frac{\max(l,l')}{2l+1} \left| R_{nl}^{n'l'} \right|^2 , \tag{8.30}$$

so that the oscillator strength for an electron is expressed as

$$\begin{aligned} f &= \frac{2m}{3\hbar} \omega_{nl}^{n'l'} \sum_{m'=-l'}^{l'} \left| r_{nlm}^{n'l'm'} \right|^2 \\ &= \frac{1}{3} \frac{\max(l,l')}{2l+1} \frac{E_{nl} - E_{n'l'}}{Re} \frac{|R_{nl}^{n'l'}|^2}{a_0^2} , \end{aligned} \tag{8.31}$$

where a_0 is the Bohr radius of hydrogen.

According to (8.14) and (8.31), the probability per second of X-ray emission for atoms due to electron transfer from the nl orbital to the $n'l'$ orbital is given by [13]

$$P_{n'l',nl} = 8.0 \times 10^9 \left[\frac{E_{nl} - E_{n'l'}}{\text{Re}} \right]^2 f_{n'l',nl} . \tag{8.32}$$

The oscillator strength for each atom is expressed as

$$f_{n'l',nl} = \frac{N_{nl}}{3} \frac{\max(l,l')}{2l+1} \frac{E_{nl} - E_{n'l'}}{\text{Re}} \frac{|R_{nl}^{n'l'}|^2}{a_0^2} , \tag{8.33}$$

where N_{nl} is the number of electrons in the initial nl shell. This result was used by Manson and Kennedy to prepare their tables of X-ray emission rates for atoms [14].

Similarly, the X-ray absorption cross-section for atoms is given by

$$\sigma(E) = \frac{4\pi^2}{3} N_{nl} \alpha E \frac{\max(l, l')}{2l+1} \left| R_{nl}^{n'l'} \right|^2 . \tag{8.34}$$

8.4 Molecular X-Ray Emission

It is typical to approximate solids as a molecular cluster with a certain symmetry and calculate the X-ray transition probability for that cluster. In the case of molecules, the X-ray emission rate is obtained using the molecular wave functions in (8.12). However, it should be noted that the dipole matrix element is evaluated with multi-center integration. When we are interested only in the relative intensity ratio or shape of the X-ray spectra, the X-ray intensity corresponding to the electron transition between ith and jth molecular states in the dipole approximation can be written as

$$I_{ij} \propto E_{ij}^3 \boldsymbol{D}_{ij}^2 , \tag{8.35}$$

where $E_{ij} = E_j - E_i$ is the transition energy of the electron between the ith and jth molecular states, and $\boldsymbol{D}_{ij}$ is the dipole matrix element corresponding to this transition.

The dipole matrix element is expressed as

$$\boldsymbol{D}_{ij} = \langle \Psi_i | \boldsymbol{r} | \Psi_j \rangle , \tag{8.36}$$

where Ψ_i and Ψ_j represent the wave functions of the ith and jth molecular orbitals, respectively. Here we omit e for simplicity and use D to denote the dipole matrix element for molecules. We assume that the initial vacancy is produced in an inner shell i. Then the wave function Ψ_i is approximated by the atomic orbital ϕ_i. On the other hand, the wave function Ψ_j is expressed by the sum of atomic orbitals in the LCAO (linear combination of atomic orbitals) scheme:

$$\Psi_j = \sum_k c_{jk} \phi_k , \tag{8.37}$$

where ϕ_k is the atomic orbital, and c_{jk} is the expansion coefficient. Using this wave function, the X-ray intensity is written as

$$I_{ij} \propto E_{ij}^3 \left| \sum_k c_{jk} \langle \phi_k | \boldsymbol{r} | \phi_i \rangle \right|^2 . \tag{8.38}$$

As described above, it is not easy to calculate the dipole matrix element $\boldsymbol{D}_{ij}$ for molecules because of the computational demands of multi-center integration. To avoid this difficulty, several approximations have been proposed to evaluate molecular X-ray emission rates. For carbon K X-rays in carbon compounds, Manne [15] neglected the contributions from interatomic transitions, sometimes called "crossover transitions", and used the single-center approximation. In this case, (8.38) reduces to

$$I_{ij} \propto E_{ij}^3 \left| {\sum_k}' c_{jk} \langle \phi_k | \boldsymbol{r} | \phi_i \rangle \right|^2 , \tag{8.39}$$

where $\sum'$ represents the summation over only the atomic orbitals corresponding to the atom in which an inner-shell vacancy is present. The calculations of the dipole matrix elements between two atomic orbitals can be performed as a single-center integral in a manner similar to the case of an isolated atom described above.

When we are interested in relative intensities of the transitions from valence shells to the core hole, and the dominant component of these valence shells consists of only one atomic orbital, the matrix elements can be omitted from (8.39), and the X-ray intensity is expressed as

$$I_{ij} \propto E_{ij}^3 {\sum_k}' |c_{jk}|^2 . \tag{8.40}$$

Manne [15] made a further assumption that the energy-dependence of the transition probability is small. Then (8.40) is simplified to

$$I_{ij} \propto {\sum_k}' |c_{jk}|^2 . \tag{8.41}$$

The validity of the single-center approximation has been tested by Støgård [16] and by Ågren and Nordgren [17] for K X-rays from simple chemical compounds of light elements. Their results indicate that this approximation can give the general trend of experimental X-ray spectra. However, more realistic calculations are needed to study X-ray spectra in detail.

The theoretical calculations of molecular X-ray intensities based on (8.38) have been performed by Ågren et al. [18, 19] and Larkins [20]. They used an ab initio molecular orbital method and calculated the dipole matrix element by multi-center integration. However, their results were limited only to simple molecules of light elements. For heavy elements and large molecules or solids with complex symmetry, the number of basis functions becomes large and the multi-center integration becomes tedious and time-consuming.

On the other hand, in the DV-Xα method, the matrix elements in the secular equation are calculated with the DV integration method [6] and the complicated multi-center integration is bypassed. For calculations of the dipole matrix elements between molecular orbital wave functions, it is also possible to apply the DV integration scheme. Using this method, the dipole matrix element can be written as

$$\langle \psi_i | \boldsymbol{r} | \psi_j \rangle \sim \sum_{k=1}^{N} w(\boldsymbol{r}_k) \psi_i(\boldsymbol{r}_k) \boldsymbol{r}_k \psi_j(\boldsymbol{r}_k) , \tag{8.42}$$

where N is the number of sampling points, $\boldsymbol{r}_k$ is the kth randomly selected sampling point, $w(\boldsymbol{r}_k)$ is the weight at the point $\boldsymbol{r}_k$, and $\psi_i(\boldsymbol{r}_k)$ is the value of the ith molecular wave function at $\boldsymbol{r}_k$. This means that the three-dimensional multi-center integral is replaced by the weighted sum of the integrand at sampling points. An advantage of the DV integration method is that (8.38) can be easily calculated, and there is no difficulty in applying this method to heavy elements and complex geometries.

8.5 Test for X-Ray Emission Rates

In the calculations of X-ray emission rates in molecules with the DV-Xα method, we use various approximations and numerical techniques. We examine here several factors that influence the molecular X-ray emission rates, i.e., the validity of the DV integration method for dipole matrix elements, the electronic relaxation effect, and the contributions from the interatomic transitions.

8.5.1 Validity of the DV Integration Method

The validity of the DV integration method has been tested by calculating the overlap integrals and dipole matrix elements for isolated atoms [21].

For the wave functions obtained with the DV-Xα method, the overlap integral can be expressed as

$$S_{\lambda\mu} = \sum_{i,j} c_{i\lambda} c_{j\mu} \langle \phi_j | \phi_j \rangle \, . \tag{8.43}$$

In the case of a single isolated atom, the molecular wave functions reduce to the atomic wave functions ($\lambda = i, \mu = j$), and $S_{\lambda\mu} = 1$ for $\lambda = \mu$ and 0 for $\lambda \neq \mu$. The calculation has been performed for a Cl atom. The result of the square of the overlap integral, $|S_{\lambda\mu}|^2$ is unity for $\lambda = \mu$ within an accuracy of better than 10^{-3} and less than 10^{-5} for $\lambda \neq \mu$ for all possible combinations of atomic orbitals. From this result, it can be said that the DV integration method can reproduce the orthonormality of atomic wave functions quite well.

The square of the dipole matrix element, $|D_{ij}|^2$, according to (8.36), for Cl ($Z = 17$) and Mn ($Z = 25$) atoms was calculated with the DV integration method for all possible dipole transitions. The obtained results are shown in Table 8.1. For comparison, the nonrelativistic Hartree–Fock–Slater (HFS) calculations were performed using the computer code of Herman and Skillman (HS) [22], and the corresponding dipole matrix elements were evaluated by the conventional numerical integration method for a single center. The calculated values are also listed in Table 8.1 and compared with the DV integration values obtained above.

However, it should be noted that in the original HS code, the exchange scaling constant α is set to unity, and the Latter tail correction for the atomic potential [23] is included. On the other hand, to compare with the DV-Xα

Table 8.1. Comparison between the squares of dipole matrix elements (atomic units.) obtained by the DV integration method and by numerical integration of HS wave functions

	Cl ($\times 10^{-2}$)		Mn ($\times 10^{-3}$)	
	HS	DV	HS	DV
$2p \to 1s$	0.417	0.417	2.181	2.181
$2p \to 2s$	17.50	17.52	66.37	66.15
$3p \to 2p$	1.316	1.328	3.681	3.687
$3p \to 1s$	0.025	0.025	0.194	0.195
$3p \to 2s$	1.150	1.169	8.701	8.684

calculations, the HS code was modified with $\alpha = 0.7$ and without the Latter correction. Accordingly, the X-ray emission rates calculated from the values in Table 8.1 are different from the tabulated values of Manson and Kennedy [14] with the HS wave functions.

It is clear from the table that the numerical values obtained with the DV integration method are in good agreement with those calculated with the numerical integration of the HS wave functions and that the DV integration method is useful in calculating the X-ray emission rates.

8.5.2 Electronic Relaxation Effect

The X-ray emission process in atoms and molecules is described by three stages. In the first stage, all of the electrons are in the lowest energy state of a neutral atom – the ground state. Then, an inner-shell vacancy is created, which produces an excited state of the positive ion called the initial state for X-ray emission. After some period, the vacancy in the inner shell is filled by one of the outer-shell electrons, and an X-ray photon is emitted. In the final state, a vacancy remains in one of the outer shells of the positive ion.

In the Xα method, the dipole matrix element is often calculated using the Slater's *transition state* (TS) concept [24]. The TS for X-ray emission corresponds to the state where the electron concerned with the transition stays halfway in the initial state and halfway in the final state. The electron configurations for the ground state, the initial state, the transition state, and the final state are shown in Fig. 8.1.

Most theoretical calculations of X-ray emission rates in atoms and molecules have been performed using the *frozen-orbital* (FR) approximation, where the same atomic or molecular potential is used before and after transition. It is usual to use the ground state (GS) configuration for this purpose. This approximation is convenient because we need only one atomic or molecular potential and the wave function of the initial state is orthogonal to that of the

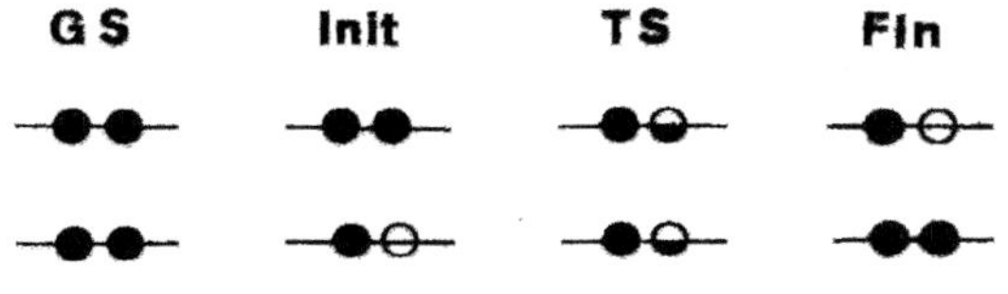

Fig. 8.1. Electron configurations for the ground state (GS), initial state (Init), transition state (TS), and final state (Fin)

final state. However, the presence of a vacancy is not taken into account. The TS method also has a similar advantage to the FR method, but the presence of the vacancy is taken into consideration. However, Mukoyama and Adachi have shown that while the TS method is useful in predicting X-ray transition energies, it is not so good at approximating the absolute X-ray transition probabilities in some cases [25].

It is more realistic to calculate transition matrix elements between the initial and final states, taking into account the electron relaxation. In this case, called the *relaxed-orbital* (RX) approximation, we must calculate wave functions separately for the initial state with an inner-shell vacancy and for the final state with a vacancy in an outer shell. The wave functions for the initial state are not orthogonal to those in the final state because of the difference in potentials. A comparison was made between the K X-ray emission rates of a CO molecule in the FR approximation with the GS configuration, the TS approach, and the RX approximation [26].

The X-ray emission rates with a $1s$ vacancy in an O atom (1σ) for all possible MO states are listed in Table 8.2. The X-ray transition energies were calculated with the TS method, because the energy eigenvalues in the Xα method do not correspond to the binding energy of the electron [24]. It can be seen that the FR approximation gives the largest values for the transition from 3σ and 4σ, while the RX values are larger for 1π and 5σ. The TS approach always gives the smallest transition probabilities.

In Table 8.3, similar results for the 1s vacancy in C (2σ) are given. In this case, the FR values are largest except for the $4\sigma \to 2\sigma$ transition. The TS

Table 8.2. X-ray energy (eV) and emission rates (arbitrary units) for the O-1s vacancy in CO

		Emission rate		
Level	Energy	FR	TS	RX
3σ	512	30.63	22.09	26.60
4σ	529	174.9	144.6	168.7
1π	531	472.5	423.7	490.4
5σ	534	52.23	37.74	64.66

Table 8.3. X-ray energy (eV) and emission rates (arbitrary units) for the C-1s vacancy in CO

		Emission rate		
Level	Energy	FR	TS	RX
3σ	265	10.71	8.04	8.45
4σ	281	1.08	1.35	3.44
1π	283	52.69	48.11	44.03
5σ	287	51.30	31.26	39.37

gives the smallest values except for the transition from 4σ. The RX value for 4σ is largest and others are between the FR and TS values.

In order to compare with other theoretical calculations and the experimental data, the relative intensity ratios of X-ray emission rates of the strongest component from the 1π orbital are calculated. The results for the O-1s (1σ) vacancy are listed in Table 8.4 and compared with the theoretical values of Ågren and Nordgren [17] and of Phillips and Larkins [27], as well as the experimental data acquired by Werme et al. [28]. Ågren and Nordgren [17] performed ab initio calculations in the FR and RX approximations, but their FR calculations were made for a state with a 1σ vacancy. Theoretical values of Phillips and Larkins [27] were also calculated with an ab initio method. It should be noted, however, that their RX calculations include the effect of exchange and overlap due to nonorthogonality of the initial and final wave functions. The similar comparison is made for the case of the C-1s (2σ) vacancy in Table 8.5.

It is clear from Tables 8.4 and 8.5 that, although the absolute X-ray emission rates depend on the approximation used, the relative intensity ratios are almost same for all theoretical calculations. The reason for the discrepancy

Table 8.4. Comparison of K X-ray energy intensity ratios for the oxygen atom in CO

	DV-Xα[a]			Ab initio[b]		Ab initio[c]		
Level	FR	TS	RX	FR[d]	RX	FR	RX	Experiment[e]
3σ	0.06	0.05	0.05	0.06	0.08	0.05	0.04	
4σ	0.37	0.34	0.34	0.37	0.51	0.42	0.20	0.38
1π	1.00	1.00	1.00	1.00	1.00	1.00	1.00	1.00
5σ	0.11	0.09	0.13	0.08	0.00	0.11	0.06	0.28

[a] Refer to [26]. [b] Refer to [17]. [c] Refer to [27]. [d] $1\sigma^{-1}$ state. [e] Refer to [28]

Table 8.5. Comparison of K X-ray energy intensity ratios for the carbon atom in CO

	DV-Xα[a]			Ab initio[b]		Ab initio[c]		
Level	FR	TS	RX	FR[d]	RX	FR	RX	Experiment[e]
3σ	0.20	0.17	0.19	0.25	0.24	0.34	0.13	
4σ	0.02	0.03	0.08	0.04	0.03	0.10	0.00	
1π	1.00	1.00	1.00	1.00	1.00	1.00	1.00	1.00
5σ	0.97	0.65	0.89	0.91	0.84	1.11	0.91	0.59

[a] Refer to [26]. [b] Refer to [17]. [c] Refer to [27]. [d] $1\sigma^{-1}$ state. [e] Refer to [28]

between the theoretical and experimental values can be attributed to the satellites due to multiple ionization, which are not included in the theoretical values [17].

It can be said that, when we are interested only in relative intensities for major transitions or the spectral shapes of X-ray emission, all of the theoretical models can give similar results.

8.5.3 Contributions from Interatomic Transitions

Most of the old theoretical calculations for molecular X-ray emission rates have been performed using the single-center approximation, and contributions from the interatomic transitions have been neglected. On the other hand, Taniguchi and Adachi [29] and Larkins [20] pointed out that in some cases, the contributions from the interatomic transitions are appreciable.

The X-ray emission rate in molecules is given by (8.38). On the other hand, the rate without interatomic transitions can be obtained as follows: the X-ray emission rate is computed with (8.38), but the summation over i in the dipole matrix element is restricted only to the atomic orbitals that belong to the same atom where the initial inner-shell vacancy is created. Then the X-ray emission rate is given by (8.39). This equation is different from the conventional single-center approximation, in which the square of D_λ is expressed as

$$D_{ij}^2 = \sum_k{}' c_{jk}^2 \left|\langle\phi_k|\boldsymbol{r}|\phi_i\rangle\right|^2 , \tag{8.44}$$

where k denotes the atomic orbital in the same atom as ϕ_i.

We show the examples for K X-ray emission rates for C and O atoms in the CO molecule [31]. The results with and without the interatomic transitions and the relative change in the emission rates are listed in Table 8.6. In general, the K X-ray emission rates increase by taking into account the existence of the interatomic transitions. Only one exceptional case is the $5\sigma \rightarrow 1\sigma$

Table 8.6. K X-ray emission rates for the carbon and oxygen in CO using the single-center approximation (I) and the two-center approximation (II) (arbitrary units)

	C			O		
Level	I[a]	II[b]	Δ[c]	I	II	Δ
3σ	3.218	8.157	153.5	21.44	25.66	19.2
4σ	0.484	0.834	70.4	132.9	145.0	9.1
1π	33.04	40.86	23.6	390.8	394.5	3.3
5σ	31.83	38.54	21.2	43.47	43.14	−0.8
Total	68.57	88.43	29.0	588.6	608.2	3.3

[a] Without interatomic contributions
[b] With interatomic contributions
[c] $\Delta = (\mathrm{II} - \mathrm{I})/\mathrm{I} \times 100$

transition in O atom. A similar effect has already been pointed out by Rowland and Larkins [30] in their CNDO/2 (Complete Neglet of Differential Overlap method) calculations.

The increase in the X-ray emission rates is significant for C K X-rays. In this case, the single-center approximation is inadequate to predict the C K X-ray emission process in CO. On the other hand, the interatomic transitions play a less important role in O K X-ray emission.

As examples of more complex molecules, the interatomic contributions to the K X-ray emission rates for chemical compounds of 3d elements were studied [31]. For compounds with octahedral symmetry, such as $CrCl_3$ and $MnCl_2$, the interatomic transitions play a minor role. On the other hand, in the case of compounds with tetrahedral symmetry, such as CrO_3 and $KMnO_4$, the interatomic transitions increase the X-ray emission rates for $4t_2$, $5t_2$, and $6t_2$ components of the 3d transition metals. The $4t_2$ component corresponds to the $K\beta''$ peak and $6t_2$ to the $K\beta_{2,5}$ peak, while the $5t_2$ component has not yet been observed experimentally. These components are generally very weak in comparison with the strong diagram lines. When we measure the $K\beta/K\alpha$ ratios for the chemical compounds of the $3d$ elements, the contributions from the interatomic transitions are less than 0.3% and are almost negligible.

8.6 Chemical Effect of the $K\beta/K\alpha$ Ratios for $3d$ Elements

The chemical effect of the $K\beta/K\alpha$ X-ray intensity ratios has been studied experimentally for various chemical compounds of $3d$ transition metals.

Mukoyama et al. measured the K X-ray spectra for several chemical compounds of Cr and Mn with the high-resolution crystal spectrometer and calculated the $K\beta/K\alpha$ ratios with the DV-Xα method [32]. For simplicity, it is assumed that all compounds are expressed as clusters with tetrahedral (Td) or octahedral (Oh) symmetry, in which the central metal atom is surrounded by four or six ligand atoms, respectively. The compounds and clusters used in the calculations are listed in Table 8.7 together with the crystal symmetry and the bond length between the metal and ligand atoms.

For metals, the $1s$–$4p$ atomic orbitals are used as the basis functions, while the $1s$–$2p$ orbitals are used for the oxygen atom and the $1s$–$3p$ orbitals are used for sulfur and chlorine. In all cases, the Slater statistical exchange parameter was chosen to be $\alpha = 0.7$. The calculations were made for the FR approximation with the GS configuration. The X-ray transition energies were obtained with the TS method.

In Fig. 8.2a, the experimental $K\beta$ spectrum for Cr in K_2CrO_4 [32] is compared with the calculated one. The energy scale is normalized to the experimental value at the $K\beta_{1,3}$ line. The theoretical shape is assumed to be a Lorentzian function. There are two satellite lines in the high-energy side of the $K\beta_{1,3}$ diagram line, which originates from the $3p \rightarrow 1s$ transition in the Cr atom. These satellites can be attributed to the transition from the MO levels. The $K\beta''$ line corresponds to the $4t_2 \rightarrow 1s$ transition and

Table 8.7. Compound, symmetry, cluster, and bond length used in the calculations

Compound	Symmetry	Cluster	Bond length (Å)
K_2CrO_4	Td	CrO_4^{2-}	1.65
$K_2Cr_2O_7$	Td	CrO_4^{2-}	1.70
CrO_3	Td	CrO_4^{2-}	1.80
CrO_2	Oh	CrO_6^{8-}	1.90
Cr_2O_3	Oh	CrO_6^{9-}	2.00
$CrCl_3{\cdot}6H_2O$	Oh	$CrCl_6^{3-}$	2.38
$CrCl_2$	Oh	$CrCl_6^{4-}$	2.55
$KMnO_4$	Td	MnO_4^{-}	1.59
K_2MnO_4	Td	MnO_4^{2-}	1.81
MnO_2	Oh	MnO_6^{8-}	1.89
MnO	Oh	MnO_6^{10-}	2.22
$MnCl_2{\cdot}2H_2O$	Oh	$MnCl_6^{4-}$	2.51
MnS	Oh	MnS_6^{10-}	2.61

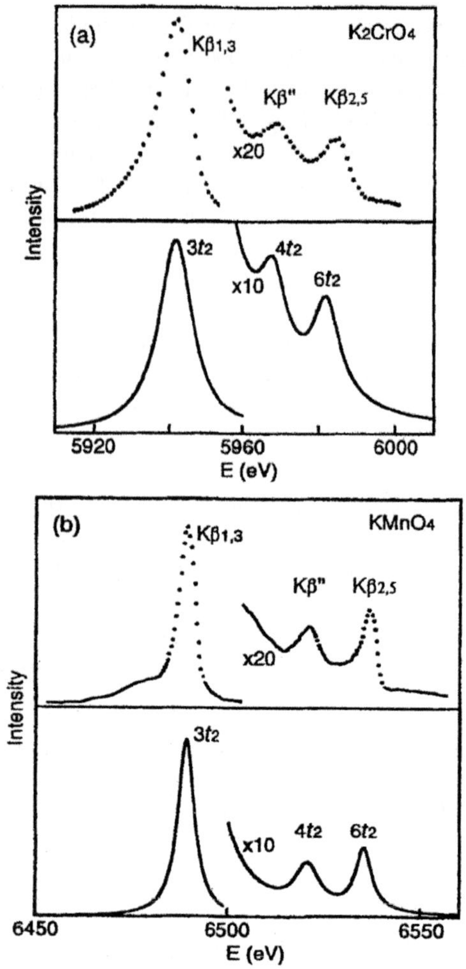

Fig. 8.2. The experimental and calculated $K\beta$ X-ray emission spectra of Cr in K_2CrO_4 and Mn in $KMnO_4$. The assignments of the molecular states are shown. The energy scale is adjusted to the experimental data at the $K\beta_{1,3}$ line

the $K\beta_{2,5}$ line is due to the transition from the $6t_2$ orbital. This assignment is same as that proposed by Best [33]. There is another line from the $5t_2$ orbital in the theoretical calculation, but this line is too weak to be observed experimentally. Fig. 8.2b shows similar experimental and theoretical spectra for Mn in $KMnO_4$ [32]. It can be seen from the figure that the experimental spectra are in good agreement with the theoretical ones.

The intensity calculations were made in the FR approximation with the GS configuration. The calculated $K\beta/K\alpha$ ratio for a single atom is 0.1144 for Cr and 0.1184 for Mn. These values are slightly smaller than the HS values of Manson and Kennedy [14] obtained with a similar model. This discrepancy is attributed to the difference in the choice of the Slater statistical scaling parameter. Manson and Kennedy [14] used the original value of the HS program, $\alpha = 1$, and a larger α value leads to a larger intensity ratio.

On the other hand, according to the relativistic calculations by Scofield [34] the $K\beta/K\alpha$ ratio is 0.1337 for Cr atom and 0.1385 for Mn. These values are in better agreement with the experimental data. He pointed out that neglecting the exchange and overlap effects underestimates the $K\beta/K\alpha$ ratios for $3d$ elements. Considering this fact, it is not favorable to compare the values calculated above directly with the experimental results.

In order to compare with the measured values, the relative $K\beta/K\alpha$ ratios with respect to a standard compound were used. As for the standard compound, K_2CrO_4 was chosen for Cr and $KMnO_4$ for Mn. The theoretical values for other compounds were divided by the theoretical value for the standard compound, while the experimental values for other compounds were also normalized with respect to the measured value for the standard compound. In Table 8.8, the calculated relative $K\beta/K\alpha$ ratios are compared with the relative measured values and calculated values by Band et al. with the multiple-scattering (MS) Xα method [35].

The calculated results are qualitatively in agreement with the experiment data, except for CrO_3. The experimental value for CrO_3 is larger than

Table 8.8. Comparison of calculated and measured $K\beta/K\alpha$ X-ray intensity ratios

		Theory		Experiment	
Compound	Symmetry	DV-Xα[a]	MS-Xα[b]	MTA[a]	
Cr (Relative to K_2CrO_4)					
$K_2Cr_2O_7$	Td	1.020		1.039	1.022±0.021[c]
CrO_3	Td	0.980		1.040	1.029±0.021[c]
CrO_2	Oh	0.949			
Cr_2O_3	Oh	0.952		0.975	0.978±0.020[c]
$CrCl_3 \cdot 6H_2O$	Oh	0.942	0.964	0.994	0.978±0.020[c]
$CrCl_2$	Oh	0.934			
Mn (Relative to $KMnO_4$)					
K_2CrO_4	Td	0.978			0.933±0.021[c]
MnO_2	Oh	0.960		0.973	1.007±0.021[c]
MnO	Oh	0.944			
$MnCl_2 \cdot H_2O$	Oh	0.937			0.971±0.020[c]
MnS	Oh	0.935	0.966		0.950±0.007[d]
					0.953±0.003[e]

[a] Refer to [32]. [b] Refer to [35]. [c] Refer to [36]. [d] PIXE, Refer to [37].
[e] Fluorescence, Refer to [37]

unity, while the calculated value is smaller. In the case of MnO_2, the theoretical value agrees well with the measured value of Mukoyama et al. [32], but is smaller than the experimental value of Tamaki et al. [36]. It should be noted that the experimental studies of Tamaki et al. [36] and of Brunner et al. [37] were performed with poor-energy-resolution Si(Li) detectors. On the other hand, a double-crystal spectrometer with high-energy resolution was employed by Mukoyama et al. [32]. In the latter experiment, the $K\beta''$ and $K\beta_{2,5}$ peaks were separately observed from the diagram line $K\beta_{1,3}$, as shown in Fig. 8.2. The MS-Xα values of Band *et al.* [35] are larger than the DV-Xα values, but only two values are available for the chemical compounds listed in Table 8.8.

From Table 8.8, it is clear that the $K\beta/K\alpha$ ratios of the compounds with Td symmetry are in general larger than those with Oh symmetry. This trend is explained as follows: as can be seen in Fig. 8.2, for compounds with Td symmetry, the satellite lines due to $4t_2$ and $6t_2$ orbitals make considerable contributions to the $K\beta$ intensity, while in the compounds with Oh symmetry, the intensity of the corresponding satellite line from the $5t_{1u}$ orbital in oxides and from the $8t_{1u}$ orbital in sulfides and chlorides is relatively weak.

8.7 Relation between $K\beta/K\alpha$ Ratios and the Number of 3*d* Electrons

The experimental values for the $K\beta/K\alpha$ X-ray intensity ratios in 3*d* elements are often plotted against the formal oxidation number of metals, and it is said that the $K\beta/K\alpha$ ratio increases with increasing oxidation number [38–40]. However, the oxidation number is not a good measure and can be used only for qualitative discussions because it is defined as an integer, and the different $K\beta/K\alpha$ ratios are observed for compounds with the same oxidation number.

On the other hand, as has been shown above, the $K\beta/K\alpha$ intensity ratios for Cr and Mn compounds depend on the crystal symmetry in which the atom of interest is located [32]. Recently Mukoyama et al. [41] have also shown that in the same chemical environment for 3*d* elements, the $K\beta/K\alpha$ ratio following electron-capture decay of radioactive sources is generally smaller than the ratio following photoionization because of an excess of 3*d* electrons in the former case.

The influence of 3*d* electrons on the $K\beta/K\alpha$ ratios has been discussed by Brunner et al. for chemical compounds of Cr, Mn, Fe, and Cu [37]. They pointed out that the influence of chemistry on the $K\beta/K\alpha$ ratio is caused by a change in the screening of 3*p* electrons due to delocalization of the 3*d* electrons and they derived a simple formula to predict the $K\beta/K\alpha$ ratio as a function of the 3*d* share of the valence-charge difference.

These facts indicate that the $K\beta/K\alpha$ X-ray intensity ratios depend on the number of 3*d* electrons as well as the crystal symmetry. However, it may also be possible that these ratios are expressed as a function of other parameters, such as bond length and the effective number of $4p$ electrons. In order to elucidate which parameter is most important, Mukoyama et al. [42] studied the dependence of the $K\beta/K\alpha$ ratio in $3d$ elements on various parameters of chemical compounds.

Unfortunately, most experimental studies on the chemical effect and the excitation-mode-dependence have been performed with solid-state detectors (SSDs). Due to the poor energy resolution of SSD, the fine structures in the $K\alpha$ and $K\beta$ lines cannot be resolved. In addition, most experimental data have been analyzed under the assumption that the $K\alpha$ and $K\beta$ lines are expressed as a single Gaussian function. As was seen above, the chemical effect is sensitive to the intensities of $K\beta_{2,5}$ and $K\beta''$ lines [32], which are less than 10% of the intensity of the $K\beta_{1,3}$ line. Without explicit observation of these peaks, the experimental uncertainty would be large. Actually, many experimental values reported for the same chemical compounds have a considerably large amount of scatter. Considering this situation, it is more favorable to use the theoretical $K\beta/K\alpha$ ratios calculated with the MO method to study the relation with various parameters.

Here, we focus our attention on the $K\beta/K\alpha$ ratios for the Cr compounds with Td and Oh symmetry and consider the case of simple clusters, where a $3d$ transition-metal atom is located at the center and is surrounded by four or six ligand atoms. The excitation mechanisms by electron-capture decay and by photoionization are considered. In the case of electron-capture decay, the atomic charge of the central atom in the cluster of the parent atoms is decreased by one, and the electronic configuration is kept unchanged [41].

The chemical compounds, the clusters used for calculations, the crystal symmetry, and the bond length a between metal and ligand are listed in Table 8.9. The symbol Cr[MnO_4^{2-}] means that the Mn atom in MnO_4^{2-} is replaced by the Cr atom, and this result corresponds to the experimental value for Cr X-rays by the K-capture decay of $^{54}MnO_4^{2-}$.

Table 8.9. Compound, cluster, symmetry, and bond length used for calculations

Compound	Cluster	Symmetry	a (Å)
CrO_4^{2-}	CrO_4^{2-}	Td	1.65
$Cr_2O_7^{2-}$	CrO_4^{2-}	Td	1.70
Cr[MnO_4^{2-}]	Cr[MnO_4^{2-}]	Td	1.81
Cr[MnO_2]	Cr[MnO_6^{8-}]	Oh	1.89
CrO_2	CrO_6^{8-}	Oh	1.90
$CrCl_3$	$CrCl_6^{3-}$	Oh	2.38
Cr[$MnCl_2$]	Cr[$MnCl_6^{4-}$]	Oh	2.51
$CrCl_2$	$CrCl_6^{4-}$	Oh	2.55

In order to find the relation between various parameters and the calculated $K\beta/K\alpha$ ratios, the method of multivariate statistical analysis was employed. For N pairs of data $(x_1, y_1), \ldots, (x_N, y_N)$, the extent of their linear relation can be measured by the correlation coefficient [43]:

$$r = \frac{N\sum_i x_i y_i - \sum_i x_i \sum_i y_i}{\sqrt{N\sum_i x_i^2 - (\sum_i x_i)^2}\sqrt{N\sum_i y_i^2 - (\sum_i y_i)^2}} . \tag{8.45}$$

The value of r ranges from 0, when there is no correlation, to $r = \pm 1$, when y_i can be expressed as a linear function of x_i. We take x_i as the parameter of the ith compound and y_i as its $K\beta/K\alpha$ ratio.

We have already shown [32] that the $K\beta/K\alpha$ ratio depends on the crystal symmetry. To confirm this, we assigned $x_i = 0$ for the compounds with Td symmetry and $x_i = 1$ for those with Oh symmetry, and calculated the correlation coefficient. The result is $r = -0.879$. This fact indicates that the $K\beta/K\alpha$ ratio is strongly correlated with the crystal symmetry, and it is more favorable to classify the compounds into two groups according to their symmetry.

From this result, the compounds were separated depending on whether they had Td or Oh symmetry, and the correlation coefficients were evaluated for other parameters, i.e., bond length, N_{3d}, N_{4s}, and N_{4p}, where N_i denotes the effective number of i-shell electrons. The values of the parameters, the $K\beta/K\alpha$ ratios, and the correlation coefficients are listed in Tables 8.10–8.11.

It can be seen from the tables that N_{3d} is linearly correlated with the $K\beta/K\alpha$ ratio for all cases and is the best parameter. The bond length is also a good parameter, but worse than N_{3d}. Other values, N_{4s} and N_{4p}, are not generally correlated with the intensity ratio.

Figure 8.3 shows the $K\beta/K\alpha$ ratios for Cr plotted against the effective number of $3d$ electrons, N_{3d}. The straight lines were drawn by the least-squares method for values of Td and Oh symmetry, respectively. It is clear from the figure that the calculated results are separated into two groups, i.e., Td and Oh symmetry, and for each symmetry, the $K\beta/K\alpha$ ratios are on a straight line as a function of N_{3d} and decrease with increasing N_{3d}.

The results indicate that $K\beta/K\alpha$ ratios for Cr are given as a linear function of N_{3d} with a negative slope, and the slope of the straight line is different

Table 8.10. $K\beta/K\alpha$ X-ray intensity ratios for Cr compounds with Td symmetry

Compound	a(Å)	N_{3d}	N_{4s}	N_{4p}	K_β/K_α ratio
CrO_4^{2-}	1.65	3.45	0.062	0.284	0.1227
$Cr_2O_7^{2-}$	1.70	3.48	0.123	0.336	0.1222
$Cr[MnO_4^{2-}]$	1.81	3.61	0.158	0.280	0.1201
r	−0.992	−1	−0.879	0.394	

Table 8.11. $K\beta/K\alpha$ X-ray intensity ratios for Cr compounds with Oh symmetry

Compound	a (Å)	N_{3d}	N_{4s}	N_{4p}	K_β/K_α ratio
CrO_2	1.90	3.56	0.251	0.077	0.1172
$Cr[MnO_2]$	1.89	3.59	0.204	0.054	0.1171
$CrCl_3$	2.38	4.21	0.429	0.232	0.1152
$CrCl_2$	2.55	4.42	0.359	0.178	0.1146
$Cr[MnCl_2]$	2.51	5.04	0.175	0.100	0.1127
r	−0.902	−1	0.012	−0.363	

for different crystal symmetries. This fact means that the $3d$ electrons play an important role for the $K\beta$ X-ray emission process, though they do not directly take part in the X-ray emission process, and the screening effect due to $3d$ electrons for $3p$ and other valence electrons is thought to be the dominant reason for the change in the $K\beta/K\alpha$ ratio.

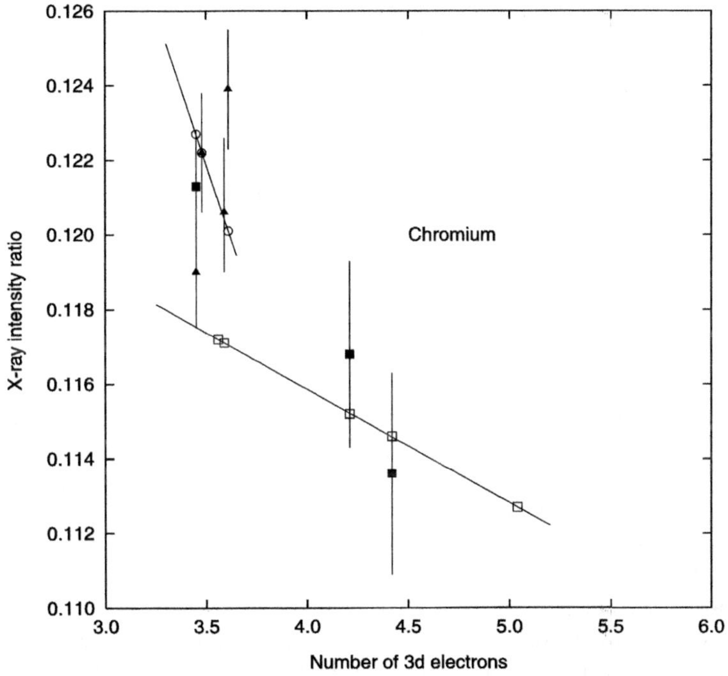

Fig. 8.3. K_β/K_α ratios for chromium compounds as a function of the effective number of $3d$ electrons. The *open circles* represent the theoretical values for Td symmetry, and the *open squares* for Oh symmetry. The experimental data are shown by the *solid triangles* [36] and the *solid squares* [44]

The above results were obtained in the frozen-orbital approximation, i.e., the same atomic potential both for the initial and the final states, and the exchange and overlap effects were not taken into consideration. It is well-known that this approximation underestimates the $K\beta/K\alpha$ ratio for a single isolated atom in the region of $3d$ elements and yields a smaller theoretical value than the experimental one. This fact suggests that the calculated values in the tables are systematically smaller than the experimental data, and direct comparison between the calculated and measured values is difficult.

Furthermore, as has been pointed out above, most of the experimental data are not very reliable because of the poor energy-resolution of SSDs and the rough method of data analysis. However, by normalizing the experimental data with respect to a certain chemical compound, in a manner similar to that outlined in the previous section, it is possible to compare the calculated results with the corresponding experimental ratios. In Fig. 8.3, the measured values of Tamaki et al. [36] and by Küçükönder et al. [44] are plotted. The experimental values were normalized with respect to $Cr_2O_7^{2-}$. Although the measured values scatter and their experimental errors are large, it can be said that the general trend of the experimental data is in agreement with the calculated results. Similar results are also found for vanadium compounds [42].

From the results obtained above, it is concluded that when the calculated $K\beta/K\alpha$ X-ray intensity ratios are plotted against the effective number of $3d$ electrons, they are on a straight line and decrease with increasing N_{3d}. The slope of the straight line is different depending on the symmetry of the cluster, and the slope for Td symmetry is more steep than that for Oh symmetry. This fact indicates that the screening effect of $3d$ electrons on $3p$ and other valence electrons plays a dominant role in the $K\beta/K\alpha$ ratio for $3d$ elements. Using these results, we can estimate the effective number of $3d$ electrons from the $K\beta/K\alpha$ ratio and vice versa.

8.8 Summary

General expressions for X-ray emission and absorption were given on the basis of the dipole approximation. The dipole matrix element for atoms was obtained, and the X-ray emission rate and absorption cross-section for atoms were given. Theoretical models for molecular X-ray emission rates were described, and the advantages of the DV-Xα method over other molecular orbital methods were discussed. Various approximations used in the calculations for molecular X-ray emission were tested with the DV-Xα method. The $K\beta/K\alpha$ X-ray intensity ratios for chemical compounds of $3d$ elements were calculated using the DV-Xα method, and the chemical effect on the $K\beta/K\alpha$ ratios was studied.

In the examples shown here, the calculations were performed only for simple small clusters and for a limited number of symmetries. However, the advantage of the DV-Xα method is that the method can be easily applied to larger clusters and to compounds with more complex geometry. This feasibility is due to the fact that numerical wave functions are used as basis functions, and the multi-center integration in the dipole matrix element calculation is done by the use of the DV integration method.

There is a problem in cluster calculations for solids: the periphery effect of clusters. In general, the clusters are not neutral, but positively or negatively charged. In addition, the outermost atoms in the cluster are not completely bonded. This effect may cause distortion of theoretical X-ray and electron spectra. In order to improve termination and electroneutrality of the cluster, the chemically complete cluster method has been developed within the framework of the DV-Xα method and successfully applied for the analysis of X-ray emission spectra and photoelectron spectra in solids [45].

Owing to recent advances in X-ray and electron spectrometers and the advent of third-generation synchrotron radiation facilities, it is possible to obtain high-resolution X-ray and electron spectra for molecules and solids with relative ease. The DV-Xα method is quite efficient for the calculation of theoretical spectra to compare with these experimental data and to obtain information about the electronic structures of materials.

References

1. A. Meisel (ed.), *Röntgenspektren und Chemische Bindung* (Karl-Marx-Universität, Leipzig, 1966)
2. R.L. Barinski, W.I. Nefedov, *Röntgenspektroskopische Bestimmung der Atomladungen in Molekülen* (Geest & Portig, Leipzig, 1969)
3. K. Siegbahn, C. Nordling, G. Johansson, J. Hedman, P.E. Hadén, K. Hamrin, U. Gelius, T. Bergmark, L.O. Werme, R. Manne, *ESCA Applied to Free Molecules* (North-Holland, Amsterdam 1969)
4. A. Meisel, G. Leonhardt, R. Szargan, *Röntgenspektren und Chemische Bindung* (Geest & Portig, Leipzig, 1977); *X-Ray Spectra and Chemical Binding* (Springer, Berlin Heidelberg New York, 1989)
5. H. Adachi, M. Tsukada, C. Satoko, J. Phys. Soc. Japan **45**, 875 (1978)
6. F.W. Averill, D.E. Ellis, J. Chem. Phys. **59**, 6412 (1973)
7. H. Adachi, Adv. Quantum Chem. **37**, 1 (2001)
8. H. Nakamatsu, T. Mukoyama, H. Adachi, J. Chem. Phys. **95**, 3167 (1991)
9. H. Nakamatsu, Chem. Phys. **200**, 49 (1995)
10. H. Adachi, Adv. Quantum Chem. **29**, 179 (1997)
11. W. Heitler, *The Quantum Theory of Radiation* (Clarendon, Oxford, 1954)
12. M.E. Rose, *Elementary Theory of Angular Momentum* (Wiley, New York, 1957)
13. H.A. Bethe, E.E. Salpeter, *Quantum Mechanics of One- and Two-Electron Atoms* (Plenum, New York, 1977)
14. S.T. Manson, D.J. Kennedy, At. Data Nucl. Data Tables **14**, 111 (1974)
15. R. Manne, J. Chem. Phys. **52**, 5733 (1970)
16. Å. Støgård, Chem. Phys. Lett. **36**, 357 (1975)
17. H. Ågren, J. Nordgren, Theor. Chim. Acta **58**, 111 (1981)
18. H. Ågren, R. Arneberg, Phys. Scr. **28**, 80 (1983)

19. H. Ågren, R. Arneberg, J. Müller, R. Manne, Chem. Phys. **83**, 53 (1984)
20. F. Larkins, In, *X-Ray and Inner-Shell Processes*, ed. by T.A. Carlson, M.O. Krause, S.T. Manson (AIP, New York 1990) p. 817
21. T. Mukoyama, K. Taniguchi, H. Adachi, Bull. Inst. Chem. Res., Kyoto Univ. **62**, 13 (1984)
22. F. Herman, S. Skillman, *Atomic Structure Calculations* (Prentice-Hall, Englewood Cliffs, 1963)
23. R. Latter, Phys. Rev. **99**, 510 (1955)
24. J.C. Slater, *Quantum Theory of Molecules and Solids*, Vol. 4 (McGraw-Hill, New York, 1974)
25. T. Mukoyama, H. Adachi, Bull. Inst. Chem. Res., Kyoto Univ. **65** 23 (1987)
26. T. Mukoyama, Bull. Soc. Discrete Variational Xα **9**, 41 (1996)
27. R.A. Phillips, F. Larkins, Aust. J. Phys. **39**, 717 (1986)
28. L.O. Werme, J. Nordgren, H. Ågren, C. Nordling, K. Siegbahn, Z. Phys. **A272**, 131 (1975)
29. K. Taniguchi, H. Adachi, J. Phys. Soc. Japan **49**, 1944 (1980)
30. T.W. Rowlands, F.P. Larkins, Theor. Chim. Acta **69**, 525 (1986)
31. T. Mukoyama, K. Taniguchi, H. Adachi, Phys. Rev. B **41**, 8118 (1990)
32. T. Mukoyama, K. Taniguchi, H. Adachi, Phys. Rev. B **34**, 3710 (1986)
33. P.E. Best, J. Chem. Phys. **44**, 3248 (1966)
34. J.H. Scofield, Phys. Rev. A **9**, 1041 (1976)
35. I.M. Band, A.P. Kovtun, M.A. Listengarten, M.B. Trghaskovskaya, J. Electron Spectrosc. Relat. Phenom. **36**, 59 (1985)
36. Y. Tamaki, T. Omori, T. Shiokawa, Radioanal. Lett. **20**, 255 (1975); *ibid.* **37**, 39 (1979); Jpn. J. Appl. Phys. **17**, S245 (1978)
37. G. Brunner, M. Nagel, E. Hartmann, E. Arndt, J. Phys. B, At. Mol. Phys. **15**, 4517 (1982)
38. B. Möser, Cryst. Res. Technol. **20**, 1503 (1985)
39. A. Küçükönder, Y. Şahin, E. Büyükkasap, J. Radioanal. Nucl. Chem. **170**, 125 (1993)
40. L. Rebohle, U. Lehnert, G. Zschornack, X-Ray Spectrom. **25**, 295 (1996)
41. T. Mukoyama, K. Taniguchi, H. Adachi, Adv. Quantum Chem. **37**, 139 (2001)
42. T. Mukoyama, K. Taniguchi, H. Adachi, X-Ray Spectrometry **41**, 8118 (2000)
43. J.R. Taylor, *An Introduction to Error Analysis* (University Science Books, Sausalito, 1997)
44. A. Küçükönder, Y. Şahin, E. Büyükkasap, A. Kopya, J. Phys. B, At. Mol. Opt. Phys. **26**, 101 (1993)
45. B. Song, H. Nakamatsu, R. Sekine, T. Mukoyama, K. Taniguchi, J. Phys., Condens. Matter **10**, 9443 (1998)

9

Response to the Creation of a Core Hole in Transition-Metal Compounds

Jun Kawai

9.1 Core-hole Spectroscopic Techniques

X-ray fluorescence spectroscopy (XRF) [1,2], X-ray photoelectron spectroscopy (XPS, ESCA) [3], Auger electron spectroscopy (AES), and X-ray absorption fine-structure (XAFS) spectroscopy or X-ray absorption near-edge structure (XANES) spectroscopy [4] are extensively used to investigate the electronic structure of materials. In these spectroscopic methods, the electronic structures are usually measured as a response to the sudden creation of a core hole. Consequently, the electronic structures observed through these core-hole spectroscopic techniques are different from those of the ground states of the materials. However, we usually approximate the observed spectra as representing the ground state electronic structure of the system. This is because the relaxation and the correlation energies cancel each other out [5]. For example, in the $1s^{-1}$ core hole state, the core orbital binding energy becomes deeper by a finite value due to the lack of screening previously provided by the photoionized electron. This energy is known as the relaxation energy. The remaining $1s$ electron, therefore, is more strongly attracted by the positive nuclear charge than before the other $1s$ electron was ionized. The correlation energy is the electron–electron repulsion energy, which becomes absent after one of the $1s$ electrons is ionized. The relaxation and correlation energies are approximately $10 - 20\,\mathrm{eV}$ and have opposite sign; therefore, we can neglect the presence of the core hole in these spectroscopic techniques, especially in the analysis of valence electronic structure. However, the effect of a core hole on the valence electron configuration is sometimes important in certain kinds of materials, and such materials have particularly useful properties due to the very effect of the change of valence electron configuration by a small perturbation such as core-hole creation.

In XPS, an X-ray photon is absorbed by the system, and according to a time-dependent perturbation, i.e., the interaction between a photon and an electron, the core electron is photoionized. We measure via XPS the kinetic energy of the ionized electron in vacuum. Since we know the incident photon energy, we can obtain information on the binding energy of the electron before photoionization, using energy conservation. The binding energy of the core electron is modified by the chemical environment of valence electrons, and the binding energy is shifted by a few eV even for the core orbitals. The observed binding energy is also modified by many-electron effects, such as relaxation and correlation effects. However, they cancel each other out, as mentioned above. This is the reason we usually assume that the binding energy measured by XPS experiments is that of the ground state, i.e., we make the one-electron approximation. The electronic binding energy information we get from an XPS experiment is almost enough to characterize the material.

The electronic system of the material as a whole changes its total energy when it loses an electron by photoionization. The exact electron binding energy is the difference of the total energy before and after photoionization. The total energy is an eigenvalue of the system, and is the sum of all of the one-electron energies (including the nucleus–electron attractive energy) and the two-electron energy. The two-electron energy is the sum of the repulsion energy and exchange energy of all of the combinations of any two electrons in the system. We usually get information on the electronic structure of materials, especially on the valence electronic structure, by the response of electrons to the sudden creation of an electron hole in a core orbital in XPS. A Green's function [6] is a response function to the sudden creation of a core hole, i.e., a response to a unit test charge in a system. Once the response function is known, the response of valence electrons of the system to any types of input of the same order or stronger can be calculated by the convolution of the Green's function and the input. XPS is, in a sense, a method to measure experimentally the Green's function of the system, because the core hole creation is approximated to be a delta function or impulse; the Green's function is defined as the response function of a system to the impulse of the delta function.

Though the meaning of the electronic structure of the ground state is easy to understand, the meaning of the core-hole state is slightly complicated. The core-hole state is, in a sense, the final state of XPS, but a more sophisticated interpretation is that it is the Green's function of the system. However, we must clarify the difference between a sudden response and an adiabatic response of valence electrons to the creation of a core hole. The sudden creation of a core hole [7] means that the time required to photoionize a core electron is infinitely short; the adiabatic response [8] means that the core electron is photoionized over an infinitely long time, and the speed of the photoelectron is slow. On the other hand, sudden core-hole creation occurs in the case of photoionization with a high enough incident photon energy and the velocity of the photoelectron is fast. Adiabatic response occurs during threshold excitation using synchrotron radiation [9–12] or electron capture decay [13–15]. Photoionization using threshold X-ray energy will reduce the probability of double-electron photoionization due to shake-off or Coster–Kronig transitions [16], but the charge-transfer effect cannot be avoided [17].

The electron capture decay is a method to annihilate the $1s$ core electron by an Fe $\rightarrow$ Mn nuclear conversion. Consequently, the potential energy felt by the valence electron before and after the conversion is mostly the same, and the Mn $K\alpha$ and $K\beta$ X-ray fluorescence spectra we observe are pure X-ray fluorescence spectra without the shake-off effect.

Also under consideration are the line shape and intensity ratio of the $L\alpha_{1,2}$ and $L\beta_1$ X-ray fluorescence lines. The $L\alpha_{1,2}$ and $L\beta_1$ X-ray fluorescence lines are due to electron transitions as follows: $L\alpha_1$ is due to the electron transition $3d_{5/2} \rightarrow 2p_{3/2}$, $L\alpha_2$ is due to $3d_{3/2} \rightarrow 2p_{3/2}$, and $L\beta_1$ is due to $3d_{3/2} \rightarrow 2p_{1/2}$ [18] as shown in Fig. 9.1. The intensity can be estimated by the degeneracy of the one-hole transition, and it is 9:1:5 if the energy interval of $2p_{1/2}$ and $2p_{3/2}$ is negligibly small. Here, we have also neglected a small contribution of the $4s \rightarrow 2p$ transition.

Chemical or solid state effects on the $L\alpha$ and $L\beta$ X-ray fluorescence lines have been known for many years [17–24]. Chemical effects are complicated because of the self-absorption effect, where the characteristic X-rays emitted from deep within a solid are absorbed by other atoms in the solid before detection. The L_2 and L_3 X-ray absorption edges are situated at the high-energy shoulder of the $L\beta$ and $L\alpha$ main peaks, respectively. Thus, the line shape and the intensity ratio change as the take-off angle of X-ray emission from the solid surface, concentration of the X-ray-emitting element, and electron acceleration potential change if the characteristic X-rays are excited by an electron beam. Though it is very difficult, once the self-absorption effect is carefully removed, we still observe several chemical effects on the shape of the spectra (e.g., in copper compounds, as is depicted in Fig. 9.2 [21]) due to the electronic structure modifications of the materials as follows:

1. The $L\beta$ intensity relative to the $L\alpha$ intensity is stronger for Cu(II) compounds than for Cu(I) compounds.

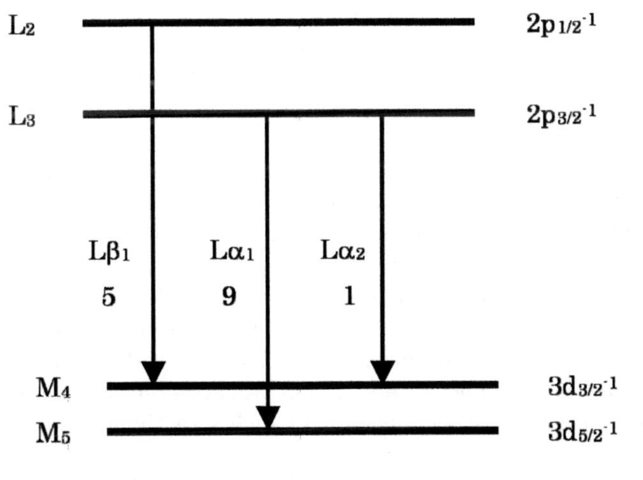

Fig. 9.1. Electron transitions associated with the $L\alpha$ and $L\beta$ X-ray fluorescence lines

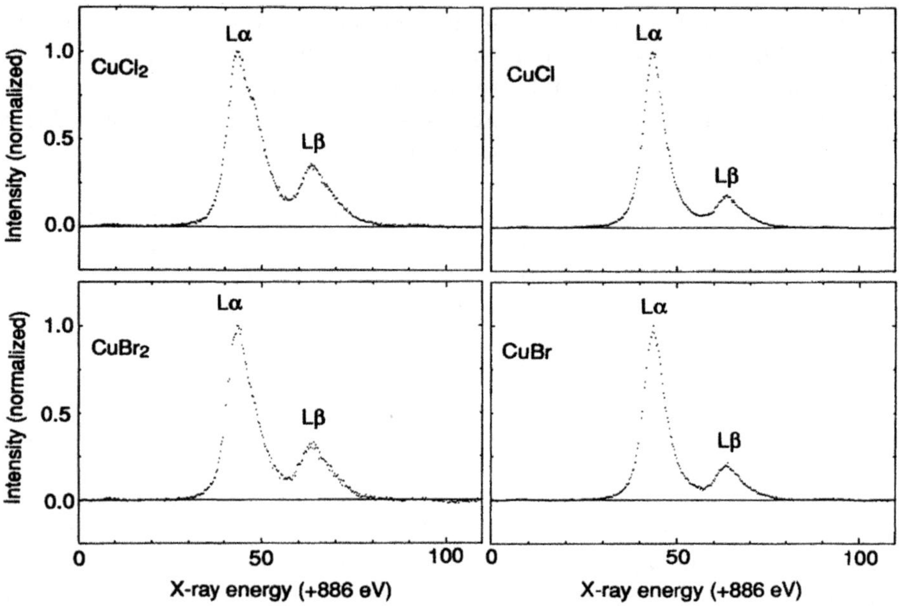

Fig. 9.2. Measured Cu $L\alpha,\beta$ spectra of divalent and monovalent copper compounds (reproduced from [21])

2. A strong hump exists at the $L\alpha$ high-energy shoulder of Cu(II) compounds but is weaker or almost absent for Cu(I) compounds. A similar tendency is observable at the high-energy shoulder of the $L\beta$ line.

The chemical effect on the high energy hump of $L\alpha$ and $L\beta$ is explained by Kawai et al. [17,18,21,25] as follows: The $2p$ XPS spectra of divalent copper compounds, which have a $3d^9$ valence electron configuration in the ground state, have a main peak, which is attributed to the $2p^{-1}3d^{10}$ well-screened final state, and have high-energy satellites, which are attributed to the $2p^{-1}3d^9$ poorly-screened state or the charge-transfer final state [26]. The L X-ray fluorescence spectral main line of divalent copper compounds corresponds to the well-screened state of XPS and the high energy hump to the poorly-screened $2p^{-1}$ hole state, as shown in Fig. 9.3 [17].

An objection was raised by Ohno [27] based on copper metal $L\alpha,\beta$ spectra at threshold excitation, which was heavily smeared by the self-absorption effect. Ohno proposed that the main source of the high-energy hump of divalent copper compounds was due to the Coster–Kronig transition. Ohno did not consider the charge-transfer effect, and he thought that the metal and divalent copper compounds were similar spectral profiles based on an atomic point of view. However Kawai et al. [28] found evidence to contradict Ohno's proposition using a DV-Xα calculation; an additional valence hole, if created by the Coster–Kronig transition or other mechanism such as a shake-off pro-

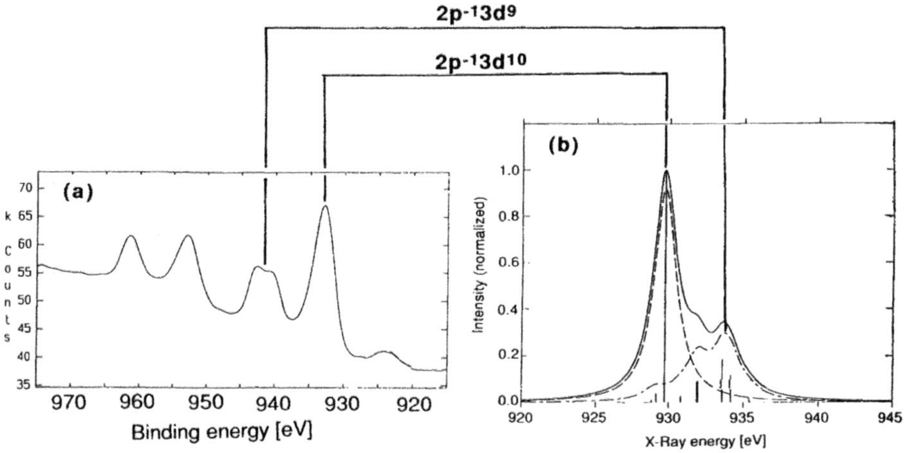

Fig. 9.3. Relation between Cu $2p$ XPS and Cu $L\alpha$ X-ray fluorescence spectra of divalent copper compounds, reproduced from [17]

cess, would be delocalized by the charge-transfer effect through the strong covalence tentatively formed by the core hole. This tentative covalence will be described later. Again, Magnuson et al. [29] raised doubts concerning the charge-transfer mechanism put forth by Kawai et al. [17, 28] based on the threshold excitation measurement of the $L\alpha$ spectra of copper metal. Magnuson et al. stated that their spectrum was not smeared by the self-absorption effect. The $L\alpha$ main line and the hump intensities observed by Magnuson et al. were 1.00 and 0.69, respectively, and the $L\beta$ main line and the hump intensities were 0.22 and 0.14, respectively, consequently the $L\beta/L\alpha$ intensity ratio observed by Magnuson et al. was 0.213. However, for metal, the $L\beta/L\alpha$ intensity ratio free from the self-absorption effect was measured by Kawai et al. [18] to be 0.27 – 0.28. Thus, the copper spectrum of Magnuson et al. was still heavily smeared by the self-absorption effect. Most importantly, Magnuson et al. should have measured the divalent copper compounds in place of copper metal, if they hoped to disprove the charge-transfer mechanism proposed by Kawai et al. We have calculated many kinds of solid systems with and without the presence of a core hole ($1s^{-1}$ hole), using the DV-Xα method and cluster approximation. Here, a cluster approximation means that an infinite solid is approximated by a central atom, where a core hole is created, and its surrounding atoms. The central and its surrounding atoms form a cluster molecule.

To calculate the electronic structure of a molecule or cluster using a molecular orbital method such as the DV-Xα method [30] requires converging the electron field into the minimum energy state. Thus, the calculated results of the self-consistent field (SCF) method is equivalent to the adiabatic limit. The XPS final state is between the sudden and adiabatic limits, but it is closer to the sudden limit; the speed of a photoelectron is fast enough in typical experimental conditions. The sudden limit is an intermediate between the ground-state SCF (but where the core electron is annihilated) and core-hole SCF results (adiabatic limit). The sudden limit wave function is thus expressed by a linear combination of the ground state (after the core electron is annihilated) and core-hole SCF result. The final state measured by XPS is sometimes closer to the ground state and sometimes closer to the core hole state.

9.2 Ionic Chemical Bond as a Perturbation of Atomic Structure

The electronic structure of an ionic solid is expressed as a perturbation of the atoms. Suppose a two-atom molecule, e.g., potassium fluoride K–F, with an ionic chemical bond, as shown in Fig. 9.4, forms two molecular orbitals, i.e., antibonding ϕ_a and bonding ϕ_b orbitals.

The results of the time-independent perturbation theory are summarized as follows:

$$\text{The 0th order energy} = E_n^0 \,. \tag{9.1}$$

$$\text{The 0th order wave function} = u_n^0 \,. \tag{9.2}$$

$$\text{The 1st order energy, } E_n' = \langle u_n^0 | H' | u_n^0 \rangle \,. \tag{9.3}$$

$$\text{The 1st order wave function, } u_n = u_n^0 + \sum_{k \neq n} \frac{\langle u_k^0 | H' | u_n^0 \rangle}{E_n^0 - E_k^0} u_k^0 \,. \tag{9.4}$$

$$\text{The 2nd order energy, } E_n^{(2)} = \sum_{k \neq n} \frac{|\langle u_k^0 | H' | u_n^0 \rangle|^2}{E_n^0 - E_k^0} \tag{9.5}$$

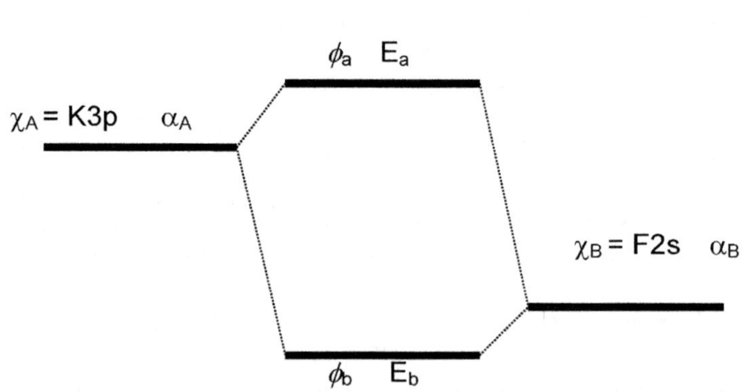

Fig. 9.4. Molecular orbital formation between potassium $3p$ and fluorine $2s$ atomic orbitals. Other atomic orbitals are omitted because they are energetically separated from these two atomic orbitals

Suppose that the molecular orbital, ϕ_a, and its energy, E_a, in Fig. 9.4, is mostly determined by the atomic orbital χ_A, which is close to the molecular orbital ϕ_a, then they are expressed using (9.1)–(9.5) as

$$E_a = \alpha_A + \frac{\beta^2}{\alpha_A - \alpha_B} \tag{9.6}$$

and

$$\phi_a = \chi_A - \frac{|\beta|}{\alpha_A - \alpha_B}\chi_B \, . \tag{9.7}$$

In the same way, for the bonding orbital ϕ_b, we obtain

$$E_b = \alpha_B - \frac{\beta^2}{\alpha_A - \alpha_B} \tag{9.8}$$

and

$$\phi_b = \chi_B + \frac{|\beta|}{\alpha_A - \alpha_B}\chi_B \, . \tag{9.9}$$

Here, β is the transfer matrix element of the Hamiltonian of the system. In these expressions, we omitted the overlap integral S, because we have supposed that the system is ionic, and thus, S is small – for instance, less than 0.1. β is usually small for an ionic system – for instance, less than 0.1 a.u. In the above derivation, we obtained (9.6)–(9.9) by an inspection. However, these expressions can be obtained in a more rigorous manner. The secular equation of the two-level molecule is

$$\begin{vmatrix} \alpha_A - E & \beta - ES \\ \beta - ES & \alpha_B - E \end{vmatrix} = 0 \, . \tag{9.10}$$

The Hamiltonian of the system is thus

$$H = \begin{pmatrix} \alpha_A & \beta - ES \\ \beta - ES & \alpha_B \end{pmatrix} .$$

If we decompose this Hamiltonian into two parts as follows,

$$H = H_0 + H' = \begin{pmatrix} \alpha_A & 0 \\ 0 & \alpha_B \end{pmatrix} + \begin{pmatrix} 0 & \beta - ES \\ \beta - ES & 0 \end{pmatrix} , \tag{9.11}$$

then

$$E_a = \alpha_A + \frac{(\beta - ES)^2}{\alpha_A - \alpha_B} = \alpha_A + \frac{(\beta - \alpha_A S)^2}{\alpha_A - \alpha_B} , \tag{9.12}$$

$$E_b = \alpha_B - \frac{(\beta - ES)^2}{\alpha_A - \alpha_B} = \alpha_B - \frac{(\beta - \alpha_B S)^2}{\alpha_A - \alpha_B} , \tag{9.13}$$

$$\phi_A = \chi_A - \frac{|\beta - \alpha_A S|}{\alpha_A - \alpha_B}\chi_B , \tag{9.14}$$

$$\phi_B = \chi_B + \frac{|\beta - \alpha_B S|}{\alpha_A - \alpha_B}\chi_A \, . \tag{9.15}$$

9.3 Covalent Bond Formation Due to a Core Hole

When a core hole is created in the potassium 1s orbital in Fig. 9.4, since the electron–electron Coulomb repulsion energy is approximately 10 – 15 eV, the one-electron energy of the K-3p orbital will become deeper by 10 – 15 eV, as shown in Fig. 9.5. However, because the energy separation between the K-3p and F-2s orbitals is less than a few eV, the energy ordering of K-3p and F-2s becomes inverted [31].

In an intermediate state between the ground state and the core hole state, it is possible that the K-3p and F-2s energies are incidentally degenerated. In this case, since the energy denominator in (9.12)–(9.15) becomes zero, a very strong covalent bond is formed between the K-3p and F-2s orbitals by resonance. After photoionization occurs, the new orbital energy and eigenfunction, α_A^* and χ_A^*, respectively, are sometimes close to α_B and χ_B.

The main component of a molecular orbital happens to change during such a level inversion. The anti-bonding orbital is mostly due to the K-3p orbital in the ground state, but it becomes F-2s in the K-1s^{-1} core hole state. The interpretation of the XPS spectra, the Auger electron spectra, and X-ray fluorescence and absorption spectra are always very complicated for such a level inversion. If both of the level-crossing orbitals are fully occupied by electrons, then the situation is not complicated, but if one of them is partially filled, then the hole in a valence orbital will be transferred to the central atom from its neighbor atoms by the level inversion.

Several examples of partially filled orbital crossings have been found to be associated with the photoionization shake-off phenomena as follows: Urch [32], Benka et al. [33], and Hartmann et al. [34,35] found in their analyses of chemical effects on the intensity of F $K\alpha$ satellites, such as the $KL-L^2$ transition due to shake-off or heavy ion collision, that the covalency between the F-2p and the ligand level-crossing is significant. Kawai et al. found several examples

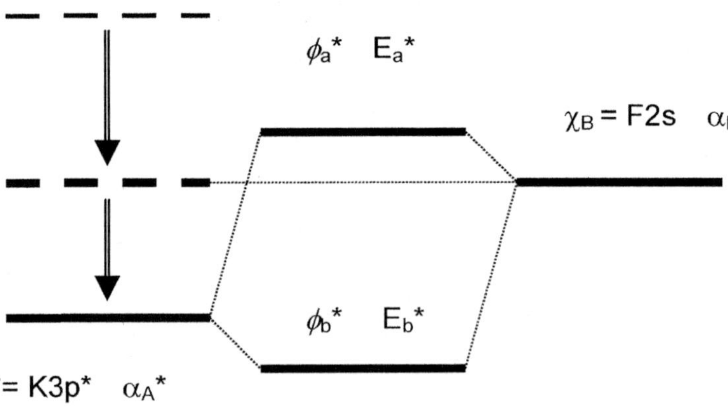

Fig. 9.5. Time-dependent deepening of the potassium 3p orbital due to the creation of a 1s core hole

of level-crossing for K $K\alpha$ [31], Cl $K\alpha$ [36,37], Ca $K\alpha$ [38], and Ti $K\alpha$ [39]. The most important feature is that such a level inversion involving a partially filled valence orbital frequently occurs in useful materials, such as superconducting copper oxides, transition-metal borides (TiB_2 [39]), transition-metal catalysts, and rare-earth compounds. To seek a level–inversim is a short cut to find a useful material. The source of useful properties for us is the presence of energetically close atomic orbitals, one of which is not fully occupied; the hole moves easily from one site to another by a small perturbation rather than a core-hole creation.

9.4 Charge Transfer Due to a Core Hole

When two levels are crossing each other, as shown in Fig. 9.6, if the central atomic orbital, say K-$3p_z$, t_{1u}, has a different symmetry block from that of the linear combination of the surrounding orbitals, then they cross but do not interact with each other. Here, the words "symmetry block" mean that within t_{1u} symmetry, even the x, y, or z directions should coincide between atoms A and B. This is because the transfer matrix element $\beta = \langle \chi_A | H' | \chi_B \rangle$ equals zero when χ_A belongs to a different symmetry block than χ_B. By contrast, when the two atomic orbitals belong to the same symmetry block, they interact each other, i.e., they form a molecular orbital. Through this molecular orbital, electrons (or complementarily, holes) are transferred to another site, if there is room. The interaction of the two atomic orbitals will split the degenerate orbital eigenvalue into two different levels. The energy gap of this interaction is 2β. The so-called "avoided crossing" takes place [40, 41]. We often encounter this kind of avoided-crossing in chemical reactions or atomic collisions. In chemical reactions, the time of crossing is long enough, but in photoionization, the time-scale of the level-crossing is very short (almost sudden) compared with adiabatic limit. Some of the electrons remain in the same atomic orbital because they have no chance to move or be rearranged during the level-crossing. In this case, the molecular orbital profile (number of holes belonging to an atom) does not change.

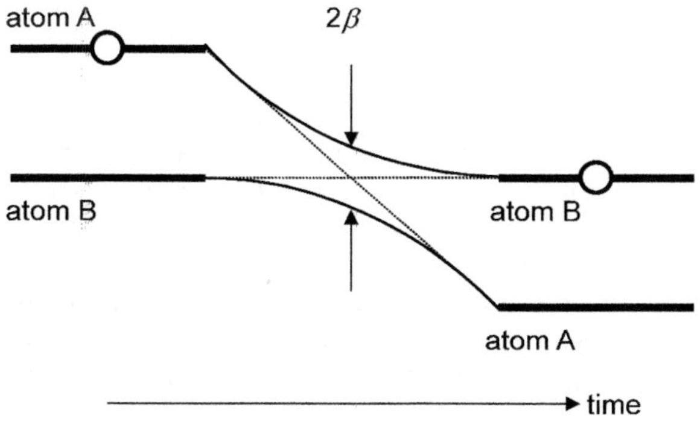

Fig. 9.6. Level-crossing of two atomic orbitals after the creation of a core hole

The problem is to determine the ratio of charge transfer to non-transfer after the photoionization. Once this ratio is known, we can estimate the responses to any input to the system, even with large perturbations, by Auger electron spectra, X-ray absorption spectra, or X-ray fluorescence spectra. Such a level crossing frequently occurs in transition-metal and rare-earth compounds [42–45] in XPS experiments. The avoided crossing and subsequent charge-transfer effect is also important for interpreting the X-ray fluorescence spectra of transition-metal compounds.

9.5 Calculation Details

The author and his colleagues have calculated the valence electron density of states for the ground state and the $1\mathrm{s}^{-1}$ core hole state for many kinds of transition-metal and rare-earth compounds. However, the DV-Xα calculations were first investigated to ascertain the impact of the following:

1. Cluster size
2. Difference between $1s\downarrow^{-1}$ and $1s\uparrow^{-1}$ hole states
3. Difference between $1s^{-1}$ and $2p^{-1}$ hole states
4. Effect of bond length difference

In what follows, these calculation details as well as the results of the calculations are described.

9.5.1 Cluster Size

The calculations for a molecule are usually not difficult because we can calculate the electronic structure of a whole molecule. However, for solids, it is sometimes difficult to calculate an electronic structure of infinite size without some approximations. The most frequently used approximation for solids is the assumption of periodic boundary conditions. In this approximation, the calculation is usually performed on a small cluster rather than on a solid of infinite size, or in Fourier space. Though this kind of calculation seems to yield quite an exact solution, it is still a crude approximation, because if we examine a gold crystal, for example, the electronic structure cannot be reproduced by only one gold atom cluster with periodic boundary conditions. The situation becomes much more complicated with an atom with a core hole, because it is considered an impurity atom that is different from the surrounding atoms in the solid. To solve this kind of an impurity problem, it is appropriate to use the cluster molecule model and to express the molecular orbital (MO) by a linear combination of atomic orbitals (LCAO). The meaning of the LCAO-MO results is straightforward, and it is easy to understand the value of the molecular orbital coefficients. However, there exists difficulties in how to cut the cluster from the infinite solid and how to define the boundary (and the corresponding electric charge, for example) of the cluster. To estimate these conditions, various kinds of clusters are calculated and compared with the X-ray fluorescence spectra [46] (A review has been published on the molecular orbital calculation of X-ray fluorescence spectra in Japanese [47]). The conclusion of these studies is summarized as follows:

1. If we need X-ray spectra of a cation in a solid, it is enough to consider a cluster made of a central cation and nearest-neighbor anions.
2. If we need X-ray spectra of an anion in a solid, one must consider a cluster made of the central anion, its nearest-neighbor cations, and the second nearest-neighbor atoms (anions). This is because the overlap integral between anions is not negligibly small, and their energy is degenerate.
3. The charge of the cluster ion should be equal to the sum of the formal charge of each atom in the cluster.
4. The boundary of the cluster is left as dangling bonds and need not to be terminated by hydrogen or other atoms.
5. If we want to calculate valence X-ray spectra such as the $K\alpha$ spectra of boron, carbon, nitrogen, or oxygen, where $K\alpha$ X-ray lines are emitted by the electron transition from the valence $2p$ to the $1s$ orbital, or the $K\beta$ spectra of Mg, Al, Si, P, S, or Cl, where the $K\beta$ spectra are emitted by the $3p \to 1s$ transition, the $2p$ or $3p$ DOS of the ground state of the cluster is enough to reproduce the X-ray spectral profiles.

However, if we need to know the response to the core-hole calculation, we must consider a cluster with a core hole state; if we need to know the transition energy, we must perform calculations on the half-electron state, or the Slater's transition state.

These criteria do not always hold, but are a good starting point to calculate the X-ray spectra of solids using the cluster molecule approximation.

9.5.2 Difference Between $1s{\downarrow}^{-1}$ and $1s{\uparrow}^{-1}$ Hole States

The core-hole state does not need to be calculated to reproduce the valence X-ray profiles. However, if we need to know the response of valence electrons to the creation of a core hole, we must calculate the core-hole state. The $1s^{-1}$ core hole has two spin configurations due to $1s{\downarrow}^{-1}$ or $1s{\uparrow}^{-1}$ electron photoionization. If the material we are studying has no extra spin in the valence orbital, then these two hole states are not different from each other. If the material has an incomplete valence shell such as in $3d$ transition-metal compounds, there is a difference between the effect of $1s{\downarrow}^{-1}$ versus $1s{\uparrow}^{-1}$ electron photoionization. We calculated the valence electron density of states of $[FeO_6]^{10-}$ and $[FeO_6]^{9-}$ for the ground state and for the $1s{\downarrow}^{-1}$ and $1s{\uparrow}^{-1}$ hole states, and the results reveal that the difference between $1s{\downarrow}^{-1}$ and $1s{\uparrow}^{-1}$ hole states is negligible, as is shown in Table 9.1 [48].

Table 9.1. Calculated effective number of valence electrons of the $[FeO_6]^{10-}$ cluster [48]

	Ground state	$1s\uparrow^{-1}$hole state	$1s\downarrow^{-1}$ hole states
$3d\uparrow$	5.00	5.00	5.00
$3d\downarrow$	1.20	1.78	1.78
$4s\uparrow$	0.00	0.07	0.07
$4s\downarrow$	–0.03	0.04	0.03
$4p\uparrow$	–0.05	0.04	0.04
$4p\downarrow$	–0.03	0.02	0.02
Sum	6.09	6.95	6.94

Therefore, in what follows, we shall mostly consider the $1s\uparrow^{-1}$ hole state.

9.5.3 Difference Between $1s^{-1}$ and $2p^{-1}$ Hole States

The $K\alpha$ X-ray fluorescence process involves the $1s$ and $2p$ hole states. Thus, we must perform calculations for both of these core hole states. However, usually the difference between these two core hole states is negligibly small. This is due to the fact that, for a $3d$ transition-metal atom, both the $1s$ and $2p$ orbitals are core orbitals and are strongly localized. For example, we calculated the difference in the electron charge between the $1s$ and $2p$ core hole states of the CuO_4 planar D_{4h} cluster, as shown in Table 9.2 [49]. The $2p^{-1}$ hole was created at the p_z orbital, which is perpendicular to the D_{4h} plane. For the two other p orbitals in the D_{4h} plane, the effect may be again negligible. However, for a $2p^{-1}$ hole, there are complications, since there are always three ways to create a $2p^{-1}$ hole. Thus, for the sake of simplicity, we focus on the creation of a $1s^{-1}$ hole in what follows.

Table 9.2. Calculated effective charges on Cu of CuO_4 clusters [49]

	Cluster	Ground state	$1s^{-1}$ hole state	$2p_z^{-1}$ hole state
Cu(II) $3d^9$	$[CuO_4]^{6-}$	$Cu^{+1.40}$	$Cu^{+0.79}$	$Cu^{+0.80}$
Cu(I) $3d^{10}$	$[CuO_4]^{7-}$	$Cu^{+0.89}$	$Cu^{+0.62}$	$Cu^{+0.64}$

9.5.4 Effect of Bond Length Difference

In calculating the electronic structure of the cluster molecule, we fixed the length of the Fe–O bond to 0.21 nm for both the ground state and the core hole state, for both FeO and Fe_2O_3. However, the equilibrium distance in

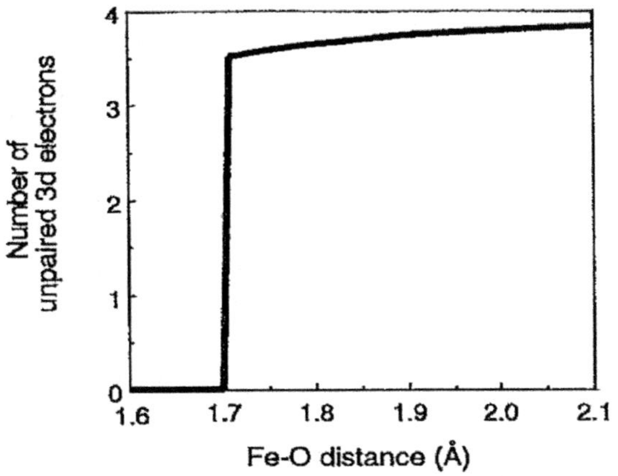

Fig. 9.7. The effect of Fe–O distance in the $[FeO_6]^{10-}$ cluster on the calculated number of unpaired $3d$ electrons

a real crystal is not always the same for all six Fe–O bond lengths in a cluster with O_h symmetry. It is not the same between FeO and Fe_2O_3, nor is it the same between the ground state and the core hole state. The creation of a core hole and the filling of the core hole occurs over a very short period (e.g., 10^{-16} sec.). By contrast, the phonon vibration period is a few orders of magnitude longer. Therefore, an observed X-ray spectrum is the sum of many X-ray spectra weighted by the phonon density of states. However, the effect of bond distance is usually negligible as is described below. Thus, the effect of bond distance difference between the core-hole and ground states is negligible. The vibration in the ground state and the statistical distribution of the bond distance are manifested in the X-ray spectra as phonon broadening. However, we have neglected this effect. This phenomenon is elaborated upon in the following references: XPS phonon broadening is described in detail in [50], XRF broadening is discussed in [51], and XANES broadening is briefly described in [4].

When the spin state or charge-transfer is sensitive to the bond length, the discussion becomes very complicated. To investigate this, we calculated the bond-length effects on the electronic structure of a cluster molecule, as shown in Fig. 9.7 [48], where the Fe–O distance in the solid is around 0.21 nm, and the spin state is conserved down to 0.17 nm.

Based on the calculation shown in Fig. 9.7, we can calculate the electronic structure of a cluster molecule with quite a rough estimation of the bond length. An error of up to 30 pm from the equilibrium distance is allowed.

9.6 Charge-Transfer Effect

The electron number of Cu in the ground state and the core hole state has been calculated for copper oxides [49]. Divalent copper has extra one electron in the core hole state, whereas the monovalent copper is unchanged in electron number. This is well described by Lenglet [52, 53]. In accord with the copper calculation, we extended the interpretation of charge transfer effect to Ni $K\beta$ spectra [54,55]. The $K\beta'$ line of a transition metal is interpreted many ways, such as exchange-splittings or plasmon satellites; many other possible explanations are described by Gamblin et al. [56]. Kawai et al. clarified that the $K\beta'$ [54] and $L\alpha$ satellites [55] of late transition-metal compounds such as NiO originates from the charge-transfer satellites based on the atomic multiplet calculations of Ni^{2+} and Ni^{+}. The $K\beta'$ is due to the multiplet structure of Ni^{+}, though the compound is Ni^{2+} in the ground state of such materials as NiO.

The charge transfer concept has extended to the interpretation of the Cu $L\alpha$ high-energy satellite [17, 57–61], although this satellite was later interpreted as a shake-up satellite by Magnusson [29], as described above. We still believe that the spectral line is due to the charge transfer satellite [28], because the satellite is more strongly observed in nickel compounds [55], as shown in Fig. 9.8.

The charge-transfer effect is a common interpretation of the chemical effect of the shake-off satellite in transition-metal compounds, as reviewed in [62, 63]. De Groot et al. [64] recalculated in a similar manner the Ni $K\beta$ line shape with the inclusion of the charge transfer effect, and their results support the assumption of the charge transfer effect on the line shape of the X-ray fluorescence spectra proposed by Kawai et al. De Groot and his colleagues studied spin-selective X-ray absorption spectroscopy using the high-resolution

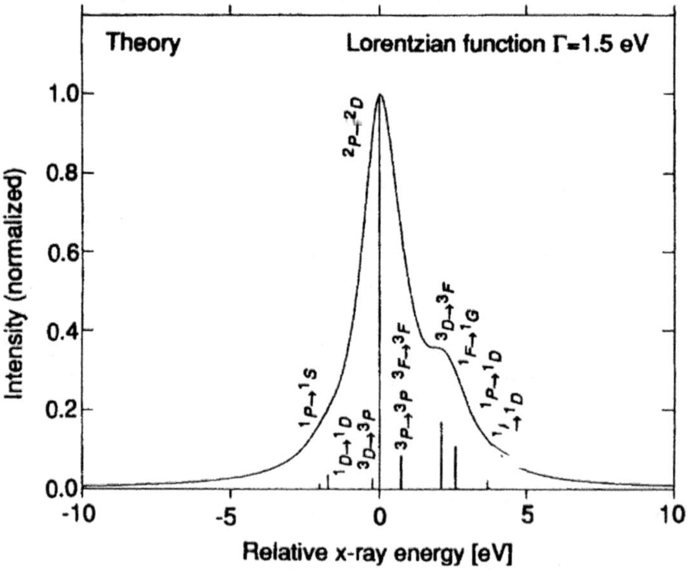

Fig. 9.8. Ni $L\alpha$ and $L\beta$ spectra of NiI_2 and a multiplet calculation of Ni^{+}, reproduced from [55]

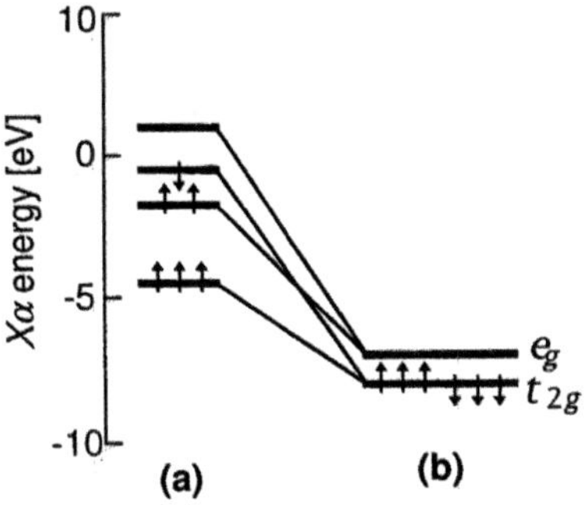

Fig. 9.9. Spin-flip of Co due to core-hole creation

X-ray fluorescence–yield method, for $4d$ transition-metal compounds [65], Fe $K\beta$-yield K-XANES [66, 67], Mn $K\beta$-yield K-XANES [68, 69], and Mn L-XANES [70]. These reports were based on the assumption that the $K\beta'$ line was due to multiplet-splitting of the same electron configuration as the ground state. Thus, if the charge-transfer effect is not negligible, the assumption of spin-selectivity fails. Here, we note that for early-transition metal compounds, the charge transfer effect is small, but for late transition-metal compounds, it is not negligible, as is described below.

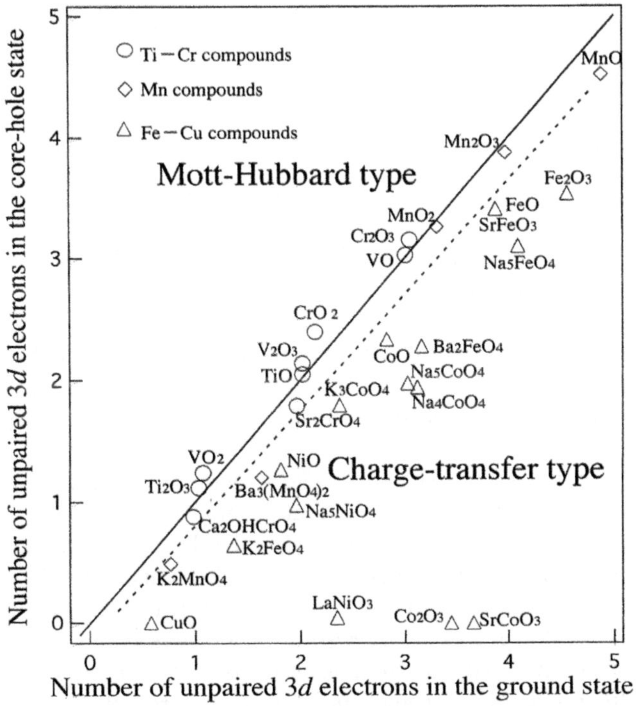

Fig. 9.10. Plot of the calculated number of unpaired $3d$ electrons in the ground state and the $1s^{-1}$ hole state, reproduced from [76]

Kawai et al. also calculated the charge transfer effect on the line shape of Fe $K\alpha$ [28]. However, when Kawai et al. extended the calculation into all $3d$ transition-metal compounds, they found a very strange effect – the spin-flip effect due to core-hole creation [71] for some kinds of cobalt oxide, as shown in Fig. 9.9.

They synthesized $LaCoO_3$ as a possible compound of spin-flip cobalt oxide and measured the line shape of Co $K\alpha$. They observed that the line shape was anomalously narrow despite the fact that the compound was a high-spin compound. This was explained as a spin-flip–type charge transfer [72].

Oku and his colleagues measured XPS and high-resolution $K\alpha$ spectra of Fe [73], Mn [74], and Ni [75]. They interpreted that several spectral features originate from the charge-transfer effect using the DV-Xα calculations.

The charge-transfer effect was summarized by Suzuki et al. [76] (Fig. 9.10), where the ground state $3d$ spin and the $1s^{-1}$ hole state 3d spin are plotted. For the early transition-metal compounds, the charge-transfer effect is negligible. This means that for compounds from Ti to Mn, the $3d$ electron number of the ground state is conserved in the $1s^{-1}$ hole state. By contrast, for the late transition-metal compounds, from Mn to Cu, the $3d$ electron number increases by one, and sometimes, the $3d$ spin becomes diminished, as is seen in Fig. 9.10.

The classification of the charge-transfer compounds and non-charge-transfer compounds is quite similar to that determined by the Zaanen–Sawatzky–Allen diagram [77], which separates charge-transfer insulators and Mott-Hubbard insulators.

9.7 Concluding Remarks

Core-level spectroscopic techniques are very useful in characterizing the electronic structure of compounds. However, we demonstrated that charge-transfer–type many-body effects are important for the interpretation of these spectra, though usually only shake-off–type many-body effects are considered from the atomic physics point of view. Though the charge-transfer effect described above was for $3d$ transition-metal compounds, Suzuki extended the calculation into La [78] and all of the $4f$ rare earth elements [79]. The charge transfer effect is again important for other compounds that have partially filled valence orbitals.

Finally, I would like to comment on the sum rule, which, mathematically, states that the sum of the diagonal matrix element is unchanged during a physical process [80]. There is an important sum rule involved in shake-up and shake-off processes [81]. Since the charge-transfer effect is a kind of shake-off or shake-up process, we can expect the existence of the same sum rule. The shake-up, shake-off, and charge-transfer processes are responses to the removal of one core electron; the sum of the intensity in the spectrum is always conserved. If the charge-transfer effect becomes strong, then extra electron ionization probability, such as shake-off, becomes small. Plasmon oscillation should be included in the responses.

References

1. A. Meisel, G. Leonhardt, R. Szargan, *X-Ray Spectra and Chemical Bonding.* Springer Series in Chemical Physics vol. 37 (Springer, Berlin Heidelberg New York, 1989)
2. F. de Groot, High-resolution x-ray emission and x-ray absorption spectroscopy, Chem. Rev., vol.101, p.1779–1808 (2001)
3. S. Hufner, *Photoelectron Spectroscopy* (Springer, Berlin Heidelberg New York, 1995)
4. J. Kawai, *Encyclopedia of Analytical Chemistry,* ed. by R.A. Meyers (Wiley, New York, 2000) pp. 13288–13315.
5. A.V. Kondratenkio, L.N. Mazalov, D.M. Neiman, Opt. Spectrosc. (USSR) **48**, 587 (1980): Opt. Spektrosc. **48**, 1072 (1980) (Russian Original)
6. M.C. Thornley, in *George Green, Mathematician & Physicist 1793–1841*, 2nd edn., ed. by D.M. Cannell (SIAM, Philadelphia, 2001) pp. 183–204
7. L.I. Schiff, *Quantum Mechanics*, 3rd edn. (McGraw-Hill, New York, 1968) pp. 292–293
8. A. Messiah, *Quantum Mechanics* (Dover, Mineola, 1999) pp. 739–761
9. R.D. Deslattes, R.E. LaVilla, P.L. Cowan, A. Henins, Phys. Rev. A **27**, 923 (1983)
10. N. Wassdahl, J.-E. Rubensson, G. Bray, P. Glans, P. Bleckert, R. Nyholm, S. Cramm, N. Mårtensson, J. Nordgren, Phys. Rev. Lett. **64**, 2807 (1990)
11. Y. Muramatsu, H. Kouzuki, T. Kaneyoshi, M. Motoyama, A. Agui, S. Shin, H. Kato, J. Kawai, Appl. Phys. A (Mat. Sci. Process.) **65**, 191 (1997)
12. H. Ishii, Y. Ishiwata, R. Eguchi, Y. Harada, M. Watanabe, A. Chainani, S. Shin, J. Phys. Soc. Japan **70**, 1813 (2001)
13. U. Bergmann, P. Glatzel, F. de Groot, S.P. Cramer, J. Am. Chem. Soc. **121**, 4926 (1999)
14. Y. Tamaki, T. Omori, T. Shiokawa, Radiochem. Radioanal. Lett. **37**, 39 (1979)
15. P. Glatzel, U. Bergmann, F.M.F. de Groot, S.P. Cramer, Phys. Rev. B **64**, 045109 (2001)
16. M. Magnuson, N. Wassdahl, J. Nordgren, Phys. Rev. B **56**, 12238 (1997)
17. J. Kawai, K. Maeda, K. Nakajima, Y. Gohshi, Phys. Rev. B 48, 8560 (1993)
18. J. Kawai, K. Nakajima, Y. Gohshi, Spectrochim. Acta B **48**, 1281 (1993)
19. J.E. Holliday, J. Appl. Phys. **33**, 3259 (1962)
20. V.R. Galakhov, E.Z. Kurmaev, V.M. Cherkashenko, Izv. Akad. Nauk, SSSR, Ser. Fiz. **49**, 1513 (1985)
21. J. Kawai, K. Nakajima, K. Maeda, Y. Gohshi, Adv. X-Ray Anal. **35**, 1107 (1992)
22. J.T. Armstrong, Anal. Chem. **71**, 2714 (1999)
23. J. Kawai, Rigaku J. **18**(1), 31 (2001)
24. J. Kawai, Adv. X-Ray Chem. Anal. **32**, 1 (2001)
25. J. Kawai, K. Maeda, Spectrochim. Acta B **46**, 1243 (1991)
26. G. van der Laan, C. Westra, C. Haas, G.A. Sawatzky, Phys. Rev. B **23**, 4369 (1981)
27. M. Ohno, Phys. Rev. B **52**, 6127 (1995)

28. J. Kawai, K. Maeda, K. Nakajima, Y. Gohshi, Phys. Rev. B **52**, 6129 (1995)
29. M. Magnuson, N. Wassdahl, J. Nordgren, Phys. Rev. **56**, 12238 (1997)
30. H. Adachi, in *Hartree-Fock-Slater Method for Materials Science*, ed. by J. Kawai. Springer Ser. Materials Science, vol. 84 (Springer, Berlin Heidelberg New York, 2005), pp. 1–18
31. J. Kawai, C. Satoko, Y. Gohshi, J. Phys. C: Solid State Phys. **20**, 69 (1987)
32. D.S. Urch, J. Chem. Soc. Chem. Commun. **526**, (1982)
33. O. Benka, R.L. Watson, K. Parthasaradhi, J.M. Sanders, R.J. Maurer, Phys. Rev. A **15**, 914 (1977)
34. E. Hartmann, C. Lauterbach, Adv. Quantum Chem. **29**, 373 (1997)
35. E. Hartmann, J. Phys. B: At. Mol. Phys. **19**, 1899 (1986)
36. J. Kawai, C. Satoko, Y. Gohshi, Phys. Rev. Lett. **57**, 988 (1986)
37. J. Kawai, C. Satoko, Y. Gohshi, Spectrochim. Acta B **42**, 1125 (1987)
38. J. Kawai, Y. Nihei, Y.-Z. Bai, K. Fujisawa, Y. Gohshi, Phys. Rev. A **39**, 3686 (1989)
39. J. Kawai, Y. Nihei, Y. Gohshi, Adv. X-Ray Chem. Anal. Japan **20**, 67 (1989)
40. L.L. Landau, E.M. Lifshitz, *Quantum Mechanics – Non-relativistic Theory*, Japanese translation by Tokyo-Tosho (1971), section 79
41. S. Glasstone, K.J. Laidler, H. Eyring, *The Theory of Rate Processes* (McGraw-Hill, New York, 1964) Chapter III
42. A. Fujimori, Phys. Rev. B **27**, 3992 (1983)
43. G.A. Sawatzky, in *Core-Level Spectroscopy in Condensed Systems*, ed. by J. Kanamori, A. Kotani (Springer, Berlin Heidelberg New York, 1988) p. 99
44. J. Kawai, M. Oku, Y. Nihei, Adv. X-Ray Chem. Anal. Japan **21**, 149 (1990)
45. J. Kawai, Adv. X-Ray Chem. Anal. Japan **22**, 145 (1991)
46. M. Sakai, S. Hayakawa, J. Kawai, Y. Gohshi, Adv. X-Ray Anal. **36**, 65 (1993)
47. J. Kawai, I. Tanaka, H. Adachi, Bunseki Kagaku **44**, 251 (1995)
48. J. Kawai, C. Suzuki, H. Adachi, T. Konishi, Y. Gohshi, Phys. Rev. B **50**, 11347 (1994)
49. J. Kawai, Y. Nihei, M. Fujinami, Y. Higashi, S. Fukushima, Y. Gohshi, Solid State Commun. **70**, 567 (1989)
50. H. Siegbahn, L. Karlsson, in *Handbuch der Physik, Band XXXI*, ed. by W. Mehlhorn (Springer, Berlin Heidelberg New York, 1982) pp. 215–467
51. J. Kawai, M. Ohta, Y. Nihei, Y. Gohshi, Spectrochim. Acta B **44**, 815 (1989)
52. M. Lenglet, Eur. J. Solid State Inorg. Chem. **34**, 405 (1997)
53. M. Lenglet, Trends Chem. Phys. **6**, 121 (1997)
54. J. Kawai, M. Takami, C. Satoko, Phys. Rev. Lett. **65**, 2193 (1990)
55. J. Kawai, K. Maeda, Y. Sasa, M. Takami, M. Uda, Riken Accel. Prog. Rep. **24**, 50 (1990)
56. S.D. Gamblin, D.S. Urch, J. Electron Spectrosc. Relat. Phenom. **113**, 179 (2001)
57. J. Kawai, K. Maeda, Spectrochim. Acta B **46**, 1243 (1991)
58. J. Kawai, K. Maeda, Physica C **185–189**, 981 (1991)
59. K. Nakajime, J. Kawai, K. Maeda, Physica C **185–189**, 983 (1991)
60. J. Kawai, K. Maeda, Riken Accel. Prog. Rep. **25**, 147 (1991)
61. J. Kawai, K. Nakajima, K. Maeda, Y. Gohshi, Adv. X-Ray Anal. **35**, 1107 (1992)
62. J. Kawai, Nucl. Instrum. Methods Phys. Res. B **75**, 3 (1993)
63. J. Kawai, Nucl. Instrum. Methods Phys. Res. B **87**, 88 (1994)
64. F.M.F. de Groot, A. Fontaine, C.C. Kao, M. Krisch, J. Phys. Condens. Matter **6**, 6875 (1994)

65. F.M.F. de Groot, Z.W. Hu, M.F. Lopez, G. Kaindl, F. Guillot, M. Tronc, J. Chem. Phys. **101**, 6570 (1994)
66. G. Peng, X. Wang, C.R. Randall, J.A. Moore, S.P. Cramer, Appl. Phys. Lett. **65**, 2527 (1994)
67. X. Wang, C.R. Randall, G. Peng, S.P. Cramer, Chem. Phys. Lett. **243**, 469 (1995)
68. G. Peng, F.M.F. de Groot, K. Hämäläinen, J.A. Moore, X. Wang, M.M. Grush, J.B. Hastings, D.P. Siddons, W.H. Armstrong, O.C. Mullins, S.P. Cramer, J. Am. Chem. Soc. **116**, 2914 (1994)
69. F.M.F. de Groot, K. Hämäläinen, A. Fontaine, Physica B **206–207**, 89 (1995)
70. M.M. Grush, J. Chen, T.L. Stemmler, S.J. George, C.Y. Ralston, R.T. Stibrany, A. Gelasco, G. Christou, S.M. Gorun, J.E. Penner-Hahn, S.P. Cramer, J. Am. Chem. Soc. **118**, 65 (1996)
71. J. Kawai, C. Suzuki, H. Adachi, J. Electron Spectrosc. Relat. Phenom. **78**, 79 (1996)
72. C. Suzuki, J. Kawai, J. Tanizawa, H. Adachi, S. Kawasaki, M. Takano, T. Mukoyama, Chem. Phys. **241**, 17 (1999)
73. M. Oku, H. Matsuta, K. Wagatsuma, T. Konishi, J. Chem. Soc. Faraday Trans. **92**, 2759 (1996)
74. M. Oku, K. Wagatsuma, T. Konishi, J. Electron Spectrosc. Relat. Phenom. **98–99**, 277 (1999)
75. T. Konishi, J. Kawai, M. Fujiwara, T. Kurisaki, H. Wakita, Y. Gohshi, X-Ray Spectrom. **28**, 470 (1999)
76. C. Suzuki, J. Kawai, H. Adachi, T. Mukoyama, Chem. Phys. **247**, 453 (1999)
77. J. Zaanen, G.A. Sawatzky, J.W. Allen, Phys. Rev. Lett. **55**, 418 (1985)
78. C. Suzuki, T. Mukoyama, J. Kawai, H. Adachi, Phys. Rev. B **57**, 9507 (1998)
79. C. Suzuki, J. Kawai, M. Takahashi, A.-M. Vlaicu, H. Adachi, T. Mukoyama, Chem. Phys. **253**, 27 (2000)
80. S. Sugano, Y. Tanabe, H. Kamimura, *Multiplets of Transition-Metal Ions in Crystals* (Academic, New York, 1970) Chapter 10
81. R. Manne, T. Åberg, Chem. Phys. Lett. **7**, 282 (1970)

[illegible] J. Chem. Phys. 101, 6676 (1994) [illegible]
[illegible] Phys. [illegible] 67, [illegible] (1991) [illegible]
[illegible] 245, [illegible]
[illegible]

10

Determining Electronic Structure from Auger Spectra in the Cluster Approximation

László Kövér

Following a short discussion of the effects of the atomic environment on Auger spectra emitted from solids, it is demonstrated how X-ray-excited Auger spectroscopy can be used to obtain information on the electronic structure near the initial core hole and on electron correlation. Local charges determined using the cluster approximation are compared to those obtained from Auger parameter analysis, while relaxation/polarization effects and satellite structures in Auger spectra are interpreted in the cluster approach. The advantages of the combined use of the X-ray induced-Auger and photoelectron spectroscopies with theoretical cluster molecular orbital models in the studies of chemical and solid state effects on Auger processes are emphasized. The power of these methods is illustrated by recent results obtained for metals, alloys, and oxides of great practical importance.

10.1 Introduction

Auger spectra of solids can contain important information on the chemical states and local electronic structure of materials [1]. The local electronic structure in the atomic environment surrounding an atom of a key element in a solid has a determining role on its macroscopic (mechanical, electronic, magnetic, optical, and chemical) properties. Auger spectra obtained from molecules, clusters, and solid surfaces reflect the effects of changes in the potential of the atomic cores and in the local density of electronic states of the core-ionized atoms. By its two-electron nature, the Auger process sensitively indicates the strength of electron correlation. In recent years, significant progress has been made in the understanding of the origin, nature, and magnitude of the effects of the atomic environment on Auger transition energies and probabilities [2,3], as a consequence of the rapid development of spectroscopic methods and theoretical models. These studies lead to a deeper insight into the Auger process itself in the case of polyatomic systems and at the same time enhance the possibilities of chemical state identification. Photon excitation of Auger spectra from solids has many advantages over inducing such spectra by charged particles [4]. In this chapter, some recent examples are provided for the combined application of high-resolution X-ray-induced photoelectron and Auger spectroscopy of metals, alloys, and insulators, with cluster MO models for interpretation of the spectroscopic data.

10.2 Effects of the Atomic Environment of Auger Spectra Excited from Solids

The transition energy $E_{i,j,k}$ of the atomic Auger process involving the levels i (with the initial vacancy), j, and k (with the final state vacancies) can be approximated by the following equation:

$$E_{i,j,k} = E_i - E_j - E_k - U_{jk} ,$$

where E_i, E_j, and E_k are the binding energies of the respective electrons, and U_{jk} denotes the repulsion or correlation energy between the final state holes. For Auger transitions in the atomic core, usually it is a good assumption that the binding energy shifts due to different atomic environments are similar, while in the case of core-valence Auger transitions, this assumption is generally invalid. Theoretical estimates of $E_{i,j,k}$ can be obtained from the difference in total energies of the initial and final states calculated using ab initio methods. However, for polyatomic molecules or clusters, the convergence of the total energy calculation is not always guaranteed [5], and determining small differences of large energy values requires high accuracy, which is sometimes very difficult to achieve. The transition probability I_{ijk} of the Auger process involving the i, j, and k atomic levels, in the simplest approximation – the frozen core independent particle approximation with the Wentzel Ansatz-approach – can be expressed as [5]

$$I_{ijk} = |\langle \Psi_f(i\varepsilon)|e^2/r_{12}|\Psi_i(jk)\rangle|^2 ,$$

where ε is the energy of the ejected Auger electron, e^2/r_{12} is the Coulomb interaction between the two electrons participating in the transition, and $\Psi_i(jk)$ and $\Psi_f(i\varepsilon)$ are the wave functions of the initial and final states, respectively. This simple picture reflects the highly local character of the Auger process. The probability of the interatomic Auger transitions (involving levels of different, neighboring atoms) is usually orders of magnitude smaller than the probability of the transitions taking place inside the atom with the initial core hole (intra-atomic transitions) [5]. Therefore, strong effects of the atomic environment on Auger transitions can be expected first of all as a consequence of environmentally induced changes in the atomic initial and final states. Due to various initial- and final-state excitation processes accompanying the Auger process (initial- and final-state shake, with spectator or participator electrons in the excited state [2,6]), satellite structures can appear in the Auger spectra. Fast Coster-Kronig processes, involving a valence electron with varying localization and preceding the decay of the initial core hole can also strongly influence the respective relative Auger intensities [7]. Different hybridizations between the XPS core-hole state and the Auger final state can also lead to Auger satellites with relative intensities depending on the atomic environment. Further changes in the Auger lineshape can appear due to phonon broadening induced by core-hole creation [9] and due to intrinsic or extrinsic plasmon loss or other inelastic energy-loss processes in the solid. Auger chemical shifts, local electronic structure, and electron-related effects on core-valence Auger lineshapes are discussed in more detail below.

10.3 X-ray Excited Auger Spectroscopy for Studying Chemical and Solid State Effects on Auger Spectra

Exciting Auger spectra of solids using photons has many advantages over excitation by electrons. In the case of photoinduced Auger spectra, the peak-to-background ratio (mainly due to the lack of primary electron-beam scattering) is usually much higher, facilitating lineshape analysis with high-energy resolution. Intensity of the peaks can be optimized using exciting X-rays with energy near the photoionization threshold, where the photoionization cross-section is high. X-ray-induced Auger spectra usually show fewer alterations of the chemical state of the surface components in the analyzed sample due to radiation effects and smaller charging effects, which makes possible the analysis of the Auger spectra of insulators. A great benefit of photon excitation is, however, the possibility of the use of continuum (synchrotron, bremsstrahlung) or wide-band radiation, which ensures efficient production of Auger spectra in a very wide energy region [10]. Excitation of high-energy (> 5 keV) Auger spectra using electrons is difficult because of the required primary beam energy, which must usually be 3–4 times higher than in the case exciting by photons. From thin films deposited on low-atomic-number substrates, good-quality spectra (suitable for lineshape analysis) can be obtained at these high transition energies [11] using X-ray excitation. When studying local electronic structures of surfaces and interfaces, the above advantages generally outweigh the higher exciting flux and spatial resolution of electron-induced Auger spectroscopy.

10.4 Local Charges in Binary Alloys

Charge transfer between constituent atoms of crystal structures as well as the local electronic configuration is determinant for the mechanical and chemical stability of these structures and the knowledge of the charge transfer process is important in the understanding of alloying and catalytic processes at alloy surfaces.

To calculate charge distributions and construct the ground-state local density of electron states (LDOS), the discrete variational Xα (DV-Xα) molecular orbital model [12] is used (non-relativistic and relativistic [13] versions), with a numerical basis set and the self-consistent charge (SCC) scheme.

The Slater exchange potential $V_{\mathrm{xc}} = -3\alpha(3\rho(r)/8\pi)^{1/3}$ is applied, where α (fixed to 0.7 for all calculations) is the exchange scaling parameter and ρ is the charge density. Within the DV-Xα model, instead of calculating multicenter integrals, the integrand is computed in discrete points, and therefore, the accuracy of the value of the integral is limited only by the capacity of the computer. The LDOS is obtained by replacing the computed MO levels with their respective energy distribution functions. Independent information on local charges can be obtained from high-accuracy experiments measuring the alloy–pure metal environmentally induced Auger parameter energy shifts and applying a simple atomic structure model to derive the transferred charge [3]. The comparison of the differences in the partial atomic charges computed by cluster molecular orbital models and the transferred charge derived from experiments is of crucial importance from the point of view of the consistency, reliability, and limits of our understanding concerning local charges in binary alloys.

Alloy-metal Auger parameter shifts can be obtained from the measured Auger kinetic and photoelectron binding energy shifts using the following definition [3]:

$$\Delta\xi(j) = \Delta E_k(ijj) - \Delta E_b(i) + 2\Delta E_b(j) \ , \tag{10.1a}$$

where $\Delta E_k(ijj)$ is the environmentally induced shift in the kinetic energy of an Auger transition involving the i and j core levels and ΔE_b is the difference in the binding energy of a core level between two atomic environments. If ΔE_b is similar for all core levels, (10.1a) corresponds to the conventional definition of the modified Auger parameter shift $\Delta\alpha' = \Delta E_k(ijk) + \Delta E_b(\mathrm{core})$ [14]. In this case, $\Delta\xi \cong \Delta\alpha' \cong 2\Delta R^{\mathrm{ea}}$ is the environmentally induced change in the final-state extraatomic relaxation energy, R^{ea}. Using a different definition for the Auger parameter,

$$\Delta\beta(j) = \Delta E_k(ijj) + \Delta E_b(i) + 2\Delta E_b(j) \approx 2\Delta V \tag{10.1b}$$

where ΔV is the change in the core potential as a consequence of the changes in the valence charge. To estimate charge transfer from an analysis of the final-state Auger parameter shifts, $\Delta\xi$, between a pure metal and its binary alloy using a model based on atomic structure parameters, it can be shown that [3]

$$\Delta\xi = \Delta\left[q(\mathrm{d}k/\mathrm{d}N) + (k - 2\mathrm{d}k/\mathrm{d}N)(\mathrm{d}q/\mathrm{d}N) + (\mathrm{d}U/\mathrm{d}N)\right] \ , \tag{10.2}$$

where q is the valence charge, k is the change in the potential in the core when a valence electron is removed, N is the occupancy of the core orbitals, and U is the contribution from the atomic environment. It is assumed that k and q depend linearly on N. In the case two conductors, because of the efficient screening, $\Delta(\mathrm{d}U/\mathrm{d}N) = 0$, and due to the on-site core hole screening $(\mathrm{d}q/\mathrm{d}N) = 1$; therefore, $\Delta(\mathrm{d}q/\mathrm{d}N) = 0$, giving

$$\Delta\xi = \Delta q(\mathrm{d}k/\mathrm{d}N) \tag{10.3}$$

for the case when the valence electrons belong to a single band. The parameter $\mathrm{d}k/\mathrm{d}N$ can be obtained from the results of atomic structure calculations for free atoms [16]. It should be noted that the core potential model is expected to provide a good approximation mainly for deep core transitions although the Auger parameter concept is extended in some cases to core-valence transitions with "atomic-like" Auger lineshape. Experimental Auger parameter shift determination can result in accurate data, because the measured Auger parameter values are not dependent on surface charging or on energy calibration of the spectrometer.

10.4.1 Experimental

Core photoelectron and Auger spectra were excited from polycrystalline Al_3Ni and $AlNi_3$, as well as from single-crystal (100) CuPd alloys and high-purity polycrystalline metal samples, using Cu $K\alpha$, Ag $L\alpha_1$ X-rays, and bremsstrahlung radiation from Cu, Ag, and Mo targets. Measurements of high-energy electron spectra were performed using a home-built, high-luminosity electron spectrometer [17] based on a 180° hemispherical analyzer with a resolution of 0.4 eV. High-resolution monochromated Al $K\alpha$ X-ray-excited core photoelectron and Auger spectra were obtained using a SCIENTA ESCA-300 spectrometer with a spectrometer energy resolution of 0.4 eV. Energy calibration of the electron spectra was performed in the XPS region with the aid of photo- and Auger lines from polycrystalline Cu and Ag samples [18, 19] and in the high-energy range by measuring the Cu $K\alpha$-excited Cu-2p lines.

10.4.2 Charge Transfer in CuPd Alloys

The nature, value and sign of the transferred charge in Cu–Pd alloy systems have recently become controversial issues [18]. Table 10.1 contains the alloy–metal Auger parameter shifts determined experimentally and the transferred charges derived using the linear potential model, in comparison with the charge derived from our relativistic DV-Xα calculations [20].

From the results in Table 10.1, it can be seen that the initial-state Auger parameter β is more sensitive to the changes in the valence charge than is the final-state parameter. On the basis of our measurements of deep core transitions (Cu KLL, Cu-2p, Pd-2s, 2p, Pd $L_{23}MM$), we can conclude that a very small, $\sim 0.05e$ overall charge is transferred from the Pd to the Cu site,

Table 10.1. CuPd Auger parameter shifts (eV) and transferred charges Δq [20]

		Δq(Cu)	Δq(Pd)
$\Delta\xi$(Cu)	0.03	−0.01	
$\Delta\beta$(Cu)	−1.30	−0.05(0.08[1]; −0.23[2])	
$\Delta\beta$(Pd)	0.70		0.05
DV-Xα cluster			
MO theory (g.s.)		−0.03[3]	0.03[3]

$\mathrm{d}k^{\mathrm{Cu}}/\mathrm{d}N = -3.30$[1]; est. err. $< |0.10|$eV k^{Pd}(average) = 11[1]
1 Refer to [19]
2 Refer to [18]
3 Refer to [21]

and the results obtained from Cu and Pd Auger parameters is consistent with the charge neutrality in the alloy. The magnitude and the direction of the transferred charge is confirmed by the DV-Xα calculations estimating ground state charges. In Pd, the increase of the *d* occupancy is overcompensated by the loss of *s* charge.

10.4.3 Charge Transfer in Al3Ni and AlNi$_3$ Alloys

Effects of alloying on electronic structure are expected to be manifested in the case where an atom of a metal can be almost entirely surrounded by atoms of the other constituent. In the case of simple metal–transition metal alloys, filling of the *d* bands of the transition metal can take place through hybridization and charge transfer. The Al–Ni alloy systems are particularly interesting from the point of view of obtaining information on these processes.

Estimating charge transfer from the measured Al alloy–metal Auger parameter shifts, an improved model (assuming k depends linearly on q and N) was used [22].

From this model,

$$\Delta\xi = \Delta q \cdot b \,, \tag{10.4}$$

where b is a model parameter obtained from atomic structure calculations [22]. To derive charge distributions from the DV-Xα calculations, the experimental lattice constants [23, 24] were used for the cubic-ordered (O_h) intermetallic compound $AlNi_3$ and for the orthorhombic (DO_{20}) structured Al_3Ni alloy, with the atomic sphere radii given by Slater [25].

Table 10.2 shows the values of the transferred charge calculated from (10.3) and (10.4) using the parameters $\mathrm{d}k/\mathrm{d}N = -2.3$ for Ni [16] and $b_p = -3.18$ for Al [22], together with the values derived from the cluster MO theory; the table indicates the charge values obtained assuming charge conservation, $\Delta q_{\mathrm{Al}} x_{\mathrm{Al}} = -\Delta q_{\mathrm{Ni}}(1 - x_{\mathrm{Al}})$, where x denotes atomic concentration.

Table 10.2. Transferred charges Δq (alloy-metal) [22]

Atom in the alloy	$AlNi_3$		Al_3Ni	
	Exp.	Theory	Exp.	Theory
Al	0.37	(0.60[2])	0.50	(0.12[2])
Ni	–0.11[1] (–0.11[2])	–0.20	–0.15[1] (–1.5[2])	–0.36

[1] average of values obtained from $\Delta\xi(3p, L_3MM)$ and $\Delta\xi(3p, L_2MM)$
[2] derived from the result for Al on the basis of charge-conservation between neighbors

It should be noted that the experimental final-state Auger parameter shifts are rather small for Ni [22], comparable to the experimental error. From the respective $\Delta\alpha'$ modified Auger parameter shifts, the assumption that a change in the valence charge has a similar effect on the core binding energies seems to be invalid for these systems.

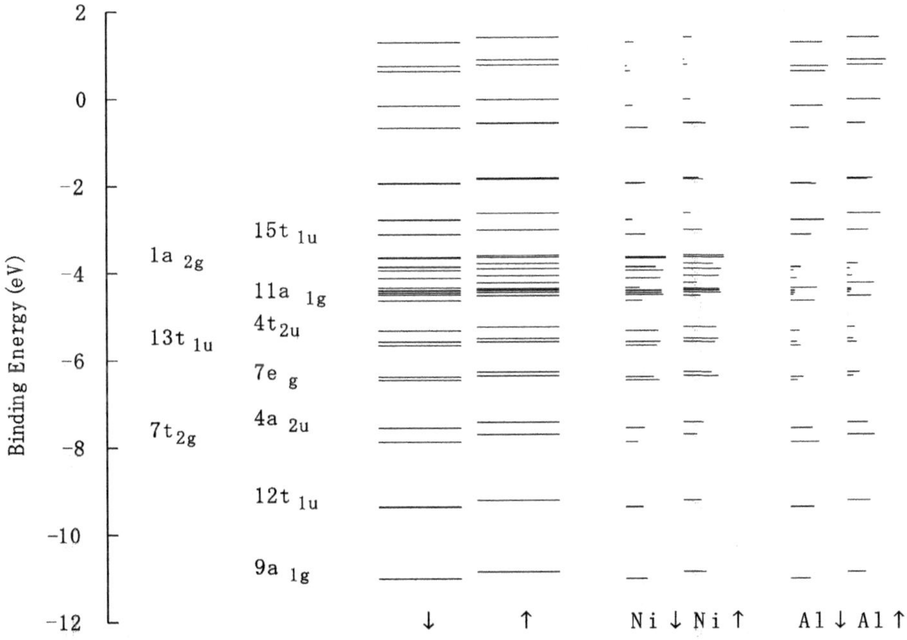

Fig. 10.1. Molecular orbital levels and spin components for $AlNi_3$ alloys [28]

In the case of $AlNi_3$, the experimental results show a small charge transfer from the Al to the Ni site (Table 10.2); the direction of this transfer agrees with the expectations on the basis of the difference in electronegativity of Al and Ni. The value determined for the charge on a Ni site in $AlNi_3$ from the Ni alloy–metal Auger parameter shifts is consistent with that obtained from the charge determined for the Al site and assuming conservation of total charge on Al and Ni neighbors. The MO theory predicts a small (although a factor of two higher) charge transfer in the same direction (Table 10.2). Large discrepancies can be seen in the case of Al_3Ni, where the experimentally derived charges are inconsistent with charge conservation, and their differences from the values predicted by the theory are also larger. There is agreement on the direction of the charge transfer, however, between the experiment and the model calculation. The calculated charge transferred from the Al-s and p orbitals to the Ni-d band shows that the charge transfer plays only a minor role in the filling of the Ni-3d band in the case of both alloys. On a Ni site, the charge transfer between d and s orbitals might be more important than charge transfer between Al and Ni [26]; furthermore, we might expect that the influence of this hybridization effect would be larger for Al_3Ni than for $AlNi_3$.

To obtain more detailed information on the particular orbitals involved in the charge-transfer process, non-relativistic spin polarized DV-Xα calculations [12, 27] were also performed for these Al–Ni systems. Using the "true" positions of the atoms in the Al_3Ni cluster (i.e., not a Cu_3Au-like structure) resulted in a much smaller (by a factor of four) derived transferred charge in the case of non-spin-polarized calculations for Al_3Ni [28]. Our spin-polarized

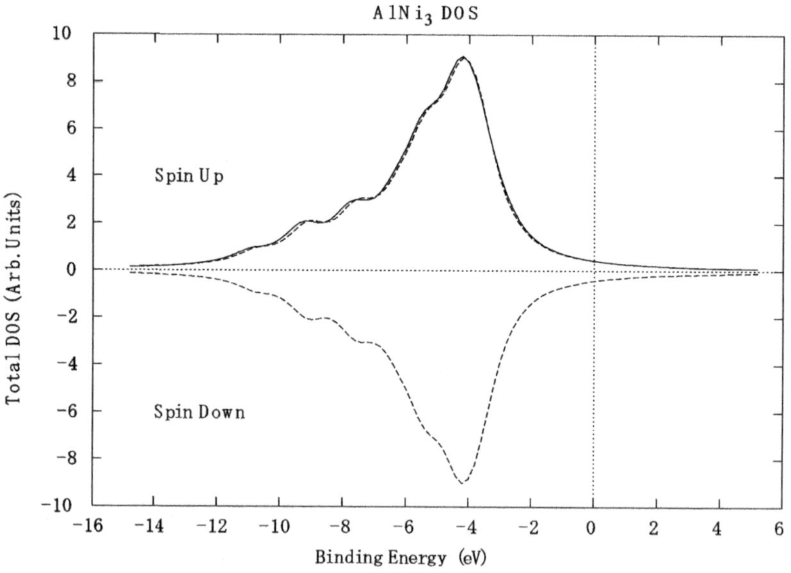

Fig. 10.2. DOS and spin components for $AlNi_3$ alloy [28]

calculations for Al_3Ni provided a zero magnetic moment, in agreement with the observation [28]. Fig. 10.1 shows the calculated level structure and spin components, and Fig. 10.2 shows the density of states for the Al_3Ni alloy. The "up" spin component in Fig. 10.2 is slightly shifted compared to the "down" spin component, resulting in a non-zero magnetic moment of the $AlNi_3$ alloy. In the case of Ni-d valence electrons, our calculated spin-magnetic moment value of $0.032\mu_B$ (where μ_B is the Bohr magneton) agrees with the results of earlier calculations [29].

10.5 Generalized Electrostatic Model for Interpreting Auger Parameter Shifts and Final State Relaxation/Polarization

For interpretation of Auger parameter environmental shifts, the most successful theoretical models include the "excited atom model," developed by Weightman et al. [30] based on parameters obtainable from atomic structure calculations (successfully applied to interpretation of metal–free atom and alloy–pure metal Auger parameter shifts [30]), and the "electrostatic model" [31,32] developed by Moretti.

Within the electrostatic model [31,32] the relaxation/polarization is described by the interaction of the electric field due to the core-ionized atom ($\boldsymbol{F}_i$) and the induced dipole on the ligands ($\boldsymbol{M}_i$). The induced dipole on the ligand i is given by $\alpha_i \boldsymbol{F}_{Li}$, where α_i is the electronic polarizability of the ligand, and $\boldsymbol{F}_{Li}$ is the total electric field on the ligand i given by the field ($\boldsymbol{F}_i$) and the field ($\boldsymbol{F}_{Dij}$) corresponding to the effect of the dipoles induced on all the other ligands:

$$\begin{aligned}\boldsymbol{F}_{Li} &= \boldsymbol{F}_i + \boldsymbol{F}_{Dij} \\ &= 1/\left(4\pi\varepsilon_0\right)\left[e\boldsymbol{R}_i/R_i^3 + \sum_{j\neq i}\left\{\left[3\left(\boldsymbol{M}_i R_{ij}\right)/R_{ij}^5\right] - \left(\boldsymbol{M}_i/R_{ij}^3\right)\right\}\right] , \quad (10.5)\end{aligned}$$

where R_i is the bond length, and R_{ij} is the minimum distance between the ligands i and j. From classical electrostatics, the extraatomic relaxation energy, R^{ea} is:

$$R^{\mathrm{ea}} = \sum_{i=1}^{n} \int_0^{F_j} M_i \mathrm{d}F_i . \quad (10.6)$$

With some strong assumptions – i.e., the use of the same ligands with same distance R from the core ionized atom, as well as high symmetry – we obtain [31,32]

$$\Delta\alpha' = 14.4n\alpha/\left(R^4 + RD\alpha\right) , \quad (10.7)$$

where n is the number of nearest neighbors, and D is a geometric factor.

The electrostatic model described above, however, is not able to deal with the local screening mechanism, when strongly localized orbitals of the core ionized atoms like d orbitals of transition metal ions are involved in the charge transfer. This, together with the strong assumptions used in obtaining (10.7), puts severe limitations on the application of the electrostatic model. These limitations can be removed by generalizing the model and using matrix formalism [33, 34]. Rearranging the N equations in (10.5), we have a matrix representing the electrostatic field on the ligands due to the central hole:

$$[F_{Li}][D_{ij}] = [F_j] \, . \tag{10.8}$$

Solving the corresponding system of linear equations, the total extraatomic relaxation energy is

$$R^{\mathrm{ea}} = \sum_i \int_0^{F_i} M_i \mathrm{d}F' = \sum \alpha_i F_i \left(F_{Li} - \left(F_i D_{ii}^{-1}/2 \right) \right) \, , \tag{10.9}$$

where D_{ii}^{-1} denotes the diagonal matrix elements of the inverse of the matrix D_{ij}, and α_i is the polarizability of the ith ligand. To calculate R^{ea}, computer codes written in C and FORTRAN languages were developed in Debrecen and Rome, respectively [33, 34]. With these programs, it is now possible to calculate Auger parameter shifts for core-ionized atoms with different nearest-neighbor ligands arranged in an arbitrary local geometry, giving the coordinates of the core-ionized atom and of its nearest-neighbor ligands, as well as the electronic polarizability of the ligands as input parameters. Alternatively, using the Auger parameter shifts as input data (e.g., the experimentally determined values), information can be obtained on the ligands and their local geometry.

The generalized electrostatic model was applied to estimate ΔR^{ea} for Si atoms located at different positions on the Si(001) surface [33]. Despite the fact that the Si(001) surface has been studied extensively, the local electronic and geometric structure of the surface and subsurface Si atoms is still under debate. In our calculation, the lattice constants and relaxed atomic coordinates were taken from [35] assuming that the electronic polarizability of Si atoms are independent of the atomic coordinates and are the same as in the free atom [36]. Taking the experimental value of ΔE_b (Si-2p), the surface-bulk Si-2p binding energy shift [37] values, and our calculated values for $\mathit{\Delta}\mathrm{R}^{\mathrm{ea}}$, the initial state energy shifts are -0.93, 0.20, $+0.06$, and -0.32 eV for the asymmetric dimer-up atom, the asymmetric dimer-down atom, the second atomic layer, and the third atomic layer, respectively. In strong contrast to the electrostatic model, the density functional calculations [38] show that ΔR^{ea} is increased at the surface and largest for the down atom. However, in the case of Si, only extended 3p orbitals should be involved in the screening and (Si being a semiconductor) the response to the presence of a localized Si core-hole can be well described by the dipoles induced on the neighboring Si atoms.

10.6 Interpretation of K-Auger Satellite Structures in 3*d* Metals and in Fluorides, Using the MO Cluster Approach

In the previous parts of this review, the importance of the accurate determination of core Auger kinetic energy shifts in deriving charge transfer in binary alloys has been demonstrated. The detailed knowledge of the respective Auger spectra is necessary to achieve the requested accuracy. Understanding initial- and final-state excitation and relaxation processes in solids is indispensable in the correct identification and interpretation of satellite structures in Auger spectra. It is well-known since the beginning of the development of X-ray photoelectron spectroscopy that satellite structures change significantly depending on the chemical environment, however, the efforts to describe these effects often lead to controversial results. Below, through examples of K-Auger spectra of the 3*d* transition metals Cu and Ni and of fluorides, it is demonstrated, that using the cluster molecular orbital model, considerable chemical effects on the satellite structures can be explained.

10.6.1 *KLL* Auger Spectra of Metallic Cu and Ni: Calculation of the Satellite Main Line Energy Separation Using the DV-Xα Cluster Molecular Orbital Model

Although earlier experimental X-ray-excited *KLL* Auger spectra of metallic Cu [39] already indicated the presence of a satellite at the lower-kinetic-energy side of the main $^1D_2$ line, attributing it to energy-loss effects, our recent studies of Cu and Ni *KLL* spectra photoexcited from metallic and alloy samples [40,41] revealed the localized intrinsic excitation origin of these satellites. In the latter experiments, Mo and Cu X-rays were used to acquire excitation *KLL* Auger spectra from metallic Cu and Ni layers of 10 nm thickness, vacuum-evaporated onto Si wafer substrates [40,41], and the spectra were obtained with a home-built electron spectrometer based on a high-luminosity hemispherical analyzer [17] using an energy resolution of 1.3 eV. In our most recent study [42], high-intensity measurements of thin-layer Cu and Ni samples of varying thickness and the use of a modified Tougaard-type correction for the background from inelastically scattered electrons helped to identify a different intrinsic satellite line at lower kinetic energies [42]. Figs. 10.3a and 10.3b show the decomposition of the background-corrected X-ray-excited Cu *KLL* and Ni *KLL* spectra, and Table 10.3 [43] contains our experimental and calculated data for the energy separation between satellite 2 and the main Cu and Ni *KLL* lines. On the basis of our Dirac–Fock–Slater-type ground-state atomic calculations, satellite 2 was attributed to a $3d_{5/2}$–$4d_{5/2}$ spectator vacancy initial-state shake-up process [41,43] However, as it can be seen from Table 10.3 [43], the free-atom unscreened calculations overestimate the experimentally derived energy separation and provide similar energy separation in the case of Cu and Ni, in sharp contradiction with the experiment. The large

Table 10.3. Energy separation between satellite 2 and the main Cu and Ni KLL Auger lines (eV) [43]

		Atomic Dirac–Fock–Slater model		
Sample	Exp.	Transition	Unscreened	Screened
Cu	11.9	$3d_{5/2} \rightarrow 4d_{5/2}$	15.8	10.5
Ni	6.4	$3d_{5/2} \rightarrow 4d_{5/2}$	15.4	6.9

	DV-Xα MO cluster model			
Sample	Transition	Uncompensated		Compensated
Cu	$8e_g \rightarrow 15e_g$	12.9	10.9[1]	10.2
	$8t_{2g} \rightarrow 16t_{2g}$	13.8		10.6
Ni	$8e_g \rightarrow 15e_g$	10.2	5.9[1]	5.7
	$8t_{2g} \rightarrow 16t_{2g}$	11.6		6.4

[1] 22 $e_g \rightarrow 42e_g$ excitation, calculated for a cluster of 43 atoms

deviations between atomic theory and experiments indicate considerable solid state effects on the shake process.

Accounting for screening, first the semiempirical "excited atom" approximation of Kleiman et al. [44], where the core holes are treated as completely screened extra protons, was used [41,43], resulting in a very good agreement with experiment (Table 10.3) and confirming the significant role of screening effects. In these metals, according to this picture, screening plays an important role, and the conduction electrons move rapidly to the attractive potential of the core hole(s) affecting the energy of the outgoing Auger electron by a change in the screening. On the basis of this model, the difference in the satellite–main Auger line energy separation for Cu and Ni metals can be explained with the occupation of the $3d$ vacancies by screening electrons ([43] and references therein). To study the phenomenon in more detail and to control the predictions of the "excited atom" model, the non-relativistic DV-Xα cluster molecular orbital calculations [12] were performed. The calculations used an fcc lattice and a cluster of 13 atoms for Cu and Ni, with lattice constants of 0.361 nm and 0.352 nm, respectively, and near-minimal basis sets: $1s$–$4d$ atomic orbitals for the central atom and $1s$–$4p$ orbitals for the 12 equivalent neighboring atoms [43]. Assuming an intraatomic $3d \rightarrow 4d$ excitation on the central atom, the energy separation between the main $KL_{23}L_{23}$ Auger line and the satellite due to the $3d$ spectator vacancy was calculated using the transition-state method [43,45]. The results of this calculation are given in Table 10.3 in the "uncompensated" column, falling between those of the screened (excited atom model) and unscreened atomic models. This indicates that the 13-atom model clusters are too small to provide enough conduction electrons to screen the holes induced by photoionization and Auger decay.

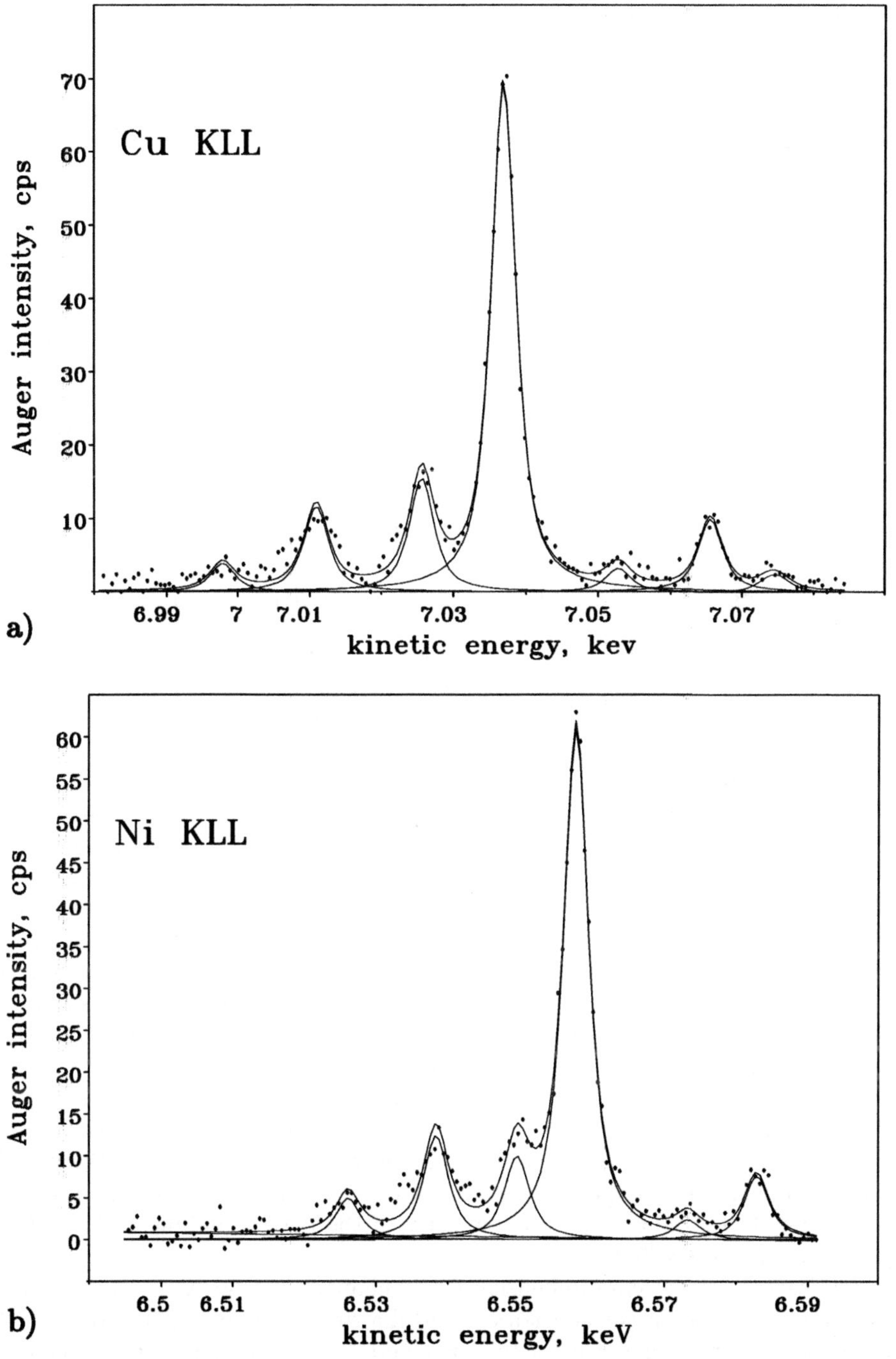

Fig. 10.3. Decomposition of the background-corrected X-ray-induced spectra (**a**) for Cu KLL, (**b**) for Ni KLL (*dots*) [42]

The lack of the flow of electrons to the cluster from its environment can be compensated by neutralizing the charge on the cluster, allowing the screening charges to adapt themselves to the local changes of the potential and to develop the charge distribution freely [43]. The results of this "compensated" cluster MO calculation, shown in Table 10.3, are in a satisfactory agreement with experiment. Using larger, 43-atom clusters, our "uncompensated" results (Table 10.3) justify the applicability of the method of charge neutralization for small clusters mentioned above. It should be noted that the larger energy separation satellite (satellite 1, Fig. 10.3), according to our present results, is likely attributable (on the basis of our Dirac–Fock atomic calculations) to final-state shake-up (e.g., $3d_{5/2}$–$4d_{5/2}$) processes [42]. In our recent work [42] very large differences were found between experiment and the sudden-approximation atomic theory concerning the intensity of the initial state shake-up satellite lines (satellite 2 in Fig. 10.3) compared to the intensity of the main KL_2L_3 ($^1D_2$) line, confirming the presence of very strong solid state effects.

10.6.2 F *KLL* Spectra in Fluorides: Determination of the Resonance Energy and Multiplet Structure Using DV-Xα Cluster Molecular Orbital Calculations

Strong variations – depending on the chemical environment – in the satellite structure of the F *KLL* Auger and F $K\alpha$ X-ray spectra of ionic fluorides excited by X-ray photons, electrons, or high-energy ions, have been reported so far [46]. Showing a very strong change in its intensity (1 – 70%, relative to the main diagram line), a new satellite line was found at the higher-kinetic-energy side of the main F $KL_{23}L_{23}$ line in the spectra of a series of alkali and alkanine-earth fluorides [47–49]. The K^1L^1 (the superscript denotes the number of vacancies in the given shell) satellite of the respective photon-excited F $K\alpha$ X-ray spectra was observed to change in the opposite way compared to the F *KLL* satellite line mentioned above [46,50,51]. Resonant Auger spectroscopy studies indicated that this Auger satellite is not induced by the photoexcitation process itself [49], while another study showed no presence of the satellite in the F *KLL* spectra of free KF molecules, although it was found to be the most intense among those of fluorides in polycrystalline phase [47,48]. These observations suggest that the process leading to the appearance of this Auger satellite is probably part of the decay or relaxation. Initial explanations [47,48] attributed this phenomenon to a resonant electron transfer between the metal and the fluorine atoms [50]; however, later it became clear that this scheme is inadequate to explain the low intensity of the corresponding K^1L^1 F $K\alpha$ X-ray satellite peak [51].

To explain more comprehensively the changes in the X-ray and Auger satellite intensities in the spectra of solid fluorine compounds, the concept of the orbital rearrangement prior to X-ray and Auger emission – the "resonant orbital rearrangement (ROR)" – has been proposed [46, 52, 53]. Here, ROR denotes a resonance between two states existing during the ionization and the decay/relaxation process; namely, between the highest-occupied molecular orbital (HOMO) of the K^0L^0 ground state and the lowest unoccupied molecular orbital (LUMO) of the K^1L^0 ionized state [46]. In the resonant transition an electron of the F^- ion in a fluoride, occupying an orbital of mainly F-2p character in the ground state (HOMO) moves to an orbital of the ionized state (LUMO) having mostly F-3s character. The ROR concept proved to be successful in interpreting the intensity changes of the new satellite in the F KLL Auger spectra of alkali fluorides [46]. In our recent work [54] we studied the structure of the Auger satellites induced from rutile-type difluorides and from NaF.

F KLL Auger spectra were excited by Al $K\alpha$ and Cu $L\alpha$ X-rays from polycrystalline powder samples of CoF_2, MgF_2, NiF_2, ZnF_2, and NaF and measured by a high-luminosity hemispherical electron spectrometer [17] using an energy resolution better than 0.6 eV [54]. The use of the Cu $L\alpha$ excitation, with a photon energy close to the F-1s ionization threshold, resulted in a very significant improvement in the peak-to-background ratio. Figure 10.4 shows the measured F KLL Auger spectrum induced by Cu $L\alpha$ radiation from the NaF sample [54]; M denotes the new satellite peak.

To remove the background due to inelastically scattered electrons, two different models were used in order to estimate the systematic error in the evaluation of the spectra [54].

The resonance energy, ΔE, assuming resonant orbital rearrangement (ROR) [46] following F-1s photoionization, was estimated as the energy

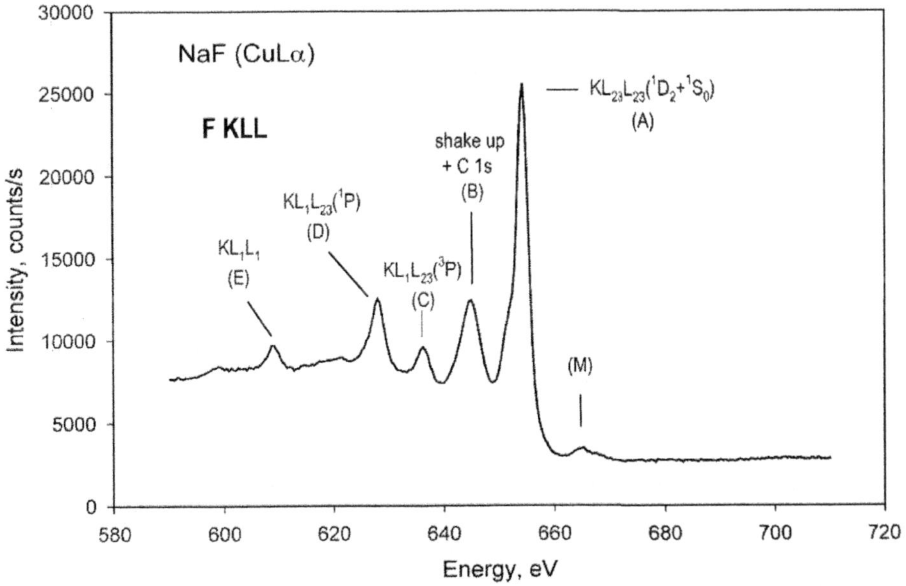

Fig. 10.4. F KLL Auger spectrum excited by Cu $L\alpha$ X-rays from the polycrystalline NaF sample [54]

difference between the highest occupied MO level having a F-$2p$ contribution, and the lowest-unoccupied MO level having a F-$3s$ contribution. To obtain ΔE, the non-relativistic DV-Xα cluster MO calculations were performed using numerical basis functions and the self-consistent charge scheme [12]. A near-minimal basis set ($1s$–$3s$ for F, $1s$–$3d$ for Mg and Na, $1s$–$4p$ for Co, Ni, and Zn atoms, respectively) was used, with the model clusters Na_6F^{5+} with O_h symmetry and M_3F^{5+} with C_{2v} symmetry, where M represents Mg, Co, Ni, and Zn [54].

Within the ROR model, the probability of the ROR process is [46]

$$P_{\mathrm{ROR}} \approx \frac{a\,(\hbar/2\tau)^2}{(\Delta E)^2 + (\hbar/2\tau)^2}\,, \tag{10.10}$$

where a is a constant, and $\tau \sim 2.4 \times 10^{-13}$ s is the lifetime of the F-$1s$ vacancy, i.e., $P_{\mathrm{ROR}} \sim 1/(\Delta E)^2$. This means that the ratio between the intensity I_M of the M satellite peak and the intensity I_A of the main diagram line A in the F KLL spectra, I_M/I_A (see Fig. 10.4), should be proportional to $(1/\Delta E)^2$ if $\Delta E \gg \hbar/2\tau$. For alkali fluorides, a very good linear correlation was found between I_M/I_A and $(1/\Delta E)^2$ [46].

From the evaluation of our measured F KLL Auger spectra we arrived at an interesting observation. Although the experimental errors are large, the energy width (FWHM, full-width at half maximum) of the M satellite

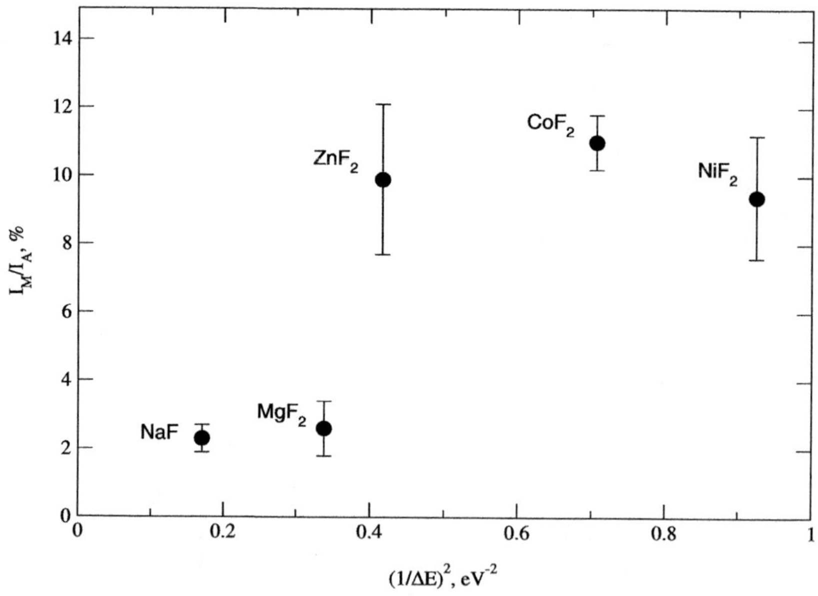

Fig. 10.5. Relative intensities I_M/I_A of the M satellite peaks as a function of $(1/\Delta E)^2$, where ΔE is the resonance energy obtained from the DV-Xα calculations [54]

peak seems to be proportional to its relative intensity, I_M/I_A (Fig. 10.5 and Table 10.1 in [54]), and the measured energy widths for the different samples differ by a factor of almost two [54]. This suggests that the intensity of the M satellite line depends on the local density of electron states available for the Auger transition, resulting in the M satellite.

Figures 10.5 and 10.6 show our experimentally derived relative intensity, I_M/I_A, and the energy width (FWHM) of the satellite peak M as a function of $(1/\Delta E)^2$ [54]. Although the results in these figures suggest some correlation between I_M/I_A or the energy width and $(1/\Delta E)^2$, the existence of a linear relationship for the rutile-type difluorides, which was obtained for the monofluorides earlier [46], cannot be definitively proved. To decrease the scatter of the data in Figs. 10.5–10.6, more accurate calculations with larger cluster sizes, a reduction of the systematic errors in the evaluation of the experimental spectra, and probably a more detailed description of the resonant process assumed to yield the M satellite line in the F KLL Auger spectra of fluorides are needed.

In the case of polycrystalline KF, the energy separation of the M satellite line from the main $^1D_2$ line is much smaller and I_M/I_A is much higher than for other fluorides [46,54]. The multiplet structure of two holes in the valence band of KF was calculated using the K_6F^{5+} cluster (initial state: K_6F^{5+}, final state:

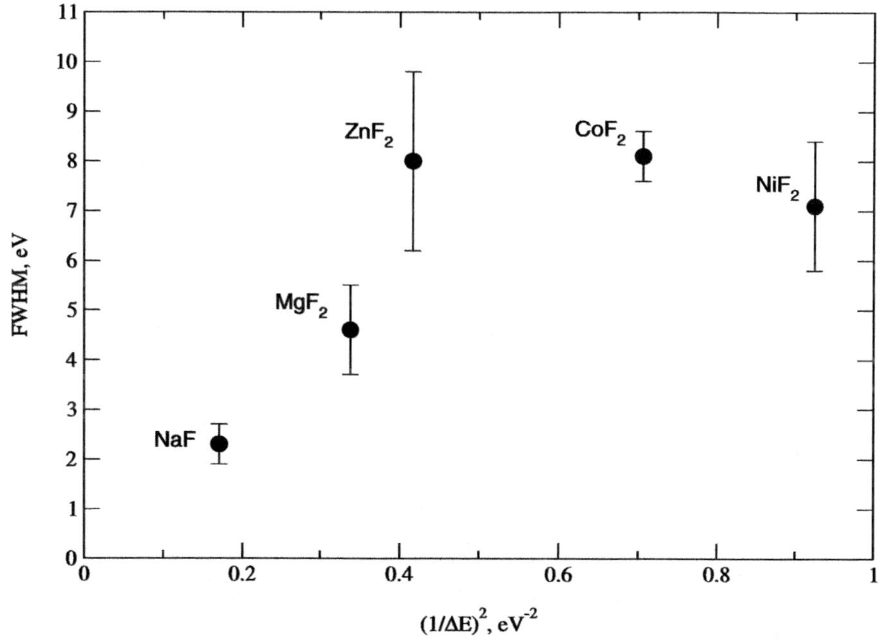

Fig. 10.6. Energy widths (FWHM) of the M satellite peaks as a function of $(1/\Delta E)^2$, where ΔE is the resonance energy obtained from the DV-Xα calculations [54]

K_6F^{7+}) in octahedral symmetry and with molecular orbitals in the ground state as well as in the final state [54]. The multiplet structure calculations were based on the configuration-interaction model [55]. In this calculation, the correlation correction was not included; its inclusion would reduce the energy splitting. From the results of the multiplet structure calculations, the predicted splitting of the ^1D state is negligible compared to the observed satellite-main line energy separation [46]; therefore, the multiplet splitting is not expected to explain the origin of the M satellite in the F KLL spectrum of KF.

10.7 Information on Local Electronic Structure and Correlation from Core-valence Auger Lineshapes

The lineshapes of core-valence Auger spectra of solids can be determined on the basis of the final-state rule, following the model of Ramaker [2]: in the absence of significant configuration mixing (localization) and shake processes, the initial state determines the relative l or ll' Auger intensities (where, for example, $l = s$ or p); the shape of each component distribution is determined by the final-state DOS. The experimental lineshapes $A(E)$ for the core-core-valence (CCV) and for the core-valence-valence (CVV) Auger processes can be described as

$$A_{\mathrm{CCV}}(E) = C_s\rho'_s(E) + C_p\rho'_p(E) \tag{10.11}$$

and

$$\begin{aligned} A_{\mathrm{CVV}}(E) \cong\; & C_{ss}R_s^2\rho_s(E)\otimes\rho_s(E) \\ & + C_{sp}R_sR_p\rho_s(E)\otimes\rho_p(E) + C_{pp}R_p^2\rho_p(E)\otimes\rho_p(E)\,, \end{aligned} \tag{10.12}$$

where $\otimes$ denotes convolution integral, $\rho_l(\rho_{l'})$ is the local LDOS (screened DOS) of the final state without (with) a core hole, C_i, C_{ij} denote atomic Auger intensities (Auger transition matrix elements) normalized per filled shell, and the R_l factors provide the ratio of local charges in the screened initial state to that in the unscreened final state of the CVV process.

Correlation effects on CVV Auger lineshapes $A_{\mathrm{CVV}}(E)$ can be approximated using the Cini-Sawatzky theory assuming on-site hole–hole interaction and completely filled bands [56, 57]:

$$A_{\mathrm{CVV}}(E) \sim \frac{\rho\otimes\rho'(E)}{[1-\Delta U I(E)]^2 + [\Delta U\pi\rho\otimes\rho'(E)]^2}\,, \tag{10.13}$$

where $I(E) = \int(\rho\otimes\rho(E)/(E-\varepsilon))\,\mathrm{d}\varepsilon$, and ΔU is the effective hole–hole correlation parameter.

Interpreting lineshapes of core-valence Auger transitions of solids, we applied the semiempirical approach outlined above using atomic Auger transition matrix elements. To calculate the LDOS distribution, the DV-Xα cluster molecular orbital method [12] was applied using the self-consistent charge scheme. The LDOS was obtained by replacing the MO levels with energy distribution functions of Lorentzian shape with fixed energy width.

10.7.1 Local Electronic Structures in Phosphorus Oxyanions

Our experimental P KLV spectra [4,5], obtained from polycrystalline Na_3PO_4, $(NaPO_3)_n$ and Li_3PO_4 using Mo X-rays for excitation and a high-luminosity electron spectrometer [17] based on a hemispherical analyzer, reflect the dominance of the effect of the common T_d symmetry. Describing the P KLV lineshapes, effects of the final state core holes were accounted for by performing the PO_4^{3+} T_d symmetry DV-Xα cluster molecular orbital calculations with the corresponding final-state (L_1 or $L_{2,3}$) vacancies [58]. A reasonably good agreement between the experimental and calculated spectra demonstrates that by using LDOS from our MO cluster calculations and atomic Auger transition probabilities, the main structure of the KLV Auger spectra of phosphates can be explained [58]. The same applies to the case of the P LVV spectra of these phosphorus oxyanions [4, 59], where both the energy separation between the two most intense (ss and pp) peaks and the lineshape of the highest-intensity line are well-reproduced by our calculations [4, 59].

10.7.2 Core-valence K-Auger Spectra of Metallic Al

The Auger spectra of phosphates show strong group localization, therefore it is interesting to see the case of a simple metal with high conductivity.

The experimental Mo X-rays excited Al KLV and KVV spectra [60] were obtained from high purity polycrystalline Al samples using our home-built electron spectrometer [17] based on a hemispherical analyzer, with the energy resolution of 0.4 eV (KLV) and 2.0 eV (KVV). High resolution Al valence band XPS spectra were excited by monochromated Al $K\alpha$ X-rays of 0.5 eV linewidth [61].

In the DV-Xα calculations O_h symmetry and a cluster of 19 atoms was used [60]. The final state core hole effects in the case of the KLV Auger peaks were accounted for in two ways: by applying the Cini-Sawatzky model and the ground state DOS, and performing the DV-Xα cluster calculations with the corresponding final state (L_1 or L_{23}) core hole, respectively [60]. Figures 10.7 and 10.8 show the comparison of the experimental and calculated Al KL_1V and Al $KL_{23}V$ spectra, respectively, including both models. For obtaining the experimental KL_1V and $KL_{23}V$ Auger lineshapes, structures due to plasmon losses were subtracted from the measured Al KLV spectra and corrections for the inelastic background were made using the Shirley method [62]. The calculated KL_1V and $KL_{23}V$ Auger lineshapes show a good agreement with the experimental spectra (Figs. 10.7 and 10.8) proving that the lineshapes reflect the screened local DOS and that the Cini-Sawatzky model is a quite good approximation for these cases [60]. It is clearly indicated in the spectra

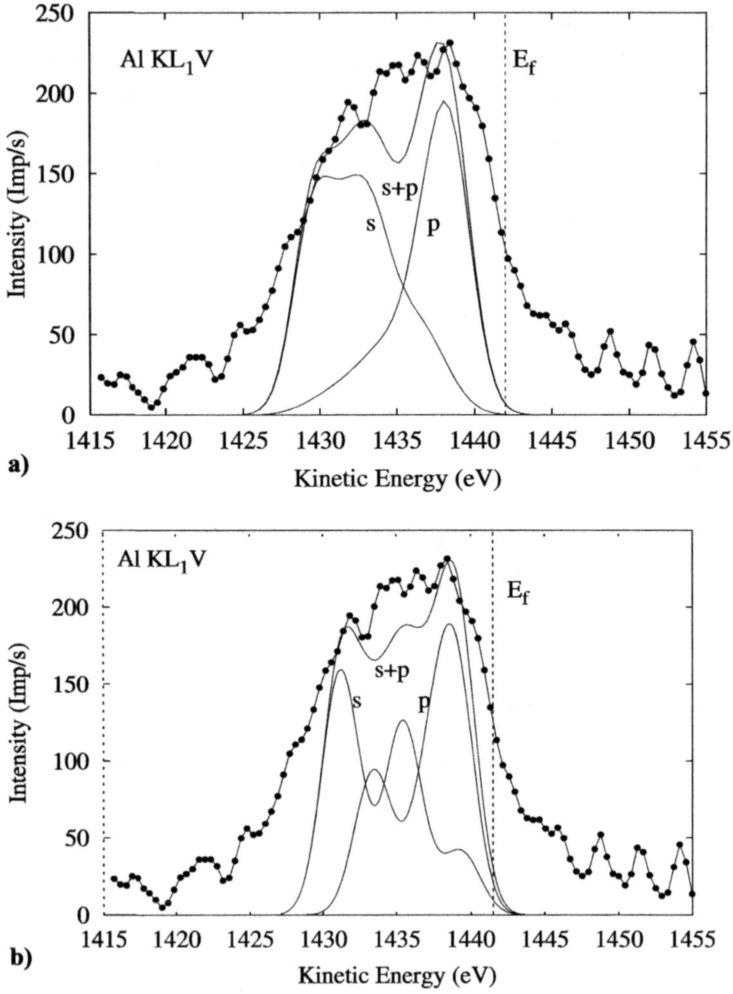

Fig. 10.7. Comparison of experimental and theoretical Al KL_1V spectra [60]. (**a**) Experimental and Cini-Sawatzky model. (**b**) Experimental and cluster MO model with a final-state L_1 hole

that the difference in the KL_1V and $KL_{23}V$ lineshapes is mainly due to the increase of the contribution from the s type partial DOS in the case of the KL_1V spectrum.

According to the final-state rule, the CVV Auger and the valence-band photoelectron spectra should reflect the ground-state DOS. In Fig. 10.9, the high-resolution Al valence-band photoelectron spectrum [61] is shown, in comparison with the result of our DV-Xα cluster MO calculations (using the photoionization cross-sections given by Scofield [63]). The agreement between theory and experiment is quite good, proving the correctness of the ground-

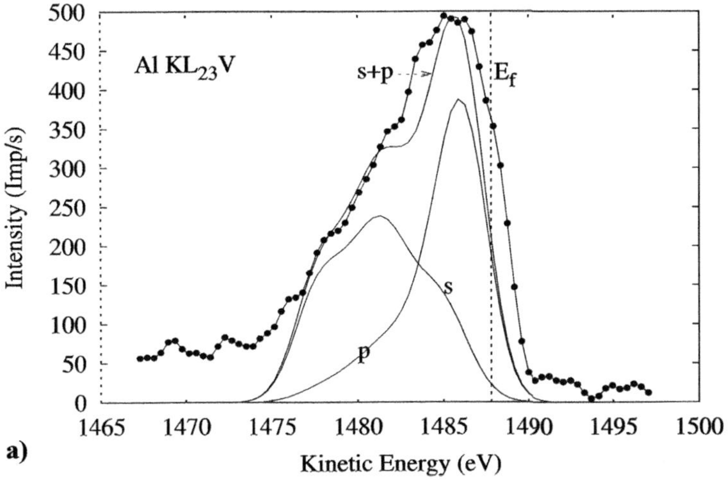

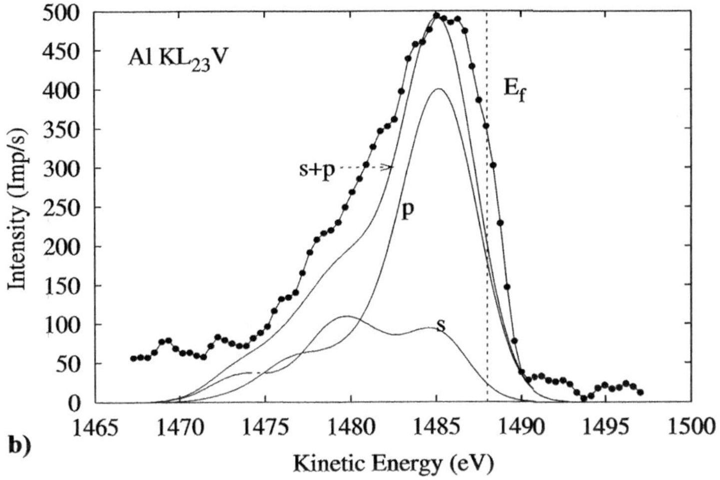

Fig. 10.8. Comparison of experimental and theoretical Al $KL_{23}V$ spectra [60]. (**a**) Experimental and Cini-Sawatzky model. (**b**) Experimental and cluster MO model with a final-state L_{23} hole

state DOS obtained by the cluster MO calculations. Figure 10.10 shows the experimental Al KVV spectrum in comparison with our theoretical spectra obtained from the cluster model calculations.

The results demonstrate that the cluster MO model describes the CVV lineshapes quite well in Al, and no correlation effects can be identified in the spectra. From further cluster MO calculations, it became clear that the strong effect of the oxygen chemisorption on the Al KVV spectra, which was observable even at normal emission, was due to the contribution of the O-2p level [4, 64].

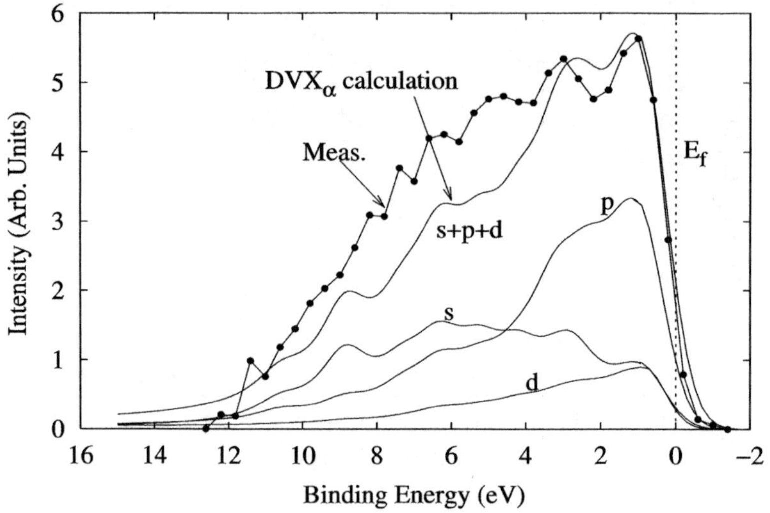

Fig. 10.9. High-resolution valence band XPS spectrum of polycrystalline Al [61], compared to the spectrum calculated using the DV-Xα cluster MO model [60]

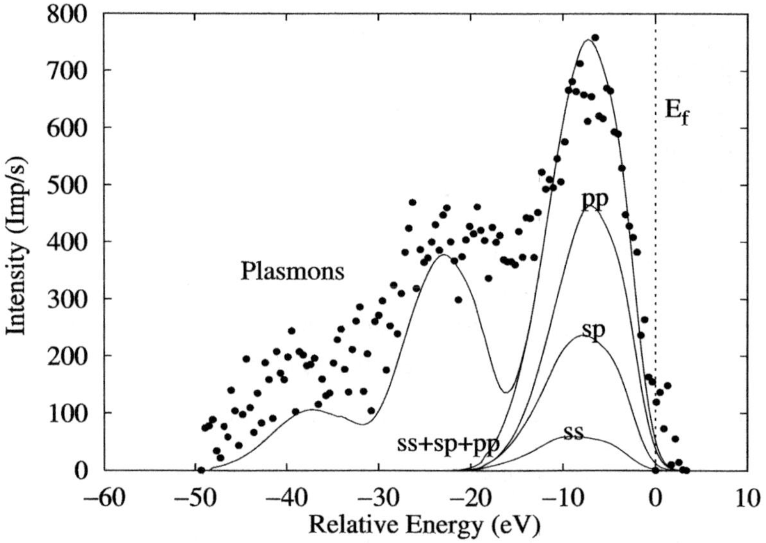

Fig. 10.10. The experimental KVV Auger spectrum of polycrystalline Al (corrected for inelastic background but adding the contribution from plasmon losses) compared to the results obtained by the DV-Xα cluster MO model [60]

10.7.3 Local Electronic Structure in Al–Ni Alloys

Studying alloy effects on the local electronic environment and on atomic transitions in alloys is very important for understanding of electronic properties of these materials. Investigations of the electronic structure of 3*d* transition metal–simple metal binary alloy systems, such as Al−Ni alloys, provide interesting information on Ni-*d* band filling and the strength of hybridization effects. Furthermore, these studies are relevant to clarify issues like the applicability of the final-state rule [65] for the interpretation of the Al core-valence Auger lineshapes in the case of Al−Ni alloys, and the effect of the transition-energy dependence of the Auger matrix ele ments on Al K-Auger spectra [66].

Al $KL_{23}V$ Auger spectra of $AlNi_3$ and Al_3Ni polycrystalline alloy samples, induced by Mo X-rays, were obtained using a home-built, high luminosity electron spectrometer [17], based on a hemispherical analyzer, with an energy resolution of 0.4 eV [67]. Valence band monochromated Al $K\alpha$ XPS spectra were measured by a SCIENTA ESCA-300 spectrometer with a spectrometer resolution of 0.4 eV [67].

To derive DOS and LDOS distributions, the DV-Xα cluster molecular orbital calculations [12] were performed using the experimental lattice constants [23,24], a 14-atom cluster and cubic O_h symmetry for $AlNi_3$, a 16-atom cluster and orthorhombic DO_{20} structure for Al_3Ni, with the outermost orbitals Al-3*d* and Ni-4*p* [67]. The DOS profile was computed by replacing each discrete energy level by a Lorentzian of 0.8 eV in width, weighted by the degeneracy of the orbitals. In the case of Al $KL_{23}V$ spectra, an additional broadening of 0.9 eV, due to the Al-1*s* and Al-2*p* energy widths [68], the contribution of the finite energy resolution and of the phonon broadening [69], was taken into account. To obtain the model XPS spectra, the atomic components of the DOS were weighted by the corresponding photoionization cross-sections as presented by Scofield [63]. The Al $KL_{23}V$ Auger lineshapes were derived using the calculated LDOS with a final-state L_{23} core hole and the final-state rule; for the ratio of the Al $KL_{23}s$ and Al $KL_{23}p$ Auger transition matrix elements, the atomic value of 0.25 [70] was used [67].

In Fig. 10.11, the comparison of the measured and calculated valence band XPS spectra can be seen, showing a satisfactory agreement between theory and experiment regarding the bandwidths and proving the correctness of the calculated ground-state DOS and the model applied for simulation of the valence-band photoelectron spectra [67]. Figure 10.11 indicates that with increasing dilution of nickel in aluminum, the dominating Ni-3*d* bands become narrower and pull away from the Fermi level [71, 72], while the d^9 correlation satellites (at about −9 eV) disappear, due to the filling of the Ni-3*d* band [73]. For Al_3Ni, our calculations predict the energy position of the maximum correctly, in agreement with experiment and with recent band-structure calculations [74], while for Al_3Ni, this maximum is shifted by 0.5 eV towards the higher binding energy compared to experiment (Fig. 10.11a). The dominating Ni-*d* contributions have different ratios in the two alloys. From our calculations, in the case of Al_3Ni, the total number of electrons around the Fermi level is in better agreement with the experimental value than the theoretical value reported earlier [67]. Our results suggest the presence of strong hybridization effects.

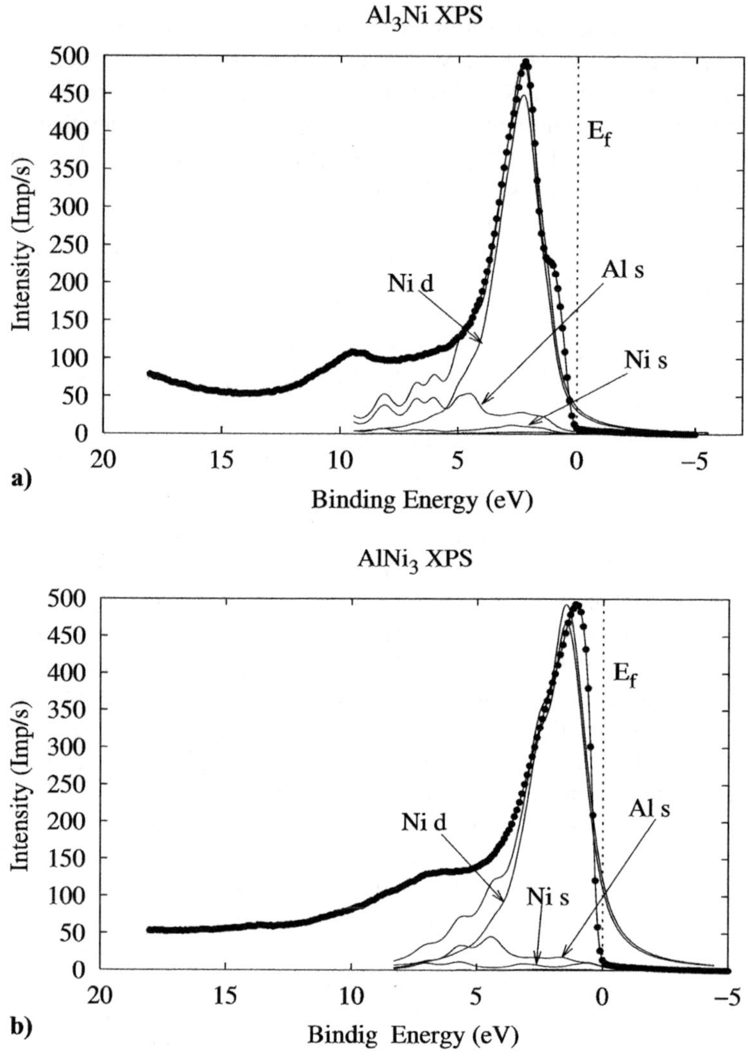

Fig. 10.11. Comparison of the measured and calculated valence band XPS spectra [67]. (**a**) Al_3Ni. (**b**) $AlNi_3$

Figure 10.12 shows the comparison of the calculated Al KL_{23} in comparison with the measured spectra, indicating the domination of the p density around the site with a core hole. The Al p/s ratios of the contributions to the $KL_{23}V$ spectra are 3.4 and 3.9 for Al_3Ni, and $AlNi_3$, respectively [67]. The very similar shape of the $KL_{23}V$ Auger peaks in the two alloys suggest weak electron–electron interaction. Due to hybridization effects, the Al DOS and the Ni-d energy position is reduced in the alloys relative to the case of the metal, causing a small energy shift.

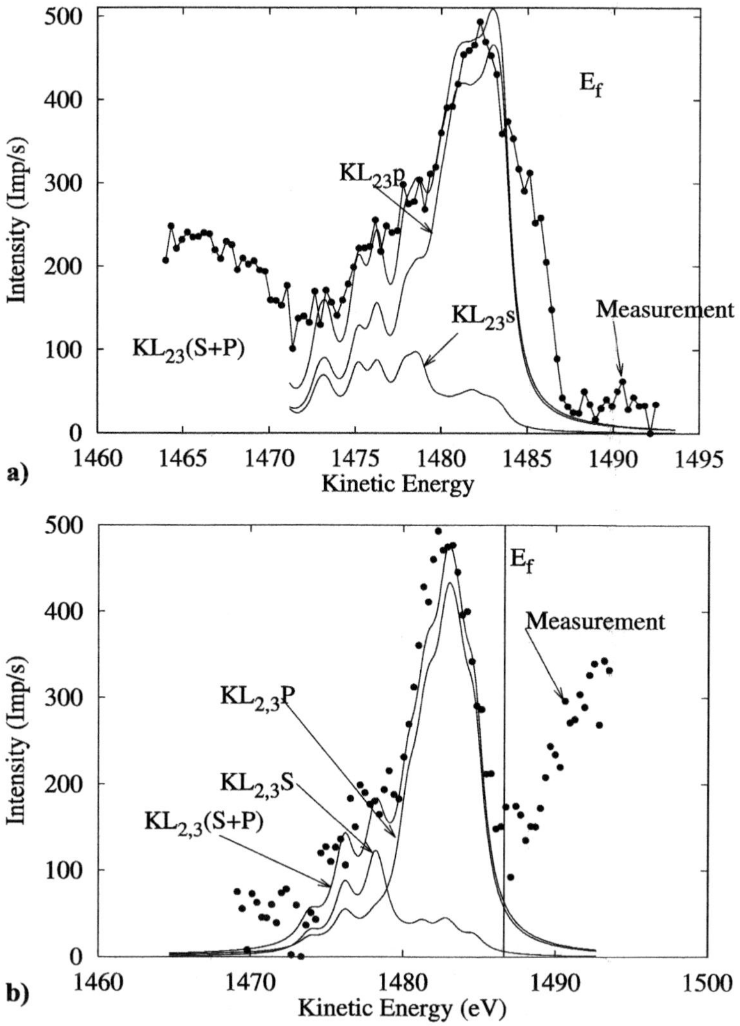

Fig. 10.12. Calculated Al $KL_{23}V$ spectra in comparison with the measured spectra [67]: (**a**) Al_3Ni; (**b**) $AlNi_3$

Comparing the calculated Al $KL_{23}V$ Auger spectra obtained for alloys to the spectra obtained for metallic Al [60], a difference occurs in the s PDOS, as it is pulled away from the Fermi level with increasing Al content. For $AlNi_3$ (Fig. 10.12b) the calculated lineshape agrees well with experiment, correctly giving the position of the maximum and the width of the energy distribution and the energy position and the relative intensity of the contributions from the s PDOS. The calculation also reproduces the slight decrease in the p PDOS as a consequence of the Ni-d–Al-p hybridization, causing a shoulder at the low- and high-kinetic-energy part of the spectrum. In the case of Al_3Ni, the calculation describes well the position of the maximum and the position of the shoulder due to the *s* PDOS; however, it gives a somewhat lower energy width compared to experiment, and it cannot account for the shoulder at the higher kinetic energy side [67]. From the results, the energy-dependence of the *KLV* Auger matrix elements seems to be negligible.

10.8 Summary

The power and advantages of the cluster approximation and the DV-Xα cluster molecular orbital model have been demonstrated through a series of recent studies. These advantages include the interpretation of photoinduced Auger spectra, the extraction of information on the electronic structure of solids from these spectra, the determination of local charges in binary alloys, the description of core-hole-induced excitations in 3*d* transition metals as well as relaxation/polarization effects, and the characterization of local electronic structure in oxyanions, metals, and alloys.

Acknowledgement. This work has been supported by the Hungarian project OTKA T038016 and by the JSPS-MTA joint project No. 59.

References

1. H.H. Madden, J. Vac. Sci. Technol. **18**, 677 (1991)
2. D.E. Ramaker, Crit. Rev. Solid State Mater. Sci. **17**, 211 (1991)
3. P. Weightman, J. Electron Spectrosc. Rel. Phenom. **68**, 127 (1994)
4. L. Kövér, Fizika A **4**, 149 (1995)
5. F.P. Larkins, J. Electron Spectros. Rel. Phenom. **51**, 115 (1990)
6. U. Becker, R. Wehlitz, Physica Scripta **T41**, 127 (1992)
7. N. Mårtensson, R. Nyholm, Phys. Rev. B **24**, 7121 (1981)
8. D.E. Ramaker, J. Electron Spectrosc. Rel. Phenom. **66**, 269 (1994)
9. J.A.D. Matthew, Phys. Rev. B **29**, 3031 (1984)
10. L. Kövér, A. Némethy, I. Cserny, D. Varga, Surf. Interface Anal. **20**, 659 (1993)
11. A. Némethy, L. Kövér, I. Cserny, D. Varga, P.B. Barna, J. Electron Spectrosc. Rel. Phenom. **70**, 183 (1995)
12. H. Adachi, M. Tsukada, C. Satoko, J. Phys. Soc. Japan **45**, 875 (1978)
13. A. Rosén, D.E. Ellis, J. Chem. Phys. **62**, 3039 (1975)
14. C.D. Wagner, L.H. Gale, R.H. Raymond, Anal. Chem. **51**, 466 (1979)
15. T.D. Thomas, P. Weightman, Phys. Rev. B **33**, 5406 (1986)
16. D.A.C. Gregory, A.D. Laine, P.S. Fowles, A. Takahashi, P. Weightman, J. Phys. **C5**, 3843 (1993)
17. L. Kövér, D. Varga, I. Cserny, J. Tóth, K. Tõkési, Surf. Interface Anal. **19**, 9 (1992)
18. R.J. Cole, N.J. Brooks, P. Weightman, S.M. Francis, M. Bowker, Surf. Rev. Lett. **3**, 1763 (1996)

19. R.J. Cole, P. Weightman, in Metallic Alloys: Experimental and Theoretical Perspectives, eds.: J.S. Faulkner, R.G. Jordan (Kluwer, 1994)
20. L. Kövér, Zs. Kovács, P. Weightman, R. Sanjinés, D. Varga, G. Margaritondo, J. Pálinkás, M. Abon, in *Proceedings of the 7th European Conference on Application of Surface and Interface Analysis*, ed. by I. Olefjord, L. Nyborg, D. Briggs. ECASIA-97, 16–20 June 1997, Göteborg (Wiley, New York, 1997) p. 185
21. L. Kövér, Adv. Quantum Chem. **42**, 331 (2003)
22. L. Kövér, Zs. Kovács, R. Sanjinés, P. Weightman, G. Margaritondo, D. Varga, J. Pálinkás, H. Adachi, in *Proceedings of the 6th European Conference on Applications of Surface and Interface Analysis*, ed. by H.-J. Mathieu, B. Reihl, D. Briggs. ECASIA-95, 1995, Montreux (Wiley, New York, 1996) p. 448
23. K.H. Hellwege, Landolt-Börnstein, Structure Data of Elements and Intermetallic Phases New Series, Group III, Vol. 6 (Springer, Berlin Heidelberg New York, 1971)
24. P. Villars, L.D. Calvert, *Pearson's Crystallographic Data for Intermetallic Phases* (American Society of Metals Park, OH, 1985)
25. J.C. Slater, *Quantum Theory of Molecules and Solids*, vol. 2, (Mcgraw-Hill, New York, 1974)
26. P. Weightman, R.J. Cole, N.J. Brooks, J.M.C. Thornton, Nuclear Instrum. and Meth. in Phys. Res. B **97**, 472 (1995)
27. H. Adachi, T. Tanabe, S. Imoto, Japan. J. Appl. Phys. **17**, 49 (1978)
28. Zs. Kovács, L. Kövér, H. Adachi, Adv. Quantum Chem. **29**, 421 (1997)
29. D. Hackenbracht, J. Kübler, J. Phys. F **10**, 427 (1980)
30. R.J. Cole, D.A.C. Gregory, P. Weightman, Phys. Rev. B **49**, 5657 (1994)
31. G. Moretti, Surf. Interface Anal. **16**, 159 (1990)
32. G. Moretti, Surf. Interface Anal. **17**, 352 (1991)
33. Zs. Kovács, L. Kövér, F. Filippone, G. Moretti, J. Pálinkás, in *Proceedings of the 7th European Conference on Applications of Surface and Interface Analysis*, ed. by I. Olefjord, L. Nyborg, D. Briggs. ECASIA-97, 16–20 June 1997, Göteborg (Wiley, New York, 1997) p. 848
34. G. Moretti, J. Electron Spectrosc. Relat. Phenom. **47**, 283 (1998)
35. N. Roberts, R.J. Needs, Surf. Sci. **236**, 112 (1990)
36. D.R. Lide (ed.), *Handbook of Chemistry and Physics*, 72nd edn. (CRC Press, Boca Raton, 1991)
37. E. Landemark, C.J. Karlsson, Y-C. Chao, R.I.G. Uhrberg, Phys. Rev. Lett. **69**, 1588 (1992)
38. E. Pehlke, M. Scheffler, Phys. Rev. Lett. **71**, 2338 (1993)
39. E. Sokolowski, C. Nordling, Arkiv för Fysik **14**, 557 (1958)
40. L. Kövér, D. Varga, I. Cserny, J. Tóth, Zs. Kovács, J. Surf. Anal. **51**, 77 (1999)
41. L. Kövér, Zs. Kovács, J. Tóth, I. Cserny, D. Varga, P. Weightman, S. Thurgate, Surf. Sci. **433–435**, 833 (1999)
42. L. Kövér, I. Cserny, J. Tóth, D. Varga, T. Mukoyama, J. Electron Spectrosc. Rel. Phenom. **114–116**, 55 (2001)
43. I. Cserny, L. Kövér, H. Nakamatsu, T. Mukoyama, Surf. Interface Anal. **30**, 199 (2000)
44. G.G. Kleiman, S.G.C. de Castro, R. Landers, Phys. Rev. B **49**, 2753 (1994)
45. J.C. Slater, *The Calculation of Molecular Orbitals* (Wiley, New York, 1979)
46. M. Uda, T. Yamamoto, T. Takenaga, Adv. Quantum Chem. **29**, 389 (1997)
47. O. Benka, M. Uda, Phys. Rev. Lett. **56**, 1667 (1986)

48. M. Uda, K. Maeda, A. Koyama, Y. Sasa, Phys. Rev. A **29**, 1258 (1984)
49. H. Aksela, E. Kukk, S. Aksela, A. Kikas, E. Nomniste, A. Ausmees, M. Elango, Phys. Rev. B **49**, 3116 (1994)
50. O. Benka, R.L. Watson, R.A. Kenefick, Phys. Rev. Lett. **47**, 1202 (1981)
51. M. Tachibana, H. Adachi, T. Mukoyama, Y. Hibino, Y. Sasa, K. Fuwa, M. Uda, Nucl. Instrum. Meth. Phys. Res. B **49**, 15 (1990)
52. M. Uda, T. Yamamoto, Nucl. Instrum. Meth. Phys. Res. B **150**, 1 (1999)
53. M. Uda, T. Yamamoto, T. Takenaga, M. Mochizuki, T. Mukoyama, Nucl. Instrum. Meth. Phys. Res. B **150**, 50 (1999)
54. L. Kövér, M. Uda, I. Cserny, J. Tóth, J. Végh, D. Varga, K. Ogasawara, H. Adachi, J. Vac. Sci. Technol. **A19**, 1143 (2001)
55. K. Ogasawara, T. Ishii, I. Tanaka, H. Adachi, Phys. Rev. B **61**, 143 (2000)
56. M. Cini, Solid State Commun. **20**, 655 (1976)
57. G.A. Sawatzky, Phys. Rev. Lett. **39**, 504 (1977)
58. L. Kövér, A. Némethy, I. Cserny, A. Nisawa, Y. Ito, H. Adachi, Surf. Interface Anal. **22**, 45 (1994)
59. L. Kövér, I. Cserny, A. Némethy, H. Adachi, ATOMKI Ann. Rep. 51 (1994)
60. Zs. Kovács, L. Kövér, D. Varga, P. Weightman, J. Pálinkás, H. Adachi, J. Electron Spectrosc. Relat. Phenom. **72**, 157 (1995)
61. P.H. Hannah, PhD Dissertation, University of Liverpool, 1985
62. D.A. Shirley, Phys. Rev. **55**, 4709 (1972)
63. J.H. Scofield, J. Electron Spectrosc. Rel. Phenom. **8**, 129 (1976)
64. Zs. Kovács, L. Kövér, D. Varga, P. Weightman, R. Sanjinés, G. Margaritondo, J. Pálinkás, H. Adachi, ATOMKI Ann. Rep. 48 (1994)
65. P.H. Hannah, P. Weightman, P.T. Andrews, Phys. Rev. B **31**, 6238 (1985)
66. Yu.N. Kucherenko, J. Electron Spectrosc. Rel. Phenom. **68**, 341 (1994)
67. Zs. Kovács, L. Kövér, P. Weightman, D. Varga, R. Sanjinés, J. Pálinkás, G. Margaritondo, H. Adachi, Phys. Rev. B **54**, 8501 (1996)
68. T. Papp, J.L. Campbell, X-ray Spectros. **24**, 307 (1995)
69. G.K. Wertheim, P.H. Citrin, Y. Baer, Phys. Rev. B **16**, 4256 (1977)
70. D.E. Ramaker, N.H. Turner, F.L. Hutson, W.N. Mei, Phys. Rev. B **33**, 2574 (1986)
71. J.C. Fuggle, F.U. Hillebrecht, R. Zeller, P.A. Bennett, Z. Zolnierek, Ch. Freiburg, Phys. Rev. B **27**, 2145 (1983)
72. P.A. Bennett, J.C. Fuggle, R. Zeller, U. Hillebrecht, A. Lenselink, G.A. Sawatzky, Phys. Rev. B **27**, 2194 (1983)
73. F.U. Hillebrecht, J.C. Fuggle, P.A. Bennett, Z. Zolnierek, C. Freiburg, Phys. Rev. B **27**, 2179 (1983)
74. G. Cubiotti, E.E. Krasovskii, O.U. Slobodyan, Y.N. Kucherenko, V.N. Antonov, J. Phys. Condens. Matter **7**, 4865 (1995)

Index

GPSR Compliance

The European Union's (EU) General Product Safety Regulation (GPSR) is a set of rules that requires consumer products to be safe and our obligations to ensure this.

If you have any concerns about our products, you can contact us on ProductSafety@springernature.com

In case Publisher is established outside the EU, the EU authorized representative is:

Springer Nature Customer Service Center GmbH
Europaplatz 3
69115 Heidelberg, Germany

Batch number: 10755246

Printed by Printforce, the Netherlands